Reintroduction Biology

Conservation Science and Practice Series

Published in association with the Zoological Society of London

Wiley-Blackwell and the Zoological Society of London are proud to present our *Conservation Science and Practice* series. Each book in the series reviews a key issue in conservation today. We are particularly keen to publish books that address the multidisciplinary aspects of conservation, looking at how biological scientists and ecologists are interacting with social scientists to effect long-term, sustainable conservation measures.

Books in the series can be single or multi-authored and proposals should be sent to:
  Ward Cooper, Senior Commissioning Editor, Wiley-Blackwell, John Wiley & Sons,
  9600 Garsington Road, Oxford OX4 2DQ, UK
  Email: ward.cooper@wiley.com

Each book proposal will be assessed by independent academic referees, as well as our Series Editorial Panel. Members of the Panel include:
  Richard Cowling, Nelson Mandela Metropolitan University, Port Elizabeth, South Africa
  John Gittleman, Institute of Ecology, University of Georgia, USA
  Andrew Knight, University of Stellenbosch, South Africa
  Georgina Mace, Imperial College London, Silwood Park, UK
  Daniel Pauly, University of British Columbia, Canada
  Stuart Pimm, Duke University, USA
  Hugh Possingham, University of Queensland, Australia
  Peter Raven,Missouri Botanical Gardens, USA
  Helen Regan, University of California, Riverside, USA
  Alex Rogers, Institute of Zoology, London, UK
  Michael Samways, University of Stellenbosch, South Africa
  Nigel Stork, University of Melbourne, Australia

Previously published

Trade-offs in Conservation: Deciding What to Save
Edited by Nigel Leader-Williams, William M. Adams and Robert J. Smith
ISBN: 978-1-4051-9383-2 Paperback; ISBN: 978-1-4051-9384-9 Hardcover; 398 pages; September 2010

Urban Biodiversity and Design
Edited by Norbert Müller, Peter Werner and John G. Kelcey
ISBN: 978-1-4443-3267-4 Paperback; ISBN 978-1-4443-3266-7 Hardcover; 640 pages; April 2010

Wild Rangelands: Conserving Wildlife While Maintaining Livestock in Semi-Arid Ecosystems
Edited by Johan T. du Toit, Richard Kock and James C. Deutsch
ISBN: 978-1-4051-7785-6 Paperback; ISBN 978-1-4051-9488-4 Hardcover; 424 pages; January 2010

Reintroduction of Top-Order Predators
Edited by Matt W. Hayward and Michael J. Somers
ISBN: 978-1-4051-7680-4 Paperback; ISBN: 978-1-4051-9273-6 Hardcover; 480 pages; April 2009

Recreational Hunting, Conservation and Rural Livelihoods: Science and Practice
Edited by Barney Dickson, Jonathan Hutton and Bill Adams
ISBN: 978-1-4051-6785-7 Paperback; ISBN: 978-1-4051-9142-5 Hardcover; 384 pages; March 2009

Participatory Research in Conservation and Rural Livelihoods: Doing Science Together
Edited by Louise Fortmann
ISBN: 978-1-4051-7679-8 Paperback; 316 pages; October 2008

Bushmeat and Livelihoods: Wildlife Management and Poverty Reduction
Edited by Glyn Davies and David Brown
ISBN: 978-1-4051-6779-6 Paperback; 288 pages; December 2007

Managing and Designing Landscapes for Conservation: Moving from Perspectives to Principles
Edited by David Lindenmayer and Richard Hobbs
ISBN: 978-1-4051-5914-2 Paperback; 608 pages; December 2007

Conservation Science and Practice Series

Reintroduction Biology: Integrating Science and Management

Edited by

John G. Ewen, Doug P. Armstrong, Kevin A. Parker and Philip J. Seddon

WILEY-BLACKWELL

A John Wiley & Sons, Inc., Publication

Library of Congress Cataloguing-in-Publication Data has been applied for.

A catalogue record for this book is available from the British Library.

This book is published in the following electronic formats: ePDF 9781444355802; Wiley Online Library 9781444355833; ePub 9781444355819; Mobi 9781444355826

Set in 10.5/12.5 pt Minion by Laserwords Private Limited, Chennai, India
Printed and bound in Malaysia by Vivar Printing Sdn Bhd

1   2012

Contents

Contributors

Karina Acevedo-Whitehouse, Institute of Zoology, Zoological Society of London, Regents Park, London NW1 4RY, United Kingdom and Facultad de Ciencias Naturales, Universidad Autonoma de Queretaro, Av. de las Ciencias S/N, Juriquilla, Queretaro 76 230, Mexico. Email: karina.acevedo. whitehouse@uaq.mx

Maurice R. Alley, Institute of Veterinary, Animal and Biomedical Sciences, Massey University, PB 11 222, Palmerston North, New Zealand. Email: M.R.Alley@massey.ac.nz

Doug P. Armstrong, Ecology Building 624, Massey University, PB 11 222, Palmerston North, New Zealand. Email: D.P.Armstrong@massey.ac.nz

Iris Biebach, Institute of Evolutionary Biology and Environmental Studies, University of Zürich, Winterthurerstrasse 190, CH-8057 Zürich, Switzerland. Email: iris.biebach@ieu.uzh.ch

Rachel Bristol, Durrell Institute of Conservation and Ecology, University of Kent, Kent CT2 7NR, United Kingdom and Nature Seychelles, PO Box 1310, Victoria, Mahé, Republic of Seychelles. Email: rmb33@kent.ac.uk

Claudia Carraro, Facoltà di Medicina Veterinaria, Dipartimento di Scienze Sperimentali Veterinarie, Università degli Studi di Padova, Padova, Italia. Email: claudia.carraro@gmail.com

Rohan H. Clarke, School of Biological Sciences, Monash University, Melbourne, Victoria 3800, Australia. Email: rohan.clarke@monash.edu

Molly J. Dickens, GIGA Neurosciences, University of Liège, 1 avenue de l'Hôpital (Bat. B36), B-4000 Liège, Belgium. Email: molly.dickens@ ulg.ac.be

John G. Ewen, Institute of Zoology, Zoological Society of London, Regents Park, London NW1 4RY, United Kingdom. Email: john.ewen@ioz.ac.uk

Steven R. Ewing, Royal Society for the Protection of Birds, Sandy, Bedfordshire SG19 2DL, United Kingdom. Email: steven.ewing@rspb.org.uk

Jim J. Groombridge, Durrell Institute of Conservation and Ecology, University of Kent, Kent CT2 7NR, United Kingdom. Email: J.Groombridge@ kent.ac.uk

Paquita E.A. Hoeck, Institute of Evolutionary Biology and Environmental Studies, University of Zürich, Winterthurerstrasse 190, CH-8057 Zürich, Switzerland and Institute for Conservation Research, San Diego Zoo Global, 15600 San Pasqual Valley Road, Escondido, CA 92027. Email: paquita.hoeck@ieu.uzh.ch

John Innes, Landcare Research New Zealand Ltd, Private Bag 3127, Hamilton 3240, New Zealand. Email: InnesJ@landcareresearch.co.nz

Ian G. Jamieson, Department of Zoology, University of Otago, PO Box 56, Dunedin 9016, New Zealand. Email: ian.jamieson@otago.ac.nz

Carl G. Jones, Durrell Wildlife Conservation Trust, Les Augrès Manor, La Profonde Rue, Trinity, Jersey, Channel Islands JE3 5BP and Mauritian Wildlife Foundation, Grannum Road, Vacoas, Mauritius. Email: carlgjones@btinternet.com

Lukas F. Keller, Institute of Evolutionary Biology and Environmental Studies, University of Zürich, Winterthurerstrasse 190, CH-8057 Zürich, Switzerland. Email: lukas.keller@ieu.uzh.ch

Robert C. Lacy, Department of Conservation Science, Chicago Zoological Society, Brookfield, Illinois 60 513, United States of America. Email: rlacy@ix.netcom.com

Pascaline Le Gouar, NIOO-KNAW, Dutch Centre for Avian Migration & Demography, Postbus 40, 6666 ZG Heteren, the Netherlands and Université Pierre et Marie Curie, Paris, France and Université Rennes 1, UMR 6553 UR1-CNRS 'EcoBio', station biologique, 35 380 Paimpont, France. Email: pascaline.legouar@univ-rennes1.fr

Tim G. Lovegrove, Auckland Council, Auckland, New Zealand. Email: Tim.Lovegrove@aucklandcouncil.govt.nz

Michael A. McCarthy, The Australian Research Centre for Urban Ecology (ARCUE), c/- The School of Botany, The University of Melbourne, Parkville, Victoria 3010, Australia. Email: mamcca@unimelb.edu.au

Don V. Merton, deceased.

Jean-Baptiste Mihoub, Université Pierre et Marie Curie, UMR 7204 MNHN-CNRS-UPMC 'Conservation des espèces, restauration et suivi des populations', 61 rue Buffon, 75 005 Paris, France. Email: mihoub@mnhn.fr

James D. Nichols, Patuxent Wildlife Research Center, U.S. Geological Survey, 12 100 Beech Forest Road, Laurel, MD 20 708-4017, United States of America. Email: jnichols@usgs.gov

Patrick E. Osborne, School of Civil Engineering and the Environment, University of Southhampton, Highfield, Southhampton SO17 1BJ, United Kingdom. Email: P.E.Osborne@soton.ac.uk

Kevin A. Parker, Ecology and Conservation Group, Institute of Natural Sciences, Massey University, PB 102904, NSMC, Auckland 0745, New Zealand. Email: k.parker@massey.ac.nz

Claire Raisin, Durrell Institute of Conservation and Ecology, University of Kent, Kent CT2 7NR, United Kingdom. Email: cr200@kent.ac.uk

Michelle H. Reynolds, USGS Pacific Island Ecosystems Research Center, Kilauea Field Station, PO Box 44, Hawaii National Park, Hawaii, HI 96 718, United States of America. Email: MReynolds@usgs.gov

David S. Richardson, Nature Seychelles, Victoria, Mahé, Republic of Seychelles and School of Biological Sciences, University of East Anglia, Norwich Research Park, Norwich NR4 7TJ, United Kingdom. Email: David.Richardson@uea.ac.uk

Michael C. Runge, Patuxent Wildlife Research Center, U.S. Geological Survey, 12 100 Beech Forest Road, Laurel, MD 20 708-4017, United States of America. Email: mrunge@usgs.gov

Anthony W. Sainsbury, Institute of Zoology, Zoological Society of London, Regents Park, London NW1 4RY, United Kingdom. Email: tony.sainsbury@ioz.ac.uk

François Sarrazin, Université Pierre et Marie Curie, UMR 7204 MNHN-CNRS-UPMC 'Conservation des espèces, restauration et suivi des populations', 61 rue Buffon, 75 005 Paris, France. Email: sarrazin@mnhn.fr

Philip J. Seddon, Department of Zoology, University of Otago, PO Box 56, Dunedin 9016, New Zealand. Email: philip.seddon@otago.ac.nz

W. Maartin Strauss, Department of Environmental Sciences, UNISA, South Africa. Email: wm_strauss@hotmail.com

Kirsty Swinnerton, Island Conservation, Center for Ocean Health, 100 Shaffer Road, Santa Cruz, CA 95 060, United States of America. Email: kirsty.swinnerton@islandconservation.org

Rosie Woodroffe, Institute of Zoology, Zoological Society of London, Regents Park, London NW1 4RY, United Kingdom. Email: rosie.woodroffe@ioz.ac.uk

Memorium of Don Merton

Donald Vincent Merton (1939–2011), a reintroduction pioneer

Don Merton died on 10 April 2011, and although many of us had known for a while that he was fighting terminal cancer the news of his passing still came as a blow, since Don, or 'Mertie' to his friends and colleagues, had been an important figure in bird restoration for nearly fifty years and he had been a mentor to many of us. Don was a consummate field man and many of the techniques in bird management that he pioneered have now become accepted practice. The conservation programmes he developed and drove in New Zealand and on Indian Ocean islands have become iconic case studies of what can be achieved.

Don played a leading role in saving many New Zealand species and is best known for his work with the North Island saddleback *Philesturnus rufusater*, South Island saddleback *P. carunculatus*, Chatham Island black robin *Petroica traversi* and the kakapo *Strigops habroptilus*, but he also helped, advised and inspired work with a range of other species. Elsewhere he worked with the noisy scrub bird *Atrichornis clamosus* in Western Australia, the echo parakeet *Psittacula eques* in Mauritius and the Seychelles magpie robin *Copsychus sechellarum*.

The first significant bird conservation work that Don was involved with was the translocation of saddlebacks in the 1960s and this continued in subsequent decades. The North Island saddleback had been reduced to just one population on Hen Island due to predation by rats, cats and stoats. Don headed up a translocation programme for this species and successfully established new populations on predator-free islands. This work became a prelude to work on the South Island saddleback, which had become doomed following the invasion of its last island home, Big South Cape Island, by rats. Translocations to other islands were largely successful and set the scene for further translocations of this, and other species, to predator-free islands. The North Island saddleback has a population of about 6000 birds on at least 14 predator-free islands and the South Island saddleback has a population of about 2000 on 17 islands (Ballance & Merton, 2007).

This work demonstrated the value of translocating species on to predator-free islands and Don was eager to communicate the results. He gave an important paper at the 1972 conference on 'Breeding Endangered Species in Captivity' and put over the idea, novel at the time, that instead of having to breed highly threatened species in captivity they could be marooned on islands where they may be able to establish viable populations (Merton, 1975).

As a result of attending this conference Don became aware that there were many techniques being developed in Europe and North America that could be applied to managing bird populations and he wanted to discover how these could be applied to the threatened birds of New Zealand. In 1973 he spent three months in Hawaii, mainland USA and Europe visiting conservation programmes. Among these he visited the Endangered Wildlife Unit of the United States Bureau of Sports Fisheries and Wildlife at Patuxent in Maryland, the Peregrine Fund at Cornell University, the Wildfowl and Wetland Trust in Slimbridge and the Durrell Wildlife Conservation Trust in Jersey. He was able to see first hand the work being conducted on endangered species such as the Hawaiian goose *Branta sandvicensis*, whooping crane *Grus americana* and peregrine falcon *Falco peregrinus*. It was on this trip that Don realized the great contribution aviculture techniques could play; techniques that bird breeders used to encourage their birds to breed could be applied to free living birds. As a result of his trip Don was able to take back to New Zealand much important information on bird management. This included ideas on the use of captive breeding and the value of using related species as 'analogues' on which to develop skills. He was able to compile information on more sophisticated techniques for trapping, handling and transporting wild birds. Don also learned about captive diets, soft-release protocols and the new technology of radio telemetry. Perhaps most usefully Don was able to gather information on avian paediatrics such as egg harvesting and incubation, double clutching, the fostering and cross-fostering of eggs and young, and hand-rearing techniques. These techniques Don would use repeatedly in his work on the Chatham Island black robin, kakapo and other species.

In 1980 Don started the recovery of the Chatham Island black robin. The species had declined to five individuals and extinction seemed imminent. The black robin was restored by using a range of intensive management techniques including clutch and brood manipulations, close guarding and supplemental feeding (see Jones & Merton, this volume, Chapter 2). This is the species for which Don is most famous and the story has been well documented (Morris

& Smith, 1988; Butler & Merton, 1992; Ballance & Merton, 2007). There were about 200 black robins by 2011.

The species that he had the longest and most continuous involvement with was the kakapo, the 'big budgie' as he affectionately called it. He had been involved with this species for five decades and played a huge role in developing this programme and providing the vision for the species recovery, and saw it increase from 51 individuals in 1996 to 100 in 2009. The honour of naming the hundredth bird was given to Don, and he named it Te Atapo, meaning 'the dawn after the night', because of its significance to the recovery effort, and he went on to point out 'the 100th bird symbolizes a very significant milestone in the ongoing struggle spanning more than a century of literally blood, sweat and tears by countless dedicated individuals to save one of New Zealand's – and the world's – most remarkable and iconic birds'.

All the kakapo now exist on predator-free islands outside their natural range since their historic habitats are now badly degraded and inhabited by exotic mammalian predators, making them largely unsuitable. Don worked with others to develop the techniques for clearing islands of exotic predators and then restoring them so they would be suitable for endangered birds like the kakapo (Butler, 1989; Ballance & Merton, 2007; Ballance, 2010).

The use of predator-free islands is a recurring theme in New Zealand conservation and has become important because of the appreciation of the damage that exotic mammals can cause. Don had witnessed the devastating impact black rats *Rattus rattus* had on Big South Cape Island in 1964–1965 when they caused the extinction of the Stewart Island snipe *Coenocorypha iredalei*, Stead's bush wren *Xenicus longipes variabilis* and the greater short-tailed bat *Mystacina robusta*. This story is told in Don's (auto)biography (Ballance & Merton, 2007) and in the book *Wild South* (Morris & Smith, 1988). The experience of seeing the collapse of the bird and other communities on the island had a profound influence on Don, and he often talked about it. He was incredulous that some scientists were myopic regarding the negative impact that rats (and other exotic mammals) were having on endemic island faunas (Merton, 1977). Together with Ian Atkinson and Brian Bell they were the first to realize just how destructive rats were to bird populations, and all three became major campaigners for the eradication of exotic rats from islands. Don led rat eradication projects on Mauritius and Seychelles as well as contributing to a range of exotic mammal eradications in New Zealand and elsewhere. On Mauritius he and a small team of helpers eradicated rabbits off Round

Island, which was at the time the largest island from which rabbits had been eradicated (Merton, 1987). Later Don went on to develop the management plan for the island (Merton *et al.*, 1989), and this plan has been successfully implemented.

A field trip with Don was an education. He planned it with meticulous attention to detail, building in contingencies for problems that may occur. For days before the trip he would work through long lists of items that were required. All provisions were carefully quarantined and packed and sealed in labelled boxes. The field camps that Don set up on Round Island and in the forest, when working on echo parakeets, were more organized than any that I or my fellow British and Mauritian biologists had experienced. So fastidious was Don in running field camps that his fellow 'kiwis' gave him the soubriquet of 'auntie'. Don was indeed a joy to share a camp site with and when we would remark that he ran a good camp he would reply that 'anybody can rough it in the field but effective field work is a product of being organized and comfortable'.

Don always had focus and stamina; he would be up earlier that everyone else and by the time the rest of the team had arisen Don would have breakfast ready, the day planned and be raring to go. In the field he had more energy than many half his age and I have memories of trying (unsuccessfully) to keep up with him. Don was, however, no paragon and could be stubborn and introspective, but these were traits that made the man determined and focused. Often when faced with a seemingly unsolvable problem he would become quiet and withdrawn while he weighed up the situation and thought through a strategy to find a solution. It was remarkable how often Don found the answer; he was very intuitive and could also draw upon his considerable experiences to solve problems big and small.

Most field conservationists spend the early part of their careers in the field and then when they reach middle age take on more and more management and office-based responsibilities. This was not Don. He spent virtually his entire career doing field-based work and resisted promotion that would take him away from hands-on conservation, for this was where he felt most comfortable and was most effective. In all he spent more than ten years of his career living in tents or field huts, usually in remote locations. In addition to this he spent several years working on the Indian Ocean islands of Christmas, Seychelles and Mauritius.

Don was never happier than when he was tinkering with birds. He used to enjoy just being with them, watching and soaking up all types of information about how they behaved and reacted. He was empathetic and had a deep understanding that arose from years of intimate contact. I remember when I showed him some recently fledged hand-reared Mauritius fodies *Foudia rubra*; the birds were very tame and readily landed on him and climbed and fluttered up and down his arms and on to his head, inspecting his glasses and probing his ears. To most biologists they are small brown passerines but to Don they were special; 'Jeez, what beauts!' he exclaimed, 'what wonderful little birds.'

When working with birds, whether setting up a supplemental feeding station or placing a nest-box, he would pay careful attention to every little detail to make it attractive to the birds. Don's attention to detail influenced all he did, and was particularly evident in his approach to trapping. I remember clearly how, on Mauritius in 1992, we had a problem feral cat that was killing pink pigeons *Nesoenas mayeri* in our last wild population; at the time we had less than twenty birds left and the cat was making a significant dent in the numbers. We unsuccessfully tried all sorts of traps to catch it. Don visited Mauritius and I asked him if he could help us. He spent a day in the field studying the situation. When he had a good idea how the cat would be using the landscape and where it would be likely to travel, he carefully set a series of traps that were well camouflaged and baited with fish. Next morning Don got up early to go and check his traps and later greeted us at breakfast, beaming with delight and holding up the dead cat he had caught!

Don was not a trained scientist but he was nevertheless a careful and systematic worker. He realized that science was the most powerful tool that could provide knowledge to inform conservation management. Although he did not see publications as an end goal he did publish important findings and techniques, including the description of lekking in the kakapo, the only parrot that is known to do so (Merton *et al.*, 1984).

Don was a modest man who never bragged about his achievements but was, however, always eager to talk and share his knowledge with anyone who was interested. He wanted to know what others had to contribute, and travelled widely visiting bird conservation projects and attended and spoke at many international conferences. Without intending to, he was the greatest ambassador of conservation that New Zealand has ever had. A very generous man by nature he readily gave his time and helped projects outside New Zealand, including projects in Western Australia, Christmas Island, Mauritius, Seychelles and Fiji. Don did much work abroad, often at his own expense, and would usually do it during his annual leave or he would take leave without pay. On Mauritius he helped develop the echo parakeet restoration project and for over 25 years he provided advice and guidance. He was a wonderful correspondent and would always be happy to respond to any queries and kept in close touch with projects he had helped.

Don was an optimist and believed that most critically endangered species are restorable and that in the future we would be restoring whole communities and rebuilding ecosystems. He showed us the way forward and the species conservation programmes he was involved with have matured into projects restoring suites of species and their island habitats.

During his more contemplative periods he would lament on just how short our human life span was compared to how long it was going to take to restore species with long generation times, like the kakapo, and to restore whole island systems. He would urge that we had to do what we could to develop the next generation of conservation leaders to continue the work. These were not just words since Don put these views into action and on his various trips to Mauritius and Seychelles he took along young conservationists at the beginning of their careers to give them the opportunity for broader conservation experience. Subsequently, in Mauritius we have had a whole stream of 'kiwi' conservationists who have been sent by Don, helping with the work while at the same time adding to their experiences and skills.

Don of course did not do his conservation work alone, and he would have been unable to develop the ideas of bird management, predator eradication and ecosystem reconstruction were it not for the intellectual and practical input of his colleagues. Notable among these have been Ian Atkinson, Brian Bell, Dick Veitch and others who together have made New Zealand the most progressive country for endangered species management. Don openly acknowledged the support of his wife Margaret and son David who often accompanied him on field trips.

Looking back at Don's achievements, he was the person who developed the techniques of intensively managing critically endangered wild birds. Don showed us that most species are saveable, and that species work drives the rebuilding of ecosystems. He was able to take the ideas of aviculturists and field conservationists, craft and combine them, and apply them to wild populations. In the coming decades more and more species are going to have to be managed in the long term if they are to survive and we can thank Don for showing us how to do this.

References

Ballance, A. (2010) *Kakapo: rescued from the brink of extinction.* Craig Potton Publishing, Nelson, NZ.

Ballance, A. & Merton, D.V. (2007) *Don Merton: the man who saved the black robin.* Reed Publishing (NZ) Ltd, Auckland, NZ.

Butler, D. (1989) *Quest for the Kakapo. The full story of New Zealand's most remarkable bird.* Heinemann Reed, Auckland, NZ.

Butler, D. & Merton, D.V. (1992) *The Black Robin: saving the world's most endangered bird.* Oxford University Press, Auckland, NZ.

Merton, D.V. (1975) The saddleback: its status and conservation. In *Breeding Endangered Species in Captivity*, ed. R.D. Martin. Academic Press, London, UK.

Merton, D.V. (1977) Controlling introduced predators and competitors on islands. In *Endangered Birds*, ed. S.A. Temple, pp. 121–128. University of Wisconsin Press, Madison, Wisconsin, USA.

Merton, D.V. (1987) Eradication of rabbits from Round Island, Mauritius: a conservation success story. *Dodo: Journal of the Jersey Wildlife Preservation Trust*, 24, 19–44.

Merton, D.V., Atkinson, I.A.E., Strahm, W. *et al.* (1989) *A Management Plan for the Restoration of Round Island, Mauritius.* Jersey Wildlife Preservation Trust, UK.

Merton, D.V., Morris, R.B. & Atkinson, I.A.E. (1984) Lek behaviour in a parrot; the kakapo *Stigops habroptilus* of New Zealand. *Ibis*, 126, 277–283.

Morris, R. & Smith, H. (1995) *Wild South: saving New Zealand's endangered birds.* Random House, NZ.

Carl G. Jones
May 2011

Foreword

Reintroductions have come far and fast over the last thirty years. Starting with large vertebrates, often charismatic and with obvious sources of decline, the IUCN Reintroduction Specialist Group database is now witness to a host of species, plant and animal, vertebrate and invertebrate, being returned to the wild around the world. As well as the quantitative increase, the quality of well-designed and planned releases with subsequent monitoring and fine-tuning is also striking. Given that any reintroduction is a step into the unknown, and involves practical issues of the selection, handling, welfare and management of the released individuals, each effort must be a marriage between science and responsible management. As these are often poor bedfellows, this book is a timely corrective that demonstrates how each can inform the other in the interests of successful reintroductions.

Through my own experiences I have witnessed the changing face of reintroductions. Planning in the late 1970s for the return of the Arabian oryx into the deserts of Oman after an extinction of only twenty years, we virtually assumed the arid grasslands were there and adequate for a grazer with only six species of grass to choose from. Disease considerations were limited to ensuring the oryx imported from California brought no blue tongue with them. However, the lesson in this instance was that while ecological conditions were adequate for initial success, human social factors in the form of envy-based capture for sale reversed the situation.

Fast-forwarding many years, I have been fortunate to be involved, only in an oversight role, with the magnificent population restorations on Mauritius and its islets. Chapter 2 covers the dramatic events there, involving habitat restoration, the use of model and analogue species, the intensive care unit approach to bird species with numbers in the tens, marooning on islets and the restoration of ecological functions: these activities are prominent weapons in the armoury for small population management today. The level of individual management used here, based on a mixture of natural history observation, and a profound understanding of species biology and psychology has led to blurring the boundary between conventional in situ and ex situ conservation. This is a welcome trend as more and more of the world's species will require some level of management support, of which reintroductions will play a significant role. In view of the pre-eminent expertise in New Zealand for both the removal of exotic and invasive species and the restoration and reintroduction of native fauna and flora, it is pleasingly appropriate that this book was conceived in that country and makes such good use of its experiences.

The evolution of reintroduction practice covers both our attention to its components and also to our perception, based on ecological understanding. The former is evidenced in the attention to genetic aspects given here (Chapters 11 to 13). The role of disease and parasites, covering not just the undesirable species whose release must be avoided but others as part of a released individual's personal community and as natural ecological factors, is important (Chapters 9 and 10). So, too, is the greater attention to the stresses faced by animals before, during and after release, with our increasing understanding that these factors can impact the performance in the wild of both individuals and populations, and may indeed be critical factors in a successful reintroduction (Chapter 4).

Our perceptions have also changed with greater ecological understanding. The simplistic view that a niche remains vacant and with fixed borders following a species' extinction is hopelessly naive. Nature abhors a vacuum, and any returning species has to fight to develop its own new niche. How can we predict what this niche will be? This is why analysis of habitat, used here in the broad sense of both a species place and landscape and also its biotic interactions, is so critical. Chapter 3 deals comprehensively with habitat suitability and selection (and these considerations for invertebrates with complex life cycles will be far more onerous), and should persuade anyone – manager or the public – that the last place a species occupied may rarely be the best place to return it to first.

A constant message across these chapters is that reintroduction has to face uncertainty and expect and deal with change. Therefore, the step into the unknown that is a reintroduction has to be based on techniques of dealing with ecological ignorance, identifying and assessing relative risks, specifying alternative outcomes in advance with indicators, to be followed by adaptively tweaked management. This is a major focus of Chapters 5 to 8.

A responsible reintroduction is already a tall order under current conditions. What for the future of reintroduction? We all know that biodiversity faces acute challenges over much of the globe: species are under ecological siege through loss of habitat and the pressures of invasives, and many species are already responding to the impacts of climate change in various ways. Chapter 1 mentions the concept of assisted colonization as a form of conservation introduction. Underlying such translocations would be uncertainties in the face of major change, and they might be seen as heretical through defying the convention that species should not be released beyond an inferred historic range; yet they must be explored and considered seriously, especially as a large proportion of the earth's surface will enjoy novel climates for which novel ecosystems and communities of species must either assemble or be constructed deliberately through management. To ensure our biodiversity survives, we may have to shuffle the cards in the pack, but we must still keep all 52.

The best reintroduction practice is ideally placed to make major contributions here. By combining cutting edge science and responsible management, this book should be a beacon for biodiversity conservation in the future.

Mark R. Stanley Price
Senior Research Fellow,
The University of Oxford, WILDCRU,
The Recanati-Kaplan Centre, Tubney House,
Abingdon Road, Tubney OX13 5QL, UK
and
Conservation Fellow, Al Ain Wildlife Park & Resort,
PO Box 1204, Al Ain,
Abu Dhabi, UAE

Preface

Like the biodiversity we are trying to conserve, this book is a product of evolution. In New Zealand the current century has seen a transition from reintroductions being largely run by government conservation managers to being often run by community conservation groups. The origin of this book came from discussions in 2005 about how to best facilitate these community-led reintroductions. The main problems being grappled with were: (1) how to decide which species are suitable for what sites; (2) how to plan and undertake the translocation process; (3) what sort of disease screening is needed; (4) what post-release monitoring is needed; and (5) how to avoid genetic problems. The community groups needed to address these issues in the proposals required by the New Zealand Department of Conservation, so were seeking advice not only on what to do but also on the basic theory underlying such advice. This presented a challenge to experienced reintroduction practitioners, but also an opportunity to think more deeply about the basis for the decisions being made.

Our original idea was to hold a workshop for community groups, but this idea morphed into an international symposium on Reintroduction Biology that was held in London in May 2008. Although the nature and location of the meeting represented a major shift from the original idea of liaison with local community groups, we felt the issues in New Zealand were not unique to the region and that an international perspective would produce benefits for reintroduction practice globally. We took pains to ensure that the symposium was attended by practitioners working at the coalface of reintroduction, as well as by researchers from a range of relevant disciplines.

This book in turn has evolved from the 2008 Symposium, but it is definitely not a conference proceedings. The topics have expanded to encompass key and emerging issues, and as a consequence many new authors have come on board, and a further three years of literature are covered. The general themes remain the same as those identified in 2005. However, because we aimed to provide a clear theoretical underpinning for decisions involved in reintroduction programmes, issues are discussed within a broad context. For example, offering sensible monitoring advice requires an understanding of how the data will be used to meet defined objectives; hence the monitoring chapter is presented as part of a broader section including population modelling and adaptive management. The organization of the book is issues-based, so we specifically avoided having chapters based on case studies or focusing on particular taxa. The one exception is Chapter 2, which summarizes insights from bird reintroductions from two highly successful and pioneering reintroduction practitioners known for taking aggressive interventions based on clear logic. We acknowledge that our largely 'taxonomically neutral' approach means the examples given are a reflection of the dominance of bird and mammal studies in reintroduction research, with relatively few plant examples. It is timely, therefore, that the publication of *Plant Reintroduction in a Changing Climate: promises and perils*, eds. J. Maschinski & K. Haskins, Island Press) coincides closely with that of this book, as the two volumes should be complementary.

The first aim of the book is to further advance the field of reintroduction biology beyond the considerable progress made since the formation of the IUCN/SSC Reintroduction Specialist Group (RSG) in 1988. The main impetus for the RSG's formation was to facilitate the planning and monitoring of reintroduction projects; hence this is the focus of the 1998 *IUCN Guidelines for Reintroductions*. While the need for planning and monitoring is as important as ever, we propose raising the bar on reintroduction practice by advocating a strategic approach where all actions (including monitoring) are guided by explicit theoretical frameworks based on clearly defined objectives.

The second aim is to break down the perceived dichotomy between research and management, which can also be represented as a dichotomy between theory and practice. We cannot see how it is in any way practical to manage anything in the absence of explicit theory. We also believe that conservation research is best done in conjunction with management, so do not see these as distinct activities. Most of the authors of this book have considerable experience in planning and undertaking reintroductions as well as producing research papers, and their contributions reflect this. Similarly, we expect that this book will most appeal to people wanting to bridge the research–management gap, such as conservation managers wanting to expand their thinking about reintroduction-related decisions or researchers who seek to make useful applied contributions to reintroductions rather than simply publishing papers.

This entire project could not have occurred without the support and contribution of many. First, we thank all the reintroduction biologists whose research the book has drawn on. We hope that your work has been captured accurately and apologize for inevitably missing some relevant publications. The burgeoning of the field is a healthy sign, but also challenging to encapsulate in a single volume. Second, we thank the IUCN/SSC Reintroduction Specialist Group and the Zoological Society of London for helping fund the London Symposium, and thanks also to all those who attended and spoke. Third, we thank the contributors to this volume. Wiley-Blackwell has been very patient and supportive, and we particularly thank Ward Cooper and Kelvin Mathews. This project was undertaken while JGE was funded on an RCUK fellowship.

Finally, we wish to acknowledge the contribution of two giants who have sadly left us recently. Devra Kleiman (1942–2010) and Don Merton (1939–2011) were true pioneers in the field of reintroduction biology. Devra was an academically trained biologist who creatively applied her research skills to solving practical problems in captive breeding and reintroduction. Don was a wildlife manager whose ability to combine on-the-ground experience with theoretical acumen allowed him to play an instrumental role in saving three species. They both managed to combine careful thinking with rapid action, facilitated by an ability to engage with and inspire the people they worked with. They are missed. We are humbled to have Don among the authors of this volume, and dedicate this book to his memory.

John G. Ewen
Doug P. Armstrong
Kevin A. Parker
Philip J. Seddon
London, Palmerston North, Auckland and Dunedin
May 2011

1

Animal Translocations: What Are They and Why Do We Do Them?

Philip J. Seddon[1], W. Maartin Strauss[2] and John Innes[3]

[1]Department of Zoology, University of Otago, New Zealand
[2]Department of Environmental Sciences, UNISA, South Africa
[3]Landcare Research, Hamilton, New Zealand

'Translocation is now well entrenched as a conservation tool, with the numbers of animals being released in reintroduction and re-enforcement projects increasing almost exponentially each year.'

Page 23

Introduction

For as long as people have been moving from one place to another, which is as long as humans have been 'human', animals and plants have been moved with them, often hidden, unnoticed or ignored, but also as valued cargo. These so-called 'ethnotramps' include economically and culturally favoured species such as deer, macaque, civets, wallabies, cassowaries and wild-caught songbirds that were commonly carried around with humans (Heinsohn, 2001).

The variety of animals shown to have been translocated by prehistoric human colonists has been described as 'astonishing', with archaeological evidence of numerous and widespread human-mediated introductions as far back as tens of millennia, during the Pleistocene (Grayson, 2001). For example, it has been shown that people moved wild animals from the New Guinea mainland

Reintroduction Biology: Integrating Science and Management. First Edition.
Edited by John G. Ewen, Doug P. Armstrong, Kevin A. Parker and Philip J. Seddon.
© 2012 Blackwell Publishing Ltd. Published 2012 by Blackwell Publishing Ltd.

to and between islands to the east and west over at least the past 20 000 years, for food and trade items as humans expanded their distribution and sought to retain access to animals whose habits were already known to them (White, 2004). It was during the Holocene (from ~11 000 years before the present), however, that the translocation of non-domesticated animals into novel habitats became one of the most significant human impacts on native animal populations (Kirch, 2005).

Clearly there are many reasons to translocate animals and some broad-scale classifications have been proposed, for example to distinguish between conservation translocations and those for commercial or amenity values (Hodder & Bullock, 1997), and along the way the terminology relating to translocations has become confused, contradictory and ambiguous. In this chapter we provide a framework for classifying the different motivations for animal translocation. We propose a simple decision tree that will enable conservation managers to categorize easily the different types of translocation, from reintroductions to assisted colonizations, and standardize the terminology applied in the species restoration literature. Throughout this chapter terms given in *italics* are defined in Box 1.1.

Box 1.1 **Glossary and definitions**

Analogue species   Closely related form that could be used as an *ecological replacement* for an extinct species (Parker *et al.*, 2010)

Assisted colonization *Translocation* of species beyond their natural range to protect them from human-induced threats, such as climate change (Ricciardi & Simberlof, 2009a)

Assisted migration   Synonym for *assisted colonization*

Augmentation   Synonym for *re-enforcement*

Benign introduction Synonym for *conservation introduction*

Biological control   Intentional use of parasitoid, predator, pathogen, antagonist or competitor to suppress a pest population (Hoddle, 2004)

Classical biocontrol The introduction of exotic natural enemies to control exotic pests (Thomas & Willis, 1998)

Conservation introduction	An attempt to *establish* a species, for the purposes of conservation, outside its recorded distribution but within an appropriate habitat and ecogeographical area (IUCN, 1998)
Ecological replacement	*Conservation introduction* of the most suitable extant form to fill the ecological niche left vacant by the extinction of a species (Seddon & Soorae, 1999)
Ecological restoration	The process of assisting the recovery of an ecosystem that has been degraded, damaged or destroyed (SER, 2004)
Establishment	Survival and successful breeding by founder individuals and their offspring; this is a prerequisite for, but not a guarantee of, population *persistence*
Follow-up translocation	Where one or more additional translocations are conducted to supplement an initial population established by *reintroduction* (Armstrong & Ewen, 2001)
Introduction	Intentional or accidental dispersal by a human agency of a living organism outside its historically known native range (IUCN, 1987)
Managed relocation	Synonym for *assisted colonization*
Marooning	*Translocation* to a predator-free offshore island
Persistence	The likelihood of population decline or extinction over some appropriate taxon-specific time frame
Re-enforcement	Addition of individuals to an existing population of conspecifics (IUCN, 1998)
Re-establishment	Synonym for *reintroduction* that implies the *reintroduction* has resulted in *establishment* (IUCN, 1998)

Rehabilitation	The managed process whereby a displaced, sick, injured or orphaned wild animal regains the health and skills it requires to function normally and live self-sufficiently (IWRC, 2009)
Reintroduction	Intentional movement of an organism into a part of its native range from which it has disappeared or become extirpated in historic times (IUCN, 1987)
Reintroduction biology	Research undertaken to improve the outcomes of *reintroductions* and other *translocations* (Armstrong & Seddon, 2008)
Relocation	Synonym for *translocation*
Restocking	Synonym for *re-enforcement*
Restoration ecology	The science upon which the practice of *ecological restoration* is based (SER, 2004)
Species restoration	The application of any of a wide range of management tools, including *translocation*, that aim to improve the conservation status of wild populations
Subspecific substitution	A subset of *ecological replacement* where the replacement taxon is a subspecies (Seddon & Soorae, 1999)
Supplementation	Synonym for *re-enforcement*
Translocation	Movement of living organisms from one area with free release in another (IUCN, 1987)
Transplantation	Synonym for *translocation*

The translocation spectrum

Seddon (2010) defined a conservation *translocation* spectrum, ranging from *reintroductions* through to forms of *conservation introduction*. Figure 1.1 broadens the scope and provides a framework for considering all motivations for moving wild animals. The first, simple, bifurcation divides movements into those that are accidental or incidental and those that are intentional

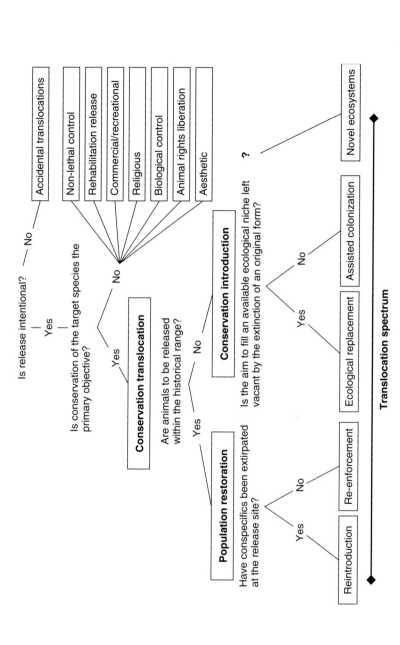

Is release intentional?

— No — Accidental translocations

— Yes —

Is conservation of the target species the primary objective?

— No —
- Non-lethal control
- Rehabilitation release
- Commercial/recreational
- Religious
- Biological control
- Animal rights liberation
- Aesthetic

— Yes —

Conservation translocation

Are animals to be released within the historical range?

— Yes — **Population restoration**

Have conspecifics been extirpated at the release site?

— Yes — Reintroduction
— No — Re-enforcement

— No — **Conservation introduction**

Is the aim to fill an available ecological niche left vacant by the extinction of an original form?

— Yes — Ecological replacement
— No — Assisted colonization

? — Novel ecosystems

Reintroduction | Re-enforcement | Ecological replacement | Assisted colonization | Novel ecosystems

Translocation spectrum

Figure 1.1 **The translocation spectrum.**

(Figure 1.1). Strictly speaking, accidental movements of wild animals are not translocations in the sense intended by the 1987 definition (IUCN, 1987). This IUCN definition, however, lacks mention of intent contained within a later, but confusing, redefinition (IUCN, 1998) that erroneously appears synonymous with re-enforcement and wild-to-wild movements. We take it that *translocations* are the deliberate and mediated (IUCN, 1998) movement of organisms, from any source, captive or wild, from one area to free release in another (IUCN, 1987). Thus *translocation* is the overarching term.

Not all translocations relate to the conservation of the species being moved. The next division in our framework therefore asks the question: is conservation of the target species the primary concern? (Figure 1.1). A split between conservation and non-conservation is in some senses simplistic and naive. Multiple, sometimes indirect, conservation benefits may accrue through translocations for, for example, recreational, commercial or wildlife rehabilitation motivations, not the least being opportunities for increased public engagement with nature and the enhanced public support for conservation measures that can arise from this engagement. Nevertheless, it is useful to make the distinction around primary concerns as many translocations may have multiple objectives and not uncommonly enhancement of the conservation status of the species may exist as a secondary goal.

Non-conservation translocations

There are at least seven types of translocation for which conservation is not the primary aim (note that species conservation may be an associated aim and protection of individual animals of endangered species may be a primary aim): non-lethal management of problem animals, commercial and recreational, biological control, aesthetic, religious, wildlife rehabilitation and animal rights activism. One of the characteristics of many of the non-conservation translocations is that they are introductions, with the sometime exception of non-lethal management and wildlife rehabilitation.

Non-lethal management of problem animals

As urban, suburban and agricultural landscapes spread, and where natural populations of wildlife species recover or expand, the potential for human–wildlife conflicts increases. In the United States, for instance, many states have Nuisance Wildlife Control Operator (NWCO) programmes in response to increased complaints of urban wildlife conflicts (O'Donnell & DeNicola, 2006). The most common forms of conflict are predation of livestock (Bradley *et al.*, 2005), attacks on humans and their domestic pets (Goodrich & Miquelle, 2005) and damage to property (Gammons *et al.*, 2009; Herr *et al.*, 2008). In large part due to public attitudes, lethal management of so-called problem wildlife is not a favoured option and instead a standard method of dealing with problem individuals is to capture and translocate them away from the focal point of conflict. The numbers of animals involved can be significant; for example, in 1994 in Illinois alone NWCO permittees moved > 18 000 animals, including > 13 000 raccoons (*Procyon lotor*), squirrels (*Sciurus* spp.) and bats, and throughout the US some hundreds of thousands of animals are moved to mitigate conflicts (Craven *et al.*, 1998). By far the majority of problem animals are mammals, with the larger species most commonly carnivores including ursids (*Ursus* spp.), felids (*Panthera* spp., *Felis concolor*, *Lynx lynx*), wolves (*Canis lupus*), and mustelids (*Mustela* spp.), but translocation of raptors, including the golden eagle (*Aquilia chrysaetos*), black eagle (*A. verreauxii*), crowned eagle (*Stephanoetus coronatus*) and martial eagle (*Polemaetus bellicosus*), has taken place in the United States and South Africa to reduce livestock predation (reviewed in Linnell *et al.*, 1997). Translocation of hen harriers (*Circus cyaneus*) has been considered to mitigate the impact of harrier predation on red grouse (*Lagopus lagopus scotica*) on UK moorland managed for grouse shooting (Watson & Thirgood, 2001). The priorities for problem animal translocations are primarily solving the specific conflict and secondarily the welfare of the individual animal. It is a public perception that translocated problem animals 'live happily ever after' (Craven *et al.*, 1998). In the few instances where post-release monitoring of such translocations has taken place, the reality is very different. Translocated problem animals typically show high post-release mortality due to the stress of capture, transport and release, aggression by territorial conspecifics, poaching, disorientation, unsuitable habitat, long-distance dispersal and disease (Craven *et al.*, 1998; Fischer & Lindenmayer, 2000). Carnivores in particular show strong homing behaviour (Bradley *et al.*, 2005), with many individuals able to return to the capture site and resume nuisance behaviour from release sites between 150 km (Lenain & Warrington, 2001) and up to nearly 500 km (Linnell *et al.*, 1997) away. Even when translocated animals do not return to the focal point of conflict, they tend to settle away from the release site and may become a new problem elsewhere (van Vuren *et al.*, 1997; Le Gouar *et al.*, this volume, Chapter 5). Consequently, translocation of nuisance animals is often ineffective at achieving either of the two objectives. This, in conjunction with challenges to the assumption that problems are caused by a few problem animals (Linnell *et al.*, 1999), have led to increased efforts to manage potential conflict proactively using such things as habitat modification, exclusions and repellents. In South Africa cheetah (*Acinonyx jubatus*) are protected under legislation and consequently, instead of trapping and shooting animals that prey on livestock, there is a programme of translocation to fenced reserves, possibly for future ecotourism, although post-release survival rates can be low, especially in the presence of competing predators at release sites (Marnewick *et al.*, 2009). There are also attempts to shift public attitudes to accept lethal control by moving away from a focus on individual animal welfare and towards an appreciation of the need for population level conservation (Linnell *et al.*, 1997).

Commercial and recreational

Commercially and recreationally motivated translocations are often indistinguishable, with introductions to support recreational hunting or derived from the pet trade having a significant commercial element. One of the most significant threats to the conservation of freshwater fishes is the introduction of alien invasive sport fish, with three of the hundred 'World's Worst Invasive Alien Species' being fish introduced solely for sport (Cambray, 2003). One of the earliest examples of fish translocation for stocking a lake was in Turkey by Murat the III, Sultan of Ottoman between 1546 and 1595, and since the 1950s in Turkey a total of 25 exotic fish species have been introduced and 14 native fish species translocated to habitats outside their natural range (Innal & Erk'akan, 2006). There is a globalization of alien fish for sport; for example rainbow trout (*Onchorhynchus mykiss*) are now established in 82 countries (Cambray, 2003). Mammals too are translocated for sport across national and international boundaries. For example, there is evidence of the illegal translocation of invasive feral pigs (*Sus scrofa*) by recreational hunters in Australia to supplement existing populations and to create new populations (Spencer & Hampton, 2005). Red foxes (*Vulpes vulpes*), grey foxes (*Urocyon cinereoargenteus*) and coyotes (*Canis latrans*) are illegally translocated to stock hunting enclosures or 'fox-pens' (Davidson *et al.*, 1992), and in one year private hunting clubs in Kentucky, USA, translocated >2 300 raccoons (Nettles *et al.*, 1979). Ungulate introductions have taken place in well over 50 countries, with the USA and South Africa having the highest numbers of introduced ungulates globally (Spear & Chown, 2009a). To date, however, with only limited evidence of negative impacts, efforts to restrict ungulate introductions may be constrained in the face of the economic gains through trophy hunting and ecotourism (Spear & Chown, 2009b).

The pet trade per se does not fit the definition of translocation because the end point is not intended to be the 'free release' of animals, but it is worth acknowledging that accidental releases and intentional releases by owners can be an endpoint. It has been estimated that in the United States 38 % of exotic bird species are established or establishing from pets that have escaped (Temple, 1992), and two of three invasive bird species in Hong Kong originate from the pet trade (Shieh *et al.*, 2006). Since 1994, 290 exotic species of pet birds have been imported into Taiwan, and of these 93 species have escaped and at least 28 of these breed in the wild (Shieh *et al.*, 2006).

Biological control

Biological control, or biocontrol, uses living organisms as pest control agents. Classical or traditional biological control involves the introduction of exotic natural enemies to control exotic pests under the simple premise that an exotic organism becomes a pest partly because it has been released from population regulation by natural enemies (Hoddle, 2004). As a form of introduction, therefore, all the risks and uncertainties of potential invasive species apply (Simberloff & Stiling, 1996). Opponents of biological control point to examples where the introduced agent has had an impact on non-target native species, through competition and predation, and other more complex interactions (Simberloff & Stiling, 1996). Proponents of biological control point out that the greatest problem arose in early programmes involving the introduction of generalist vertebrate predators, such as mosquito fish (*Gambusia affinis*), now in about 70 countries worldwide, cane toads (*Bufo marinus*) to target various invertebrate crops pests in Australia, and red foxes and mustelids to control rabbits (*Oryctolagus cuniculus*) in Australia and New Zealand, and that since then increased regulation has reduced risks (Thomas & Willis, 1998; Hoddle, 2004). Increasingly, biological control is being considered for conservation as well as for agriculture (Henneman & Memmott, 2001) and there are examples of proposed development of biological control, such as genetically engineered viral and bacterial diseases, to target invasive arthropods that threaten native flora and fauna (Hoddle, 2004).

Aesthetic

Although less common now, the introduction of exotic species, mostly birds, by Western colonists in the 18th and 19th centuries was a major form of species translocation. Settlers from Europe attempted to create a huntable resource of familiar species, or sought to create new populations of exotic songbirds and other species that were well known to them for purely aesthetic reasons (Duncan *et al.*, 2003). Around 70 % of bird introductions (953 events) have been to islands, and over half to Pacific Islands (271 events) and Australasian regions (216 events) (Blackburn & Duncan, 2001). In New Zealand as many as 137 exotic bird species were introduced before 1907, with 284 releases of 43 species occurring mainly between 1861 and 1885 (Duncan, 1997); 28 (20%) of these established populations persist to the present day (Veltman *et al.*, 1996). Translocation of animals to support both consumptive and non-consumptive nature-based tourism addresses several motivations, including aesthetic, recreational, education and advocacy, and commercial, as well as conservation (Cousins *et al.*, 2008; Mbaiwa, 2008; see Box 1.2).

Box 1.2 **Game ranching in Southern Africa: commercial or conservation translocations?**

Traditionally, the conservation of species, ecosystems and their under-lying functions has been accomplished by setting land aside for conservation purposes – the so-called Yellowstone paradigm. In sub-Saharan Africa, as elsewhere, the main shortcoming of the Yellowstone paradigm was the fact that it ignored the vast majority of land that falls outside National Parks (Child, 2000). The generic term 'game ranch-ing', which here includes private game reserves, embodies what has been referred to as a 'second conservation paradigm', which is based on sustainable use, wildlife ownership and pricing (Child, 1996, in Child, 2000). Game ranching, as practised in South Africa (see Figure 1.2), Zim-babwe and Namibia, is defined as the managed, extensive production of free-living animals on large tracts of land that are fenced or unfenced for purposes of live animal sales, hunting, trophy hunting, venison pro-duction, tourism or other uses (Bothma, 2002). Officially proclaimed conservation areas in South Africa total 5.8 % of the country's surface area, with private game ranches estimated to contribute an additional 13 % (ABSA 2003, in Carruthers, 2009).

Figure 1.2   **A southern reedbuck from the Drakensberg region of Kwazulu/ Natal, South Africa. This species is commonly used in game ranching. (Photo: W. Maartin Strauss).**

The pivotal role that early ranchers in South Africa played in the conservation of ungulates such as the bontebok *Damaliscus dorcas dorcas*, the black wildebeest *Connochaetes gnou* and the Cape mountain zebra *Equus zebra zebra* is well documented. Although the game ranching industry has expanded significantly in recent decades, the conservation role that it plays today is not clear-cut. In an attempt to increase local diversity and thereby economic viability of game ranches, managers now frequently apply artificial selection in breeding indigenous ungulates. The so-called white and black springboks (*Antidorcas marsupialis*) are, for example, popular animals for translocation and at game auctions their monetary value has, on average, increased by 38.4 % and 20.3 % per year over a 10 year period (van der Merwe *et al.*, 2008). Nevertheless, there is an assumption that game ranching contributes more to biodiversity conservation than other forms of agricultural land use (Bond *et al.*, 2004), despite profit being the primary reason for their establishment (Hearne & McKenzie, 2000). There are no official figures available, but du Toit (2007) estimated that up to 70 000 animals are captured and translocated annually in South Africa, resulting in an estimated turnover, including the value of the captured animals, of between R750 million (US$101 million) and R900 million (US$121 million). Prior to the enactment of recent (2004) laws aimed at regulating and controlling the translocation and introduction of large mammals in South Africa, exotic and/or extralimital ungulate species were translocated and introduced into game ranching areas across the country, resulting in South Africa having the second highest number of introduced ungulates globally (Spear & Chown, 2009b).

Game ranching has undoubtedly contributed to the increase in the number of wild animals in South Africa during the last few decades. Conservation in South Africa is, however, at a crossroads, as the game ranching industry pushes for exemption from all nature conservation regulatory control.

Religious

Approximately 30 % of people of all religions in East Asia believe they can accrue merit by freeing captive animals during ceremonies termed 'prayer animal releases'. These ceremonies are organized by temples using local and exotic animals, mostly birds, supplied by pet stores (Severinghaus & Chi, 1999). With the practice prevalent in Taiwan, Malaysia, Thailand, Cambodia, Vietnam, Honk Kong and Korea, and with some temples organizing as many as 24 release ceremonies per year, the scale of translocations is huge. It was estimated, for example, that 128 000 birds were released in only one year in Taichen City, Taipei (Severinghaus & Chi, 1999).

Wildlife rehabilitation

Capture, care and release of wildlife is a significant and growing practice internationally; for example in Britain some 30 000 to 40 000 wild animal casualties end up in wildlife hospitals, the most common species being the European hedgehog (*Erinaceus europaeus*) (Molony *et al.*, 2006). The most frequent causes of injuries to terrestrial species include collisions with vehicles and domestic animal attacks (Hartup, 1996). For seabirds oiling has been a major threat ever since the start of large-scale transportation of petroleum products by sea, with large-scale seabird mortality due to dumping of tanker waste oil from 1917 and oil spills in 1937 off San Francisco (Carter, 2003) and during WWII (Mezat *et al.*, 2002). Following significant oil spills in the late 1960s, rehabilitation efforts for oiled seabirds developed in the United States (Carter, 2003) and South Africa (Nel *et al.*, 2003). Rehabilitation of oiled seabirds involves capture, transport, cleaning and release, and has been characterized by high failure rates, with in some cases only 1–20 % of birds surviving the first year post-release (Mead, 1997). Post-release survival rates vary, however, with the type and degree of oiling, but also with the species. The highest success rates have been achieved with African penguins (*Spheniscus demersus*), with up to 84 % of processed penguins being released (Nel *et al.*, 2003) and up to 65 % of released penguins being resighted within two years (Underhill *et al.*, 1999). Nearly comparable survival rates have been achieved for little penguins (*Eudyptula minor*) (Goldsworthy *et al.*, 2000).

Although conservation of endangered species is one of the reasons cited for the rehabilitation of marine mammals (Moore *et al.*, 2007), the greatest risks involved in the release of rehabilitated animals is that of disease transmission from captivity to wild populations (Quakenbush *et al.*, 2009). There is general agreement by authorities that the health of wild populations should be a greater concern than the welfare of an individual animal, thus euthanasia is often the best option, but one that carries a significant negative image and risks loss of public support for rehabilitation efforts (Moore *et al.*, 2007). Consequently, there may be public pressure to sustain rehabilitation efforts even for species that have healthy populations for which the rehabilitation of a single animal has no conservation value (Moore *et al.*, 2007) but which poses significant risks to wild populations (Quakenbush *et al.*, 2009). In contrast, the numbers of individual birds involved in major oil spills can be significant at a population conservation level; for example oil spills off the South African coast from the *MV Treasure* in 2000 (Parsons & Underhill, 2005) and the *Apollo Sea* in 1994 (Underhill *et al.*, 1999) resulted in the processing of $\sim$ 10 000 and nearly 20 000 oiled African penguins, respectively, from a world total population around that time of <60 000 pairs (Birdlife International, 2008).

Animal rights activism and animal liberations

While conservation biologists are rightly concerned with animal welfare and the reduction of unnecessary suffering, there is a difference between an individual animal welfare perspective that may motivate activities such as animal rehabilitation and the conservation management of wildlife populations. For the most part these differences do not create problems; for example welfare concerns will dictate that any release of captive animals must ensure that each animal has the skills and behaviours necessary for survival in the wild (Waples & Stagoll, 1997). Different perspectives become problematic, however, where animal welfare becomes animal rights. Animal rights activists are committed, *inter alia*, to the total abolition of use of animals in science, commercial agriculture and sport hunting, and consider as fundamentally wrong any system that views non-human animals as resources to be used by humans (Regan, 1983). There exists a challenging incompatibility between a conservation ethic and animal rights, which some see as a 'highly reductionist view' that focuses exclusively on individual sentient animals (Hutchins, 2008). This can lead to illegal liberations of captive animals that effectively expand the range of introduced species and have detrimental impacts on native fauna (Lewis *et al.*, 1999). Furthermore, the released animals can suffer. The liberation of captive-bred furbearers such as mink (*Mustela vison*) from fur farms provides one example. Accounts in the media indicate multiple liberations of groups of up to 6000 captive mink from farms in Canada, USA, UK, Ireland, Finland, the Netherlands and Greece over the last decade. Many of the liberations of mink are attributed or claimed to be the actions of the Animal Liberation Front (ALF). Inevitably, released mink start to die in large numbers soon after release, before survivors can be recovered. Nevertheless, liberationists claim that outside their cages mink have a fighting chance of survival. This is despite overwhelming evidence that most freed mink face a slow death in the wild versus a humane end in captivity.

Conservation translocations

Where conservation of the target species is the primary objective we can consider *conservation translocations* (Hodder & Bullock, 1997) and ask a new question: 'are animals to be released within the historical distribution range of the species?' Releases within the documented natural range may be classified as translocations for *population restoration*, the implication being that the goal is to recover populations of the species back to some past target state. Releases outside the historical range, but with population conservation as the primary objective, are termed *conservation introductions* (IUCN, 1998) (Box 1.1).

Population restorations

Reintroduction is the release of an organism into an area that was once part of its range but from which it has been extirpated (IUCN, 1987) (Box 1.1). In broadly stated terms the objective of a reintroduction is to re-establish a self-sustaining population of a species within its historic range (Griffith *et al.*, 1989), and ideally that population will have a high probability of persistence with minimal or no intervention (Seddon, 1999). Despite some early reintroduction success stories, such as Arabian oryx (*Oryx leucoryx*) in Oman (Stanley Price, 1989) and peregrine falcon (*Falco peregrinus*) in North America (Cade & Burnham, 2003), the failure of other, less well-conceived, reintroduction attempts meant that reintroduction project success rates were low (Griffith *et al.*, 1989; Wolf *et al.*, 1996). The situation was not helped by a lack of post-release monitoring, which meant that the timing and causes of failures was not known (Seddon *et al.*, 2007a). In response to the perceived problems the World Conservation Union (IUCN) Reintroduction Specialist Group (RSG) was formed in 1988 under the auspices of the Species Survival Commission (Stanley Price & Soorae, 2003). One of the first actions of the RSG was the formulation of Guidelines for Reintroductions (IUCN, 1998) in order to improve reintroduction practice; for example the guidelines place emphasis on the identification of release sites within the historic range of the species and acknowledge a need to ensure that previous causes of decline have been addressed, both factors having been shown to strongly influence project outcomes (Fischer & Lindenmayer, 2000). In part due to the actions and outputs of the RSG, improved pre-release planning, care over the selection of founders and the composition of founder groups, release site preparation and detailed post-release monitoring have improved project success rates, at least in the short term (Soorae, 2008). Although assessment of reintroduction success is not straightforward it is useful to think of any project needing to progress through two phases (Armstrong & Seddon, 2008): population establishment, which requires survival of founders, and breeding by founders and their offspring; and population persistence, which may be assessed for taxonomically relevant time frames using population modelling tools (Seddon, 1999).

Meta-analyses of factors contributing to reintroduction success indicate the importance of habitat quality at the release site and the number of individuals released (Germano & Bishop, 2009; Griffith *et al.*, 1989; Wolf *et al.*, 1996, 1998). The effects of habitat quality on reintroduction success have been confirmed in recent experimental studies (Moorhouse *et al.*, 2009). The number of animals released, however, is often correlated with several other factors that may be important prerequisites of success. For example, projects that release the most individuals are usually those that are well funded and well resourced, and that are perceived *a priori* to have the greatest chance of success. In contrast, only few founders are released in short-term projects that do not have significant institutional and community support.

Re-enforcement (IUCN, 1998), also termed *restocking* (IUCN, 1987) and *supplementation* (IUCN, 1998), or *augmentation* (Maguire & Servheen, 1992), involves the release of individuals into an existing population of conspecifics (Box 1.1), in order to increase population size and reduce the risks of genetic or demographic collapse due to stochastic effects. Translocations for re-enforcement are used to overcome barriers to natural dispersal from other free-ranging populations (e.g. Gusset *et al.*, 2009), to speed up population growth, or to enhance genetic diversity and avoid inbreeding depression (Jamieson *et al.*, 2006). In some cases ongoing re-enforcement may be required to sustain non-viable free-ranging populations until natural productivity is sufficient to support population growth and persistence. For example, kaki or black stilt (*Himantopus novaezelandiae*) are sustained in the wild through the release of captive-reared birds while habitat restoration measures are being trialled (e.g. Keedwell *et al.*, 2002).

Seddon (2010) poses the question of when does a reintroduction become re-enforcement? While seemingly trivial semantics, this question does relate to a more significant one – that of when to stop releases. There is a substantial body of literature that discusses evaluation of reintroduction success (Fischer & Lindenmayer, 2000; Griffith *et al.*, 1989; Seddon, 1999; Wolf *et al.*, 1996, 1998), and there is now widespread use of population modelling to set re-establishment goals, to define optimal reintroduction strategies and to assess population persistence (Armstrong *et al.*, 2002, 2006; Rout *et al.*, 2009; Schaub *et al.*, 2009). Pre-release target setting considers the number, size and composition of founder cohorts and the efficacy of single versus multiple releases. Nevertheless, post-release monitoring will enable refinement of pre-release models (Armstrong & Davidson, 2006; Armstrong *et al.*, 2007;

Wakamiya & Roy, 2009) and may indicate a low probability of population persistence that could be addressed through the release of more individuals. Such *post hoc* secondary releases have been termed *follow-up translocations* (Armstrong & Ewen, 2001) and could be seen as supplementation of the re-established free-ranging population, but should strictly be considered part of the original, but not yet successful, reintroduction attempt (Seddon, 2010).

Conservation introductions

Mediated movement of organisms outside their native range constitutes a species introduction (IUCN, 1987) and if the goal is the establishment of a new population explicitly and primarily for conservation, then such a translocation is regarded as a *conservation*, or *benign* (in intent at least), *introduction* (IUCN, 1998). The current IUCN guidelines consider conservation introductions to be justified 'when there is no remaining area left within a species' historic range' (IUCN, 1998). This limited rationale marks conservation introductions as a somewhat reactive, stop-gap measure, in some cases perhaps to mark time until appropriate habitat restoration can take place within the historical distribution range of the target species. However, more pro-active interventions are now being considered by natural resource managers, and we can broadly define two types of conservation introduction: *ecological replacement* and *assisted colonization*.

Ecological replacement is the release of species outside their historic range in order to fill an ecological niche left vacant by the extinction of a native species. Extinction removes the option of reintroduction through the release of either wild or captive individuals and may mean the loss of critical or otherwise desirable ecological functions. One option is therefore to restore lost ecological function through the establishment of a viable population of an ecologically similar species (Atkinson, 2001). The most readily acceptable approach will be the release of a *subspecific substitute*, such as using the North African subspecies of ostrich (*Struthio camelus camelus*) as a replacement for the extinct Arabian subspecies *S. c. syriacus* (Seddon & Soorae, 1999). The recent use of other *analogue species* includes yellow-crowned night heron (*Nycticorax violacea*) for an extinct endemic *Nyctocorax* species in Bermuda, tundra musk ox (*Ovibos moschatus*) for the extinct *O. palantis* in Sibera (review in Parker *et al.*, 2010) and North Island kokako (*Callaeas wilsoni*) for the extinct South Island form *C. cinerea* (see Box 1.3). While subspecific substitutes may be expected to be the most appropriate ecological replacements, other forms may potentially be better functional equivalents. For example, Parker *et al.*, (2010) make a case for the replacement of the extinct New Zealand quail (*Coturnix novaezelandiae*), not with its closest extant relative, the Australian stubble quail *C. pectoralis*, but with the more distantly related but ecologically better suited Australian brown quail *C. ypsilophora*. It may not be the case that the analogue species is rare or threatened in its natural range and thus its conservation may not be a primary objective of its introduction as an ecological replacement, necessitating a broader interpretation of the earlier dichotomy between conservation and non-conservation translocations to include the conservation objective beyond the target species (Figure 1.1).

Box 1.3 **North Island kokako translocation to the South Island as an example of an ecological replacement**

In October 2008, 10 North Island kokako *Callaeas wilsoni* were translocated from Mapara in the central North Island of New Zealand to 8 140 ha Secretary Island on the southwestern corner of the South Island, 1 000 km south (see Figure 1.3). South Island forests were previously occupied by a southern kokako species *Callaeas cinerea*, declared extinct in 2004. The intention of the release was to restore the ecological functions of kokako into a South Island forested ecosystem. It is inherently experimental.

IUCN guidelines (1995) and Seddon & Soorae (1999) suggested that an ecological substitute should be selected from extant subspecies or races (rather than species) to avoid fundamental differences in habitat preferences between the original and substitute taxa. However, North and South Island kokako were regarded as subspecies until recently (Holdaway *et al.*, 2001). Plumage of the two is the same although South Island birds had orange wattles (small fleshy appendages arising from the gape and lying against the throat) while North Island wattles are blue (Higgins *et al.*, 2006). Subtly different behaviours, such as more ground-feeding, may have led to the early decline and extinction of the South Island form due to predation by introduced pest mammals, but this is unclear (Clout & Hay, 1981; Holdaway & Worthy, 1997). The extinction of *C. cinerea* prevents further comparison of behaviours of the two taxa.

Figure 1.3   A North Island kokako nestling being banded for monitoring purposes on Tiritiri Matangi Island, New Zealand. (Photo: John G. Ewen).

Diet and behaviour of *C. wilsoni* are very well known (Higgins *et al.*, 2006), whereas *C. cinerea* was never studied in detail. *C. wilsoni* mainly eat leaves and fruits, and some insects (Higgins *et al.*, 2006). Kokako were very abundant in both islands before human settlement, and their ecological roles included herbivory, pollination and fruit dispersal, as well as being prey for New Zealand's original predators (raptors), some of which are also extinct. The demise of native birds has in turn impaired pollination and perhaps seed dispersal of trees and shrubs (Kelly *et al.*, 2010). Kokako are quite large (38 cm; 230 g) with 13 mm gapes, capable of dispersing fruits of several structurally important large-seeded plants (Clout & Hay, 1989; Kelly *et al.*, 2010). Reintroducing kokako to Secretary Island is a small part of the biotic restoration planned there. While this translocation meets most objectives of a reintroduction listed by IUCN (1995) – to enhance the long-term survival of a species; to re-establish a keystone species (in the ecological or cultural sense) in an ecosystem; to maintain and/or restore natural biodiversity; to provide long-term economic benefits to the local and/or national economy; to promote conservation awareness – it is primarily an attempt at ecological restoration of a lost biotic community, as first championed in New Zealand by Atkinson (1988).

Monitoring survival of some of the kokako released in October 2008 with transmitters revealed only one death, due to falcon *(Falco novaeseelandiae)* predation. The monitored birds settled in the general area of the island where they were released. A further 17 kokako released in 2009 were sourced from two additional source populations – Kaharoa and Rotoehu – to increase the genetic representation of the North Island species and minimize future inbreeding.

Assisted colonization, also referred to as *assisted migration* (McLachlan *et al.*, 2007) and *managed relocation* (Richardson *et al.*, 2009) has been best defined as 'translocation of a species to favourable habitat beyond their native range to protect them from human induced threats' (Ricciardi & Simberlof, 2009a). Recent interest in this form of conservation introduction has been driven by predicted habitat change due to rapid climate change (Hoegh-Guldberg *et al.*, 2008), but assisted colonization could be and has been used to mitigate a variety of threats, including agricultural expansion and urbanization (Ricketts & Imhoff, 2003), filling of hydroelectric reservoirs (Richard-Hansen *et al.*, 2000) and the threats posed by (other) deliberately introduced species (Vitousek *et al.*, 1997). Specific examples include the translocation of slow-worm *(Anguis fragilis)* from sites for future housing development in the UK (Platenberg & Griffiths, 1999) and giant land snails *(Powelliphanta augusta)* from sites designated for coal mining in North Westland, New Zealand (Walker *et al.*, 2008), both of which involved releases outside the species' known historical range.

The debate around assisted colonization has focused on the risk of impacts of introduced species (Mueller & Hellmann, 2008; Ricciardi & Simberlof, 2009a, 2009b; Sax *et al.*, 2009; Seddon *et al.*, 2009; Vitt *et al.*, 2009) and is assumed by many commentators to mark a major shift in conservation translocation practice. However, assisted colonization is a well-established (if previously unnamed) conservation tool in some parts of the world. For example, in New Zealand the extinction threats to endemic birds, herptiles and invertebrates posed by introduced mammalian predators have been addressed with some success through 'marooning', whereby species are translocated to predator-free offshore islands (Saunders & Norton, 2001). In many cases these islands were not within historically documented parts of the species range. In fact one of the earliest examples of assisted colonization could be the pioneering conservation work undertaken by Richard Henry in New Zealand (Jones & Merton, this volume, Chapter 2).

During the 1890s Richard Henry was caretaker of Resolution Island in remote and rugged Fiordland on the west coast of New Zealand's South Island. A keen naturalist, he noted with dismay the impact on native birds of the arrival of recently introduced stoats (*Mustela erminea*) as they invaded this last corner of New Zealand. In a desperate attempt to protect populations of the flightless kakapo (*Strigops habroptilus*) and little spotted kiwi (*Apteryx oweni*) between 1894 and 1900 he translocated hundreds of individuals from the mainland on to Resolution Island (Saunders & Norton, 2001). Unfortunately Resolution was too close to the mainland and stoats invaded in 1900 and Henry's efforts were in vain (Clout, 2006). Nevertheless, the technique of marooning vulnerable species on predator-free islands that may or may not have been occupied by the species in the past became a vital tool to avert extinctions in the face of predation by introduced mammalian predators in New Zealand (Saunders & Norton, 2001).

We can therefore envisage a simple dichotomy between ecological replacement and assisted colonization (Figure 1.1). If the aim is to fill an available ecological niche left vacant by the extinction of the original form, then the type of conservation translocation is an ecological replacement. In the case of marooning and other translocations to move members of a species outside their range to avoid some threat, the primary aim is not to fill an available ecological niche but rather to sustain a viable population, perhaps until appropriate habitat restoration has taken place within their core distribution. In the specific case of climate change mitigation, where a suitable habitat has opened up outside the historic distribution range, then assisted colonization may be used to fill a newly available niche that has not been created by an extinction and is thus similarly distinguished from the release of ecological substitutes for an extinct form (Figure 1.1). In all cases assisted colonization is the human-mediated movement of individuals of a species that would otherwise be unable to survive current or anticipated threats within its current distribution.

Human dimensions in animal translocations

It is now well understood that intensive conservation interventions, such as reintroductions, cannot hope to succeed without some level of engagement with local and national government, non-governmental agencies and professionals, and, critically, the public (Kleiman, 1989). This is seen nowhere more strongly than in the restoration of populations of large carnivores, where local community attitudes and behaviours can be significant determinants of project outcomes (Hayward *et al.*, 2007; Hunter *et al.*, 2007; Jule *et al.*, 2008; Lohr *et al.*, 1996; Meadow *et al.*, 2005; Nilsen *et al.*, 2007; Zahniser & Singh, 2004).

Public engagement with a translocation project can come at different levels and stages, including the provision of funding (and conversely through economic benefits that accrue to local communities (Kleiman, 1989) and labour, support for approvals, advocacy and lobbying for political and legislative change, and wider attitudinal changes through education and interaction (Williams *et al.*, 2002). In many cases community engagement is not just a useful part of translocation planning, nor even just a prerequisite for success; rather it is one of the desirable outcomes.

Translocation projects provide a means to engage with the public, to make them collaborators in the programme (Kleiman, 1989) and to potentially change negative or biased public views of scientists and resource managers. The opportunity to learn about conservation projects, to see wild animals up close and even to participate in their liberation into new areas provides a powerful means to counter the view that conservation is preservation and entails the locking up of resources. Meaningful and positive contact with native species and natural areas can mitigate the alienation from the natural world and the extinction of experience that is a feature of increasingly urbanized societies. It has been suggested that advocacy for the natural world may be the main role of conservation biology (Brussard & Tull, 2007). Parker (2008) proposes that we view translocations as a vehicle to enhance linkages between scientists, managers and the general public, and that meaningful community participation should be considered one of the main outputs of a translocation, along with its management and scientific objectives.

Concluding comments

Translocation is now well entrenched as a conservation tool, with the numbers of animals being released in reintroduction and re-enforcement projects increasing almost exponentially each year. The more conservation translocation activity there is, the greater the proliferation of terms and concepts being used to describe the various actions, resulting in a mass of synonyms and variations that may end up obscuring meaning. What we have tried to do in this chapter is to provide a standardized and justified terminology to describe the full spectrum of conservation translocation activities. Our hope is that these will enable practitioners and researchers to be very clear about what they are doing or what they propose to do, not only for themselves, but also in discussions or debate with others.

Resource managers have been forced to deal with critical population declines and extinctions, ecosystem degradation and changing habitat conditions due to global climate change (Sekercioglu *et al.*, 2008); consequently new forms of conservation interventions are being explored. These include ecological replacements, whereby functionally equivalent taxa fill a niche left available by the extinction of the original form, and assisted colonization, where species are moved into areas not previously occupied in order to avoid some human-induced threat. These conservation introductions are perhaps not as radical as they first appear, but they do mark the start down an interesting pathway leading away from strict reintroductions towards more controversial decisions about which species we want to have where. There is a clear convergence in thinking between the disciplines of reintroduction biology and restoration ecology (Lipsey & Child, 2007; Seddon *et al.*, 2007b), whereby historical restoration targets are seen as arbitrary and unrealistic, and increasingly there is talk of futuristic restoration (Choi, 2004) and novel (Figure 1.1) or designer ecosystems (Temperton, 2007; Seddon, 2010). The debates will no doubt rage for decades as we consider what our future natural world could or should look like.

Acknowledgements

This chapter was improved by the comments of Doug Armstrong, Tim Blackburn, John Ewen, Ian Flux, Richard Maloney, Ollie Overdyck, Kevin Parker, Francois Sarrazin, Yolanda van Heezik and Megan Willans.

References

Armstrong, D.P. & Davidson, R.S. (2006) Developing population models for guiding reintroductions of extirpated bird species back to the New Zealand mainland. *New Zealand Journal of Ecology*, 30, 73–85.

Armstrong, D.P. & Ewen, J.G. (2001) Assessing the value of follow-up translocations: a case study using New Zealand robins. *Biological Conservation*, 101, 239–247.

Armstrong, D.P. & Seddon, P.J. (2008) Directions in reintroduction biology. *Trends in Ecology and Evolution*, 23, 20–25.

Armstrong, D.P., Castro, I. & Griffiths, R. (2007) Using adaptive management to determine requirements of reintroduced populations: the case of the New Zealand hihi. *Journal of Applied Ecology*, 44, 953–962.

Armstrong, D.P., Davidson, R.S., Dimond, W.J. *et al.* (2002) Population dynamics of reintroduced forest birds on New Zealand islands. *Journal of Biogeography*, 29, 609–621.

Armstrong, D.P., Raeburn, E.H., Lewis, R.M. *et al.* (2006) Estimating the viability of a reintroduced New Zealand robin population as a function of predator control. *Journal of Wildlife Management*, 70, 1020–1027.

Atkinson, I.A.E. (1988) Presidential address: opportunities for ecological restoration. *New Zealand Journal of Ecology*, 11, 1–12.

Atkinson, I.A.E. (2001) Introduced mammals and models for restoration. *Biological Conservation*, 99, 81–96.

BirdLife International (2008) Species factsheet: *Spheniscus demersus*. Downloaded from http://www.birdlife.org on 9/9/2009.

Blackburn, T.M. & Duncan, R.P. (2001) Determinants of establishment success in introduced birds. *Nature*, 414, 195–197.

Bond, I., Child, B., Harpe, D. de la *et al.* (2004) Private land contribution to conservation in South Africa. In *Parks in Transition: biodiversity, rural development and the bottom line*, ed. B. Child. IUCN, SASUSG and Earthscan, London, Sterling, Virginia.

Bothma, J. du P. (2002) *Game Ranch Management*. Van Schaik Publishers, Pretoria, South Africa.

Bradley, E.H., Pletshcer, D.H., Bangs, E.E. *et al.* (2005) Evaluating wolf translocation as a nonlethal method to reduce livestock conflicts in the Northwestern United States. *Conservation Biology*, 19, 1498–1508.

Brussard, P.F. & Tull, J.C. (2007) Conservation biology and four types of advocacy. *Conservation Biology*, 21, 21–24.

Cade, T.J. & Burnham, W. (2003). *Return of the Peregrine*. The Peregrine Fund, Boise, Idaho.

Cambray, J.A. (2003) Impact on indigenous species biodiversity caused by the globalisation of alien recreational freshwater fisheries. *Hydrobiologia*, 500, 217–230.

Carruthers, J. (2009) 'Wilding the farm or farming the wild?' The evolution of scientific game ranching in South Africa from the 1960s to the present. *Transactions of the Royal Society of South Africa*, 63(2), 160–181.

Carter, H.R. (2003) Oil and California's seabirds: an overview. *Marine Ornithology*, 31, 1–7.

Child, B. (2000) Making wildlife pay: converting wildlife's comparative advantage into real incentives for having wildlife in African savannas, case studies from Zimbabwe and Zambia. In: *Wildlife Conservation by Sustainable Use*, eds H.H.T. Prins, J.G. Grootenhuis & T.T. Dolan, pp. 335–387. Kluwer Academic Publishers, Massachusetts.

Choi, Y.D. (2004). Theories for ecological restoration in changing environments: Toward 'futuristic' restoration. *Ecological Restoration*, 19, 75–81.

Clout, M.N. (2006) A celebration of kakapo: progress in the conservation of an enigmatic parrot. *Notornis*, 53, 1–2.

Clout, M.N. & Hay, J.R. (1981) South Island kokako (*Callaeas cinerea cinerea*) in Nothofagus forest. *Notornis*, 28, 256–259.

Clout, M.N. & Hay, J.R. (1989) The importance of birds as browsers, pollinators and seed dispersers in New Zealand forests. *New Zealand Journal of Ecology*, 12 (supplement), 27–33.

Cousins, J.A., Sadler, J.P. & Evans, J. (2008) Exploring the role of private wildlife ranching as a conservation tool in South Africa: Stakeholder perspectives. *Ecology and Society*, 13, 43 [online] URL: http://www.ecologyandsociety.org/vol13/iss2/art43/.

Craven, S., Barnes, T. & Kania, G. (1998) Toward a professional position on the translocation of problem wildlife. *Wildlife Society Bulletin*, 26, 171–177.

Davidson, W.R., Appel, M.J., Doster, G.L. *et al.* (1992) Diseases and parasites of red foxes, gray foxes, and coyotes from commercial sources selling to fox-chasing enclosures. *Journal of Wildlife Diseases*, 28, 581–589.

Duncan, R.P. (1997) The role of competition and introduction effort in the success of passeriform birds introduced to New Zealand. *American Naturalist*, 149, 903–915.

Duncan, R.P., Blackburn, T.M. & Sol, D. (2003) The ecology of bird introductions. *Annual Review of Ecology and Evolutionary Systematics*, 34, 71–98.

du Toit, J.G. (2007) Role of the private sector in the wildlife industry. Report, Tshwane, Wildlife Ranching SA/Du Toit Wilddienste. 87 pp.

Fischer, J. & Lindenmayer, D.B. (2000) An assessment of the published results of animal relocations. *Biological Conservation*, 96, 1–11.

Gammons, D.J., Mengak, M.T. & Conner, L.M. (2009) Translocation of nine-banded armadillos. *Human–Wildlife Conflicts*, 3, 64–71.

Germano, J.M. & Bishop, P.J. (2009) Suitability of amphibians and reptiles for translocation. *Conservation Biology*, 23, 7–15.

Goldsworthy, S.D., Giese, M., Gales, R.P. *et al.* (2000) Effects of the *Iron Baron* oil spill on little penguins (*Eudyptula minor*). II. Post-release survival of rehabilitated oiled birds. *Wildlife Research*, 27, 573–582.

Goodrich, J.M. & Miquelle, D.G. (2005) Translocation of problem Amur tigers *Panthera tigris altaica* to alleviate tiger–human conflicts. *Oryx*, 39, 454–457.

Grayson, D.K. (2001) The archaeological record of human impacts on animal populations. *Journal of World Prehistory*, 15, 1–68.

Griffith, B., Scott, J.M., Carpenter, J.W. *et al.* (1989) Translocation as a species conservation tool: status and strategy. *Science*, 245, 477–480.

Gusset, M., Jokoby, O., Muller, M.S. *et al.* (2009) Dogs on the catwalk: modelling re-introduction and translocation of endangered wild dogs in South Africa. *Biological Conservation*, 142, 2774–2781.

Hartup, B.K. (1996) Rehabilitation of native reptiles and amphibians in DuPage County, Illinois. *Journal of Wildlife Diseases*, 32, 109–112.

Hayward, M.W., Adendorff, J., O'Brien, J. *et al.* (2007) Practical considerations for the reintroduction of large, terrestrial, mammalian predators based on reintroductions to South Africa's Eastern Cape Province. *Open Conservation Biology Journal*, 11, 11.

Hearne, J. & McKenzie, M. (2000) Compelling reasons for game ranching in Maputaland. In *Wildlife Conservation by Sustainable Use*, eds H.H.T. Prins, J.G. Grootenhuis & T.T. Dolan, pp. 417–438. Kluwer Academic Publishers, Massachusetts.

Heinsohn, T.E. (2001) Human influences on vertebrate zoogeography: animal translocation and biological invasions across and to the east of Wallace's Line. In *Faunal and Floral Migrations and Evolution in SE Asia–Australia*, eds I. Metcalfe, M.B.J. Smith, M. Morwood & I. Davidson, pp. 153–170. Balkema, Rotterdam.

Henneman, M.L. & Memmott, J. (2001) Infiltration of a Hawaiian community by introduced biological control agents. *Science*, 293, 1314–1316.

Herr, J., Schley, L. & Roper, T.J. (2008) Fate of translocated wild-caught and captive-reared stone martens (*Martes foina*). *European Journal of Wildlife Research*, 54, 511–514.

Higgins P.J., Peter J.M. & Cowling S.J. (2006) *Handbook of Australian, New Zealand and Antarctic Birds*, vol. 7, *Boatbill to Starlings*. Oxford University Press.

Hodder, K.H. & Bullock, J.M. (1997) Translocations of native species in the UK: implications for biodiversity. *Journal of Applied Ecology*, 34, 547–565.

Hoddle, M.S. (2004). Restoring balance using exotic species to control invasive exotic species. *Conservation Biology*, 18, 38–49.

Hoegh-Guldberg, O., Hughes, L., McIntyre, S. *et al.* (2008) Assisted colonization and rapid climate change. *Science*, 321, 345–346.

Holdaway R.N. & Worthy T.H. (1997) A reappraisal of the late quaternary fossil vertebrates of Pyramid Valley Swamp, north Canterbury, New Zealand. *New Zealand Journal of Zoology*, 24, 69–121.

Holdaway R.N., Worthy T.H. & Tennyson, A. J. D. (2001) A working list of breeding bird species of the New Zealand region at first human contact. *New Zealand Journal of Zoology*, 28, 119–187.

Hunter, L.T.B., Pretorius, K., Carlisle, L.C. *et al.* (2007) Restoring lions *Panthera leo* to northern KwaZulu-Natal, South Africa: short-term biological and technical success but equivocal long-term conservation. *Oryx*, 41, 196–204.

Hutchins, M. (2008) Animal rights and conservation. *Conservation Biology*, 22, 815–816.

Innal, D. & Erk'akan, F. (2006) Effects of exotic and translocated fish species in the inland waters of Turkey. *Reviews in Fish Biology and Fisheries*, 16, 39–50.

IUCN (World Conservation Union) (1987) IUCN position statement on the transloca-tion of living organisms: introductions, re-introductions, and re-stocking. IUCN, Gland, Switzerland.

IUCN (World Conservation Union) (1995) Guidelines for re-introductions. Annex 6 to the minutes of the 41st meeting of council. Gland, Switzerland.

IUCN (World Conservation Union) (1998) Guidelines for re-introductions. IUCN/SSC Re-introduction Specialist Group, IUCN, Gland, Switzerland and Cambridge, UK.

IWRC (International Wildlife Rehabilitation Council) (2009) What is wildlife reha-bilitation? http://www.iwrc-online.org/. Retrieved 7 September 2009.

Jamieson, I.G., Wallis, G.P & Briskie, J.V. (2006) Inbreeding and endangered species management: is New Zealand out of step with the rest of the world? *Conservation Biology*, 20, 38–47.

Jones, C.G. & Merton, D.V. (2011) A tale of two islands: the rescue and recovery of endemic birds in New Zealand and Mauritius. In Reintroduction Biology: integrating science and management, eds J.G. Ewen, D.P. Armstrong, K.A. Parker & P.J. Seddon, Chapter 2. Wiley-Blackwell, Oxford, UK.

Jule, K.R., Leaver, L.A. & Lea, S.E.G. (2008) The effects of captive experience on rein-troduction survival in carnivores: a review and analysis. *Biological Conservation*, 141, 355–363.

Keedwell, R.J., Maloney, R.F. & Murray, D.P. (2002) Predator control for protect-ing kaki (*Himantopus novaezelandiae*): lessons from 20 years of management. *Biological Conservation*, 105, 369–374.

Kelly D., Ladley J.J., Robertson A.W. *et al.* (2010) Mutualisms with the wreckage of an avifauna: the status of bird pollination and fruit-dispersal in New Zealand. *New Zealand Journal of Ecology*, 34, 66–85.

Kirch, P.V. (2005) Archaeology and global change: the Holocene record. *Annual Review of Environmental Resources*, 30, 409–440.

Kleiman, D.G. (1989) Reintroduction of captive mammals for conservation. *Bioscience*, 39, 152–161

Le Gouar, P., Mihoub, J.-B. & Sarrazin, F. (2011) Dispersal and habitat selection: behavioural and spatial constraints for animal translocations. In Reintroduction Biology: integrating science and management, eds J.G. Ewen, D.P. Armstrong, K.A. Parker & P.J. Seddon. Wiley-Blackwell, Oxford, UK.

Lenain, D.M. & Warrington, S. (2001) Is translocation an effective tool to remove predatory foxes from a desert protected area? *Journal of Arid Environments*, 48, 205–209.

Lewis, J.C., Sallee, K.L. & Golightly Jr, R.T. (1999). Introduction and range expansion of nonnative red foxes (*Vulpes vulpes*) in California. *American Midland Naturalist*, 142, 372–381.

Linnell, J.D.C., Annes, R., Swenson, J.E. *et al.* (1997) Translocation of carnivores as a method for managing problem animals: a review. *Biodiversity and Conservation*, 6, 1245–1257.

Linnell, J.D.C., Odden, J., Smith, M.E. *et al.* (1999) Large carnivores that kill livestock: do 'problem individuals' really exist? *Wildlife Society Bulletin*, 27, 698–705.

Lipsey, M.K. & Child, M.F. (2007) Reintroduction biology and restoration ecology: are two disciplines better than one? *Conservation Biology*, 21, 1387–1388.

Lohr, C., Ballard, W.B. & Bath, A. (1996) Attitudes toward gray wolf reintroduction to New Brunswick. *Wildlife Society Bulletin*, 24, 414–420.

MacLachlan, J.S., Hellmann, J.J. & Schwartz, M.W. (2007) A framework for debate of assisted migration in an era of climate change. *Conservation Biology*, 21, 297–302.

Maguire, L.A. & Servheen, C. (1992) Integrating biological and social concerns in endangered species management: augmentation of grizzly bear populations. *Conservation Biology*, 6, 426–434.

Marnewick, K., Hayward, M.W., Cilliers, D. *et al.* (2009) Survival of cheetahs relocated from ranchland to fenced protected areas in South Africa. In *Reintroduction of Top-Order Predators*, eds M. Hayward & M. Somers. Wiley-Blackwell Publishing, Oxford, UK.

Mbaiwa, J.E. (2008) The success of consumptive wildlife tourism in Africa. In *Tourism and the Consumption of Wildlife*, ed. B. Lovelock, pp. 141–154. Routledge, London and New York.

Mead, C. (1997) Poor prospects for oiled birds. *Nature*, 390, 449–450.

Meadow, R., Reading, R.P., Phillips, M. *et al.* (2005) The influence of persuasive arguments on public attitudes toward a proposed wolf restoration in the southern Rockies. *Wildlife Society Bulletin*, 33, 154–163.

Mezat, J.A.K., Newman, S.H., Gilardi, K.V.K. *et al.* (2002) Advances in oiled bird emergency medicine and management. *Journal of Avian Medicine and Surgery*, 16, 146–149.

Molony, S.E., Dowding, C.V., Baker, P.J. *et al.* (2006) The effect of translocation and temporary captivity on wildlife rehabilitation success: an experimental study using European hedgehogs (*Erinaceus europaeus*). *Biological Conservation*, 130, 530–537.

Moore, M., Early, G., Touhey, K. *et al.* (2007) Rehabilitation and release of marine mammals in the United States: risks and benefits. *Marine Mammal Science*, 23, 731–750.

Moorhouse, T.P., Gelling, M. & Macdonald, D.W. (2009) Effects of habitat quality upon reintroduction success in water voles: evidence from a replicated experiment. *Biological Conservation*, 142, 53–60.

Mueller, J.M. & Hellmann, J.J. (2008) An assessment of invasion risk from assisted migration. *Conservation Biology*, 22, 562–567.

Nel, D.C., Crawford, R.J.M. & Parsons, N. (2003) The conservation status and impact of oiling on the African Penguin. In *Rehabilitation of Oiled African Penguins: a conservation success story*, eds D.C. Nel & P.A. Whittington, pp. 1–7. Birdlife South Africa and the Avian Demography Unit, Cape Town, South Africa.

Nettles, V.F., Shaddock, J.H., Sikes, R.K. *et al.* (1979) Rabies in translocated raccoons. *American Journal of Public Health*, 69, 601–602.

Nilsen, E.B., Milner-Gulland, E.J., Schofield, L. *et al.* (2007) Wolf reintroduction to Scotland: public attitudes and consequences for deer management. *Proceedings of the Royal Society B*, 274, 995–1002.

O'Donnell, M.A. & DeNicola, A.J. (2006) Den site selection of lactating female raccoons following removal and exclusion from suburban residences. *Wildlife Society Bulletin*, 34, 366–370.

Parker, K.A. (2008) Translocations: providing outcomes for wildlife, resource managers, scientists, and the human community. *Restoration Ecology*, 16, 204–209.

Parker, K.A., Seabrook-Davison, M. & Ewen, J.G. (2010) Opportunities for non-native ecological replacements in ecosystem restoration. *Restoration Ecology*, 18, 269–273.

Parsons, N.J. & Underhill, L.G. (2005) Oiled and injured African penguins *Spheniscus demersus* and other seabirds admitted for rehabilitation in the Western Cape, South Africa, 2001 and 2002. *African Journal of Marine Science*, 27, 1–8.

Platenberg, R.J. & Griffiths, R.A. (1999) Translocation of slow-worms (*Anguis fragilis*) as a migration strategy: a case study from south-east England. *Biological Conservation*, 90, 125–132.

Quakenbush, L., Beckmen, K. & Brower, C.D.N. (2009) Rehabilitation and release of marine mammals in the United States: concerns from Alaska. *Marine Mammal Science*, 25, 994–999.

Regan, T. (1983) *The Case for Animal Rights*. University of California Press, Berkeley, California.

Ricciardi, A. & Simberlof, D. (2009a) Assisted colonization is not a viable conservation strategy. *Trends in Ecology and Evolution*, 24, 248–253.

Ricciardi, A. & Simberlof, D. (2009b) Assisted colonization: good intentions and dubious risk assessment. *Trends in Ecology and Evolution*, 24, 476–477.

Richard-Hansen, C., Vie, J.-C. & de Thoisy, B. (2000) Translocation of red howler monkeys (*Alouatta seniculus*) in French Guiana. *Biological Conservation*, 93, 247–253.

Richardson, D.M., Hellmann, J.J., McLachlan, J.S. *et al.* (2009) Multidimensional evaluation of managed relocation. *Proceedings of the National Academy of Science*, 106, 9721–9724.

Ricketts, T. & Imhoff, M. (2003) Biodiversity, urban areas, and agriculture: Locating priority ecoregions for conservation. *Conservation Ecology*, 8(2), 1 [online] URL: http://www.consecol.org/vol8/iss2/art1.

Rout, T.M., Hauser, C.E. & Possingham, H.P. (2009) Optimal adaptive management for the translocation of a threatened species. *Ecological Applications*, 19, 515–526.

Saunders, A. & Norton, D.A. (2001) Ecological restoration at Mainland Islands in New Zealand. *Biological Conservation*, 99, 109–119.

Sax, D.F., Smith, K.F. & Thompson, A.R. (2009) Managed relocation: a nuanced evaluation is needed. *Trends in Ecology and Evolution*, 24, 472–473.

Schaub, M, Zink, R., Beissmann, H. *et al.* (2009) When to end releases in reintroduction programmes: demographic rates and population viability analysis of bearded vultures in the Alps. *Journal of Applied Ecology*, 46, 92–100.

Seddon, P.J. (1999) Persistence without intervention: assessing success in wildlife re-introductions. *Trends in Ecology and Evolution*, 14, 503.

Seddon, P. J. (2010). From reintroduction to assisted colonization: moving along the conservation translocation spectrum. *Restoration Ecology*, 18, 796–802.

Seddon, P.J. & Soorae, P. (1999) Guidelines for subspecific substitutions in wildlife restoration projects. *Conservation Biology*, 13, 177–184.

Seddon, P.J., Armstrong, D.P. & Maloney, R.F. (2007a) Developing the science of reintroduction biology. *Conservation Biology*, 21, 303–312.

Seddon, P.J., Armstrong, D.P. & Maloney, R.F. (2007b) Combining the fields of reintroduction biology and restoration ecology. *Conservation Biology*, 21, 1388–1390.

Seddon, P.J., Armstrong, D.P., Soorae, P. *et al.* (2009) The risks of assisted colonization. *Conservation Biology*, 23, 788–789.

Sekercioglu, C.H., Schneider, S.H., Pay, J.P. *et al.* (2008) Climate change, elevational range shifts and bird extinctions. *Conservation Biology*, 22, 140–150.

SER (Society for Ecological Restoration) International Science and Policy Working Group (2004) *The SER International Primer on Ecological Restoration*. Society for Ecological Restoration International, Tucson. www.ser.org.

Severinghaus, L.L. & Chi, L. (1999) Prayer animal release in Taiwan. *Biological Conservation*, 89, 301–304.

Shieh, B.-S., Lin, Y.-H., Lee, T.-W. *et al.* (2006) Pet trade as sources of introduced bird species in Taiwan. *Taiwania*, 51, 81–86.

Simberloff, D. & Stiling, P. (1996) Risks of species introduced for biological control. *Biological Conservation*, 78, 185–192.

Soorae, P.S. (ed.) (2008) Global re-introduction perspectives: re-introduction case studies from around the globe. IUCN/SSC Re-introduction Specialist Group, Abu Dhabi, UAE.

Spear, D. & Chown, S.L. (2009a) Non-indigenous ungulates as a threat to biodiversity. *Journal of Zoology, London*, 279, 1–17.

Spear, D. & Chown, S.L. (2009b) The extent and impacts of ungulate translocations: South Africa in a global context. *Biological Conservation*, 142, 353–363.

Spencer, P.B.S. & Hampton, J.O. (2005) Illegal translocation and genetic structure of feral pigs in Western Australia. *Journal of Wildlife Management*, 69, 377–384.

Stanley Price, M.R. (1989) *Animal re-introductions: the Arabian oryx in Oman.* Cambridge University Press, Cambridge, UK.

Stanley Price, M.R. & Soorae, P. (2003) Re-introductions: whence and wither? *International Zoo Yearbook*, 38, 61–75.

Temperton, V.M. (2007) The recent double paradigm shift in restoration ecology. *Restoration Ecology*, 15, 344–347.

Temple, S.A. (1992) Exotic birds: a growing problem with no easy solution. *Auk*, 109, 395–397.

Thomas, M.B. & Willis, A.J. (1998) Biocontrol – risky but necessary? *Trends in Ecology and Evolution*, 13, 325–329.

Underhill, L.G., Bartlett, P.A., Baumann, L. *et al.* (1999) Mortality and survival of African Penguins *Spheniscus demersus* involved in the Apollo Sea oil spill: an evaluation of rehabilitation efforts. *Ibis*, 141, 29–37.

van der Merwe, P., Saayman, M. & Krugell, W. (2008) Factors that determine the price of game. *Koedoe*, 47(2), 105–113.

van Vuren, D., Kuenzi, A.J., Loredo, I. *et al.* (1997) Translocation as a non-lethal alternative for managing California ground squirrels. *Journal of Wildlife Management*, 61, 351–359.

Veltman, C.J., Nee, S. & Crawley, M. (1996) Correlates of introduction success in exotic New Zealand birds. *American Naturalist*, 147, 542–557.

Vitousek, P.M., D'Antonio, C.M., Loope, L.L. Rejmanek, M. & Westbrooks, R. (1997) Introduced species: a significant component of human-induced global change. *New Zealand Journal of Ecology*, 21, 1–16.

Vitt, P. Havens, K. & Hoegh-Guldberg, O. (2009) Assisted migration: part of an integrated conservation strategy. *Trends in Ecology and Evolution*, 24, 473–474.

Wakamiya, S.M. & Roy, C.L. (2009) Use of monitoring data and population viability analysis to inform reintroduction decisions: Peregrine falcons in the Midwestern United States. *Biological Conservation*, 142, 1767–1776.

Walker, K.J., Trewick, S.A. & Barker, G.M. (2008) *Powelliphanta augusta*, a new species of land snail, with a description of its former habitat, Stockton coal plateau. *Journal of the Royal Society of New Zealand*, 38, 163–186.

Waples, K.A. & Stagoll, C.S. (1997) Ethical issues in the release of animals from captivity. *Bioscience*, 47, 115–121.

Watson, M. & Thirgood, S. (2001) Could translocation aid hen harrier conservation in the UK? *Animal Conservation*, 4, 37–43.

White, J.P. (2004) Where the wild things are: prehistoric animal translocations in the New Guinea Archipelago. In *Voyages of Discovery: the archaeology of islands*, ed. S.M. FitzPatrick, pp. 147–164. Praeger Publishers, Westport, Connecticut.

Williams, C.K., Ericsson, G. & Heberlein, T.A. (2002) A quantitative summary of attitudes toward wolves and their reintroduction (1972–2000). *Wildlife Society Bulletin*, 30, 575–584.

Wolf, C.M., Garland, T. & Griffith, B. (1998) Predictors of avian and mammalian translocation success: reanalysis with phylogenetically independent contrasts. *Biological Conservation*, 86, 243–255.

Wolf, C.M., Griffith, B., Reed, C. *et al.* (1996) Avian and mammalian translocations: update and reanalysis of 1987 survey data. *Conservation Biology*, 10, 1142–1154.

Zahniser, A. & Singh, A. (2004) Return of the wolves to Yellowstone National Park, USA: a model of ecosystem restoration. *Biodiversity*, 5, 3–7.

A Tale of Two Islands: The Rescue and Recovery of Endemic Birds in New Zealand and Mauritius

Carl G. Jones[1,2] *and Don V. Merton*[†]

[1]Durrell Wildlife Conservation Trust, Jersey, Channel Islands
[2]Mauritian Wildlife Foundation, Grannum Road, Vacoas, Mauritius

'Most extinctions are avoidable – especially as we learn more about the implications of different threats, and as wildlife managers become more skilled and confident in adapting and applying effective conservation management techniques.'

Page 35

Introduction

The world is facing a biodiversity crisis as a result of human activities, with current predictions suggesting we may lose one quarter of all vertebrate species within the next century (Ballie *et al.*, 2010). These declines are most obvious on oceanic islands, where native and often highly endemic faunas continue to be susceptible to changes brought by successive waves of human colonists. In New Zealand, for example, around 50 bird species have been driven to extinction by hunting, ecosystem loss and fragmentation, and predation and competition from exotic species (Craig *et al.*, 2000). Many surviving species have declined and are often preserved only as small, reintroduced

Reintroduction Biology: Integrating Science and Management. First Edition.
Edited by John G. Ewen, Doug P. Armstrong, Kevin A. Parker and Philip J. Seddon.
© 2012 Blackwell Publishing Ltd. Published 2012 by Blackwell Publishing Ltd.

populations on predator-free island refugia (Craig *et al.*, 2000; Jamieson *et al.*, 2006). Mauritius has lost 19 out of 43 native birds and 11 of these are globally extinct (Cheke & Hume, 2008). These extinction drivers make up an 'evil quartet' of habitat loss and degradation, overexploitation, invasive alien species and cascading community effects resulting from these disturbances (Diamond, 1989).

Understanding the causes of decline is central to any attempts at restoration. While some causes of decline, such as habitat destruction, appear obvious, others, such as the impact of various exotic organisms or environmental contaminants have not always been so. The impact of exotic rats (*Rattus* spp.) on some island bird populations was not realized worldwide until comparatively recently with the publication of the conference proceedings on *The Ecology and Control of Rodents in New Zealand Nature Reserves* (Dingwall *et al.*, 1978). This document had limited circulation but was followed by a landmark paper by Atkinson (1985), further raising awareness of the impacts of exotic rodents. New Zealanders have, however, appreciated the damage that rats could do since the invasion of black rats, *Rattus rattus*, on to Big South Cape Island (Taukihepa) in 1962 (Bell, 1978). Within three years of invasion the rats had exterminated five species of native bird and one bat. This tragedy served to convince even the most sceptical that rats are capable of inducing extinction cascades. Until this time few New Zealand biologists believed that rats posed any significant threat to island faunas. This event has had a massive and enduring impact in shaping conservation policy and practice in New Zealand, and on islands around the world (Atkinson, 1985).

Unsurprisingly, similar patterns are seen in Mauritius. There are records of exotic rat predation on the nests of Mauritius fody, *Foudia rubra*, Mauritius kestrel, *Falco punctatus*, pink pigeon, *Columba mayeri*, and echo parakeet, *Psittacula eques* (Jones, 2008). For example, up to 48% of unprotected pink pigeon nests have failed due to predators with exotic crab-eating macaques, *Macaca fascicularis*, and black rats being identified as major nest predators (Swinnerton, 2001; Jones, 2008). Feral cats, *Felis catus*, and the exotic small Indian mongoose, *Herpestes javanicus*, have been identified as the main predators of fledgling Mauritius kestrels (Jones *et al.*, 1994) and, where not controlled, cats are considered the main threat to pink pigeon survival (Jones, 2008).

Unfortunately invasive alien mammalian predators are only one of the threats typically faced by island species. Other invasive alien vertebrate predators and competitors, as well as invertebrates, pathogens and invasive alien plants ('weeds'), are also contributing to the rapid decline of island biotas and to the widespread degradation of island ecosystems. The impacts of invasive alien species are additive to other anthropogenic effects such as habitat destruction and excessive hunting. As a consequence ecological interactions on islands, virtually everywhere, have been seriously disrupted by extirpations, extinctions and introductions. It is against this background that the rescue and recovery of endemic birds in New Zealand and Mauritius has taken place. Here, we present personal perspectives based on our collective observations from more than 70 years of wildlife management in New Zealand and Mauritius. We summarize examples of bird conservation activities in both countries where different approaches and techniques have been applied to avert extinction, recover species and create opportunities for ecological restoration. In both New Zealand and Mauritius, species-focused management has driven the restoration of ecosystems on satellite islands and at main island sites.

While much of the conservation work on New Zealand and Mauritius is pioneering, we acknowledge the early influences of the work on endangered species management by the Peregrine Fund (Cade, 1980–2000; Cade & Burnham, 2003), the Patuxent Wildlife Research Centre (Erickson, 1980), the Wildfowl and Wetlands Trust (Kear, 1975a–1986; Kear & Berger, 1980), the Durrell Wildlife Conservation Trust (Durrell, 1990; Durrell & Durrell, 1980) and others (e.g. Martin, 1975; Temple, 1978; Olney *et al.*, 1994; D'Elia, 2010).

A range of management options

Most extinctions are avoidable – especially as we learn more about the implications of different threats and as wildlife managers become more skilled and confident in adapting and applying effective conservation management techniques. We summarize the approaches that have been used in various combinations to avert extinctions and to recover endemic bird populations in New Zealand and Mauritius. These are:

- re-enforcement and reintroduction,
- eradication or control of invasive alien species and
- intensive management (also referred to as 'close order management').

Monitoring and research

While not technically a management action, research to underpin management and monitoring to measure and interpret results are both key activities. With any focal species there is a need to have an understanding of distribution, numbers and population trend, and to be able to establish a need for conservation before starting any form of management. For evaluating causes of population decline it is essential to collect data systematically on productivity and survival and the factors influencing these. To be able to interpret such data and to be able to manipulate productivity and survival effectively, a thorough knowledge of the life history and ecology of the focal species is necessary.

A major advance in species-focused conservation occurred in the 1940s–1950s when a coherent theory of what factors limited populations emerged (Lack, 1954, 1966; Newton, 1998). This gave bird managers the information on which to build a framework for their work. Most bird populations are naturally regulated by relatively few variables, of which food, availability of safe nest sites, predation, competition and disease are among the most important (Newton, 1998). It can be assumed that one variable, or an interaction of these variables, will be limiting the population of most critically endangered bird populations. A threatened species may respond to a broad approach addressing several of the more likely limiting factors (see Table 2.1 for a summary of corrective measures). In declining or very small populations productivity and/or survival may be enhanced by management without fully understanding the causes of population failure. For example, recovery programmes for the Chatham Island black robin, *Petroica traversi*, pink pigeon and the echo parakeet addressed the species' extreme rarity by providing supplemental feeding, enhancing and protecting nest sites, controlling/excluding predators and competitors around nest sites and controlling parasites at nest sites (Butler & Merton, 1992; Jones *et al.*, 1992, 1998, 1999; Jones & Duffy, 1993; Jones & Swinnerton, 1997; Jones, 2004) (see Tables 2.2 and 2.3).

These management actions resulted in population increases for all species even though it was not definitively known at the time what was restricting the populations or which management actions were critical for achieving recovery. In these cases, where several management approaches were attempted at once, it was not initially clear how each was affecting the population, but once populations had increased the level of management was relaxed or adjusted, reducing management practices one at a time. By doing this it was possible to see the impacts of the different management scenarios. Improving nest

Table 2.1   Causes of population limitation and corrective measures.

Causes of population limitation	Corrective measures
Food shortage	Supplementary feeding
Interspecies competition	
• Nest site competition	Nest site design, excluding competitors
• Competition for food	Feeding hoppers that favour the focal species
Predation	
• Predation upon adults	Predator control, eradication or exclusion
• Predation at nest	Predator exclusion, improved nest site design, egg and brood manipulations
Nest site limitation	Nest site enhancement, nest boxes
Parasites and pathogens	
• Nest site parasites	Treatment of nest sites with insecticides
• Disease transmission at supplemental feeding sites	Improvement of feeding protocols including better hygiene
• Disease transmission between populations	Disease screening and quarantine of birds being translocated

sites led to improved productivity in the echo parakeet, suggesting that the shortage of high-quality nest sites had previously been limiting, whereas with the pink pigeons the provision of supplemental food boosted productivity. In ideal scenarios management approaches can be trialled and implemented before species reach such a perilous status.

Translocation and reintroduction

Translocations for re-enforcement and reintroduction are important tools used by conservation managers in both New Zealand and Mauritius (see Seddon *et al.*, this volume, Chapter 1 for definitions). We note, however, that translocating animals entails inherent risks. Wherever possible the preferred option should be to manage species where they naturally exist. However, on islands such as New Zealand and Mauritius many unmanaged populations will continue to decline to certain extinction without management interventions. Therefore, translocation to suitable safe areas will remain an essential tool for recovery.

Table 2.2 **Summary of the previous status, recovery and main management initiatives, and the current status of key bird species examples from Don Merton's work in New Zealand.**

Problem	Solution	Outcome
North Island saddleback (c. 500 birds in 1964; current population estimate >10 000 birds in 2011)		
Range restricted (one island population)	Translocation to islands where exotic predators have been eradicated	13 island populations and two mainland populations established
Nest and roost site shortage	Provision of artificial nest and roost boxes	Successful
Predation by exotic mammals	Behavioural manipulation to encourage use of safe roosting and nesting sites	Unsuccessful
South Island saddleback (minimum of 36 birds in 1964; current population estimate >1 200 birds in 2011)		
Range restricted (one island population)	Translocation to islands where exotic predators have been eradicated	15 island populations established
Chatham Island black robin (minimum of 5 birds in 1979; current population estimate of 200 in 2011)		
Range restricted (one island population)	Translocation	Two island populations established
Deteriorating habitat	Translocation	Successful establishment of two island populations. There are currently no birds on the original island
Low productivity	Cross-fostering to Chatham Island warblers and Chatham Island tits	Increased productivity. The warblers were not suitable foster parents but tits successfully reared robin chicks. Imprinting problems with tit-reared robins were resolved through intensive management of nests and switching between tit and robin foster parents

Table 2.2   (*Cont'd*)

Problem	Solution	Outcome
Poor nest sites and seabird collision	Nest site repair and modification and nest boxes provided	Increased productivity and reduced nest losses
Nest site competition with exotic starlings	Protective mesh over nest boxes to exclude starlings	Increased productivity and reduced nest losses
Nest mite infestation	Replacement of nests and control with miticides	Increased productivity and reduced nest losses
Kakapo (minimum of 50 birds in 1995; current population of 122 in 2011)		
Exotic predators	Translocation to predator free islands	Predation stopped
Range restriction	Translocation	Ongoing translocation of birds to meet breeding targets and maximize genetic diversity
Infrequent breeding	Supplementary feeding	
Poor chick survival	24/7 nest minding, heat pads applied during female absences, hand rearing	Increased productivity and reduced nestling losses
Harassment of chicks by rodents	24/7 monitoring of nests, rat aversion devices and rat trapping at nest sites	Increased productivity and reduced nestling losses
Poor genetic diversity	Management of mate access and collection of sperm for artificial insemination	

Examples where people have deliberately reintroduced or introduced (released outside the historical range) free-living animal populations are found throughout history and in most cultures, some dating back hundreds, if not thousands, of years. Birds have been moved within and beyond natural biogeographic ranges for reasons as diverse as for food, recreation and commerce, as well as for cultural and aesthetic purposes (Long, 1981; Lever, 2005). Translocating threatened species to aid their conservation has been a relatively recent innovation.

Table 2.3 **Summary of the previous status, recovery and main management initiatives, and the current status of key bird species examples from Carl Jones's work in Mauritius.**

Problem	Solution	Outcome
Mauritius kestrel (minimum of 4 birds in 1974; current population estimate in 2011 of 500)		
Small population	Captive breeding Captive rearing and release by fostering or hacking	Produced 110 young that fledged between 1985 and 1993, with 331 birds released over 10 years, 32% by fostering of young between 5 and 18 days old to pairs on eggs (normally incubated for ≈ 2 weeks) (Cade & Jones, 1994; Jones *et al.*, 1994)
Nest predation	Nest guarding and predator trapping	Predation by black rats, feral cats, crab-eating macaques and lesser Indian mongooses minimized
Predation of recent fledged young	Trapping of cats and mongooses around nest sites	Rates of predation minimised (Jones *et al.*, 1991)
Low productivity	Induce double clutching by removing eggs	266 eggs harvested over 10 seasons and re-nesting occurred 95% of the time. Fertility rate of 70% for harvested eggs although fertility declines as more clutches are induced (by fourth clutch all infertile). Success of repeat clutches lower (34% fledged) compared to unmanipulated first clutches (68% fledged) (Cade & Jones, 1994; Jones *et al.*, 1994).
	Supplemental feeding	Lay more clutches and raise more young (Jones *et al.*, 1991)
	Foster captive reared young to wild pairs	Increases productivity (Jones *et al.*, 1991, 1994)

Table 2.3   (*Cont'd*)

Problem	Solution	Outcome
Poor nest sites (overheating, vulnerable to predation)	Nest modification. Rocks removed, sun shade and perches added, substrate replaced. Poor sites blocked and new sites added including monkey proof nest boxes	Improved nesting success and increased number of birds breeding
Echo parakeet (minimum of 8–12 birds in 1987; current population estimate in in 2011 of 570)		
Small population	Initiated captive breeding	Captive breeding successful, over 60 young produced 1993–2008
	Release of captive-bred and young reared from downsizing and rescuing from wild	Released 139 captive reared birds (84 from wild sourced eggs or young and 55 captive bred) between 1997 and 2005 (Cheke & Hume, 2008)
Poor rearing rate in wild	Rescuing eggs and young from failing nests	Young successfully reared and used for captive breeding or released (Jones *et al.*, 1998)
	Cross-fostering rescued young to reliable rose-ringed parakeet parents in captivity	All young reared by rose-ringed parakeets maintained in captivity for captive breeding because of possible mal-imprinting
	Testing quality of echo parakeet parents by giving rose-ringed parakeet eggs and young to rear.	Identification of reliable echo parakeet parents. Technique used in wild and captivity young removed before fledging from wild nests because they would imprint to the wrong species
Infertile eggs	Replacement with fertile eggs	Trials successful and young fledged

(*Continued*)

Table 2.3   (*Cont'd*)

Problem	Solution	Outcome
Low productivity	Induce double clutching by removing eggs	Rate of re-nesting variable (64% of time) so not widely used
	Supplementary feeding	Mostly used by captive-reared birds, their partners and progeny. Parakeets taking supplemental feeding have larger broods than wild breeders. About 50% of population take supplemental food in 2011
Poor nest sites	Destruction of poor sites. Drainage holes, weather guards added. Substrate changed before and during nesting. Entrance holes reduced to exclude competitors. Predator guards added	Increased nesting success
Lack of nest sites	Nest boxes	Modification until a design was accepted. Now widely used, especially by released birds and their progeny. In the 2009–2010 season 33 of 72 (46%) known breeding pairs used artificial nest boxes
Nestling mortality, wild pairs rarely able to fledge more than 1 or 2 young	Downsizing broods to 1 by removing chicks at 5–8 days old. Chicks fostered to wild pairs or captive reared	Increased productivity and a higher percentage of pairs rearing young
Juvenile mortality due to PBFD	Strict protocols when examining nests to avoid spreading virus between sites	Reduced incidence of disease (Raisin, 2010)

Table 2.3   (*Cont'd*)

Problem	Solution	Outcome
Loss of genetic diversity	Maximizing productivity from all breeding pairs	Preserving what genetic diversity still exists in the population (Raisin, 2010)
Pink pigeon (minimum of 9–10 birds in 1990; current population estimate in 2011 of 400)		
Low population levels	Captive breeding and release	256 captive reared birds released between 1987 and 1997. Higher survival post-release if younger than 150 days (Swinnerton, 2001)
Small distribution	Pigeons released in additional sites	In 2011 a total of six subpopulations
Poor parenting skills in captivity	Cross-fostering eggs to Barbary doves	669 fertile eggs, most hatched under Barbary doves, 440 eggs hatched and 272 reared to fledging with no mal imprinting (1977–1993)
	Give pink pigeons Barbary dove nestlings	Improvement in the parenting ability of the pigeons
Nestling mortality	Squab rescue and captive rearing	58 rescued <25 days old. 36 survived to 30 days and 31 released
Low productivity	Supplementary feeding maize and wheat	Widely used by wild birds. Increase productivity and probably survival
Releasing captive reared young	Replacement fostering to wild females on infertile eggs.	Females accept squabs only if presented 1–2 days around expected hatch date and $\leq$ 10 days old. 11 squabs fostered of which 4 fledged
Infertile eggs in wild	Replacement with fertile eggs	Trial successful but discontinued due to the need to closely synchronize laying dates. Pink pigeons did not successfully rear young if they hatch more than a day either side of the normal 14 day incubation period

(*Continued*)

Table 2.3   (*Cont'd*)

Problem	Solution	Outcome
Trichomoniasis infecting adults and killing squabs	Control of exotic dove reservoir host	Constant reinvasion has meant that culling is not effective
	Exclusion of exotic doves at feeders	Successful
	Treatment of infected adults	Early trials suggest increased breeding success
Mortality caused by non-native predators	Predator control	Improved adult survival rates.
	Rat control to reduce nest predation	Discontinued as productivity high without any rat control
Mauritius fody (minimum of 100 breeding pairs (c. 300 birds) in 2001; current estimate in 2011 of 160 breeding pairs (c. 500 birds))		
Declining population	Captive breeding	In 16 breeding attempts, 58 eggs laid, 46 hatched and 34 raised to fledging and 31 released on Ile aux Aigrettes (Cristinacce *et al.*, 2009)
Limited distribution	Establish an additional population on Ile aux Aigrettes. 93 released 2003–2006	New population established of about 160 birds and 50 breeding pairs (2011)
Poor nest success	Harvest and rescue of eggs and young	73 birds hand-reared and 62 released on Ile aux Aigrettes
Food shortage	Supplemental feeding (Ile aux Aigrettes)	Improved breeding success and survival

Perhaps the first translocation of birds for conservation purposes was in the 1890s when the New Zealand Government recognized that some native birds, particularly flightless species, were in steep decline as a result of predation by introduced mustelids (ferrets, *Mustela putorius furo*, stoats, *M. erminea*, and weasels, *M. nivalis*). Resolution Island (23 000 ha) in Dusky Sound, Fiordland, was designated a sanctuary and the government employed an Irishman, Richard Henry, as custodian (Hill & Hill, 1987). Part of Henry's job involved the transfer of kakapo *Strigops habroptilus* and kiwi *Apteryx* spp., from the South Island 'mainland' for release on the island. Several hundred flightless birds were introduced to Resolution Island. Unfortunately, stoats reached the island in 1900 and the venture failed. The first successful avian conservation translocations appear to be those in the early 1960s involving two species of New Zealand wattlebirds, the North Island and South Island saddlebacks (*Philesturnus rufusater* and *P. carunculatus*). By the early 20th century the saddlebacks had been exterminated from their mainland ranges by invasive alien mammalian predators and were each confined to a single island. Following three unsuccessful attempts in the 1920s–1950s effective translocation techniques were pioneered and self-sustaining populations of both species were established on a number of different islands (Merton, 1965, 1975; Lovegrove, 1996). Translocations have subsequently become essential for the recovery of many New Zealand birds (see the Reintroduction Specialist Group Oceania Section website for an up-to-date summary: http://rsg-oceania.squarespace.com/nz/).

Although reintroduction into part of their former range (see Seddon *et al.*, this volume, Chapter 1) is preferred, this is not possible in many situations due to the unsuitability of historic habitat. The factors making habitats unsuitable are varied but by the mid to late 20th century it was recognized that invasive alien mammalian predators were a key factor in biological declines, particularly on oceanic islands. This led to the idea of 'marooning' endangered species on predator-free islands – perhaps outside their natural ranges (i.e. a conservation introduction – rather than a reintroduction), but in habitats that were suitable for their persistence and in which they were able to be managed (Williams, 1977). A dramatic illustration of this approach was when a breeding population of kakapo was first discovered on Stewart Island only to find the population was in steep decline due to feral cat predation with about 56% of all radio-tagged birds depredated (see Powlesland *et al.*, 2006, and references therein). All 61 remaining kakapo were translocated to cat-free islands (Bell & Merton, 2002), thereby averting certain extinction.

Translocations have received a great deal of attention due to their high profile nature. While there is still a great deal to learn about how to translocate endangered species, successes are increasing as we learn more about the needs of different species and appropriate management techniques have been refined. Releases may be 'hard' or 'soft'. A hard release is when the bird is released without any preliminary conditioning to the new area and is not given any support on the assumption that it will be able to look after itself. This technique is most commonly employed when translocating self-sufficient juveniles and adult wild birds from one area to another. In soft releases birds are held at a release site in an aviary and habituated to the new area prior to their release and are provided with some support during the release process. Soft releases are the release technique of choice when releasing captive-bred or hand-reared birds. These releases vary in the level of support provided to the birds but provision of food and water is typical, and where appropriate predators are also controlled or excluded from around the release site so the birds gradually become independent of human care. A soft release typically goes through three phases:

- Pre-release training and conditioning
- The release process
- Post-release support.

Birds produced in captivity for release must be appropriately socialized, with care taken to ensure that the stimuli during early learning stages are appropriate. For most releases using captive-reared birds, it is better to have parent-reared or foster-raised birds, rather than those that are hand-reared since they will have experienced normal species imprinting and socialization. We have found that if hand-raised birds are being used it is important that they are raised with siblings to ensure socialization with conspecifics, or if they are being raised alone that they are fed with the aid of a hand puppet that mimics the head of the adult so that the young birds imprint upon an appropriate image (Wallace, 2000). Puppet rearing, for example, has been successfully used in the hand rearing of takahe, *Porphyrio mantelli*, intended for release (Eason & Willans, 2001). Attention also needs to be given to early learning for those species that need to learn nest site, habitat and feeding characteristics. In all cases, if possible, individuals should be provided with opportunities to develop physical and survival skills (Wallace, 2000).

Some of the most successful releases of captive-reared individuals have been with birds of prey. This is because their conservation management is well understood (Jones, 2004). For example, post-release survival in captive-reared Mauritius kestrels is comparable to that of wild fledged birds (Nicoll *et al.*, 2004). Young kestrels were introduced to the wild by fostering to nesting wild kestrels, or by 'hacking', a soft-release technique widely used with birds of prey (Cade & Jones, 1994; Sherrod *et al.*, 1981) (see Table 2.3). Post-release survival of captive-reared takahe equals that of wild-reared birds (Maxwell & Jamieson, 1997) while in pink pigeons, released and wild fledged young survive equally well to one year old (Swinnerton, 2001). The high survival rate of released pink pigeons is probably because of the intensive post-release support that they received (Table 2.3). It seems in most species that the earlier post-fledging that individual birds can be released, the better their survival. Presumably this is because younger birds are more adaptable and capable of learning survival skills (e.g. see the pink pigeon, Table 2.3). In social or gregarious species, such as the pink pigeon and echo parakeet, releases may be more successful if there are other birds from previous releases from which they can learn. However, in other social species the presence of conspecifics may decrease post-release survival (e.g. in the New Zealand hihi *Notiomystis cincta* (Castro *et al.*, 1994)) and there may be no benefit in releasing social species in groups or with familiar individuals (e.g. New Zealand's North Island saddleback, North Island robin *Petroica australis longipes*, whitehead *Mohoua albicilla* (Armstrong *et al.*, 1994; Armstrong, 1995; Armstrong & Craig, 1995) and hihi (Castro *et al.*, 1994)). These examples emphasize that the translocation and management of social species poses specific challenges, especially for those with complex learning traditions. For example although we advocate that species should be released at ages comparable to the natural fledging age, we recognise that in some species including takahe (Maxwell, 2001), Griffon vultures *Gyps fulvus* (Sarrazin & Legendre, 2000) and California condors *Gymnogyps californianus* (Clark *et al.*, 2007; Wallace *et al.*, 2007), they fare better if released at a year old after being appropriately socialized with older birds.

Invasive species management, particularly the eradication of suites of predators and competitors (see below), may create high-quality habitats that are suitable for many threatened species. In New Zealand this has facilitated the hard release of many translocated species with the subsequent management focus being on maintaining the predator-free status of the habitat; the species themselves often require little post-release management. In our view, however, post-release monitoring is still important to determine the success of the translocation and to identify if any further management is required. Translocations should involve a sequence of planned actions, including monitoring of numbers, productivity and survival and, where necessary, trialling management options and research focused on better understanding of the ecology of the species and fine tuning the long-term conservation management of the species.

Eradication of alien invasive mammals

The realization that many bird species could not co-exist with exotic mammalian predators led to work in New Zealand to eradicate a growing range of invasive species from increasingly large and challenging islands. More recently intensive control of suites of invasive species from habitats on the main islands of New Zealand (so-called 'Mainland Islands') has led to island-type responses in native plants and animals (Saunders, 2000). The removal of invasive predators from islands and Mainland Islands has allowed populations of native species to rebound – in some cases spectacularly. On 3083 hectare Little Barrier Island (Hauturu), for example, the eradication of feral cats resulted in a substantial increase in the numbers of hihi, which had become restricted to this single island by the late 19th century (Veitch, 1985). Eradications also create opportunities for further endangered species to be reintroduced (or, perhaps, introduced as a marooning measure); on Little Barrier Island this included the successful reintroduction of kokako *Callaeas cinerea* and the introduction of kakapo. This technique of eradicating invasive species from islands, followed in some cases by reintroductions of extirpated species as part of community-focused restoration programmes has emerged as a key biodiversity conservation strategy for the New Zealand Department of Conservation and is also a developing strategy of the Mauritian Wildlife Foundation.

Following the example set by New Zealand, efforts to clear Mauritian islands of invasive mammals were initiated in the 1970s. Important species recovery goals were achieved in the 1980s and 1990s with many projects continuing today (Jones, 2008). Following the eradication in 1991 of black rats and feral cats from Ile aux Aigrettes (25 ha) the island has been used to establish communities of Mauritian plants, reptiles and birds. Mauritius kestrel (Jones *et al.*, 1991, 1994), pink pigeon (Swinnerton, 2001), Mauritius fody (Cristinacce, *et al.*, 2009) and olive white-eye *Zosterops chloronothos* have all been translocated to the island. Whereas the Mauritius kestrels all dispersed to the mainland, where they have helped establish a population, the other three species have established breeding populations on the island, albeit with ongoing supportive management (see below). More recently young seabirds, wedge-tailed shearwater *Puffinus pacificus*, white-tailed tropicbird *Phaethon lepturus* and red-tailed tropicbird *P. rubricauda* have been harvested as well-grown nestlings from Round Island, where there are large colonies, and reared on Ile aux Aigrettes and allowed to fledge from artificial nest sites. Since seabirds are strongly philopatric it is hoped they will return and breed on Ile aux Aigrettes where they once occurred before the introduction of rats and cats.

Following the successes on Ile aux Aigrettes, we have been restoring the much larger Round Island (219 ha) for over a decade (Jones, 2008) and a trial release of Mauritius fodies has been attempted in early 2011.

Intensive management

Intensive management of wild and free-living birds was developed in the 1980s largely through work on the Chatham Island black robin in New Zealand and later on the Mauritius kestrel, pink pigeon and echo parakeet. Intensive management focuses on the individuals in the population to ensure that their survival and productivity are maximized. This is achieved by providing the necessary conditions and care (e.g. food, water, predator control, nest site enhancement, disease control) and employing clutch and brood manipulations to boost productivity as well as reacting to any problems that may occur. This approach was essential for the rescue and recovery of all four species (Butler & Merton, 1992; Cade & Jones, 1994; Jones *et al.*, 1991, 1992, 1994, 1998, 1999; Jones, 2004; Jones & Duffy, 1993; Jones and Swinnerton, 1997; see Table 2.2). The techniques for intensive management are continually being refined and developed, and utilize advanced aviculture methods (Jones, 2004) prompting Bell & Merton (2002) to talk about 'taking captive breeding into the field'. For example, intensive management has been employed extensively for wild kakapo and has included a range of egg and brood manipulations and the guarding of nesting pairs – in addition to captive management, artificial incubation, translocations and habitat management (Bell & Merton, 2002; Elliot *et al.*, 2001). Captive management and the reintroduction of captive-bred birds have played a role in the reintroduction of several New Zealand species (Merton *et al.*, 1999), albeit with a preference for *insitu* management. For example, the rescue and recovery of the black stilt *Himantopus novaezelandiae*, which declined to a population minimum of 23 adults in 1981, the takahe, which was presumed extinct until rediscovery in 1948, and several declining kiwi species has involved a range of intensive management approaches in addition to translocations and habitat management (Keedwell *et al.*, 2002; Colbourne *et al.*, 2005; Lee & Jamieson, 2001).

Nest guarding and monitoring

For many of the species we have worked with, guarding of nests and the intensive monitoring of all pairs has been a common feature. Nest guarding has been an important component in the recovery of kakapo, Chatham Island black robin, Mauritius kestrel, pink pigeon and echo parakeet with increased productivity of focal pairs (e.g. Butler & Merton, 1992; Jones & Duffy, 1993;

Jones *et al.*, 1998; Elliot *et al.*, 2001). The main purposes of nest guarding are to:

- Monitor the progress of the focal pairs to ensure maximum productivity, i.e. all viable eggs hatch and all chicks fledge.
- Build a body of knowledge on the individual birds/pairs and on the biology and behaviour of the species.
- Assess the suitability of the pair for possible clutch and brood manipulations and to plan the timing of any manipulations.
- Monitor the results and progress of manipulations.
- Provide support for the focal pairs and react to problems that threaten them and their nesting attempts (e.g. nest site enhancement, supplemental feeding, parasite and predator control and deterrence).
- Rescue clutches and broods from failing nesting attempts or, if necessary, hand-feed and re-hydrate failing chicks in the nest.

The amount of time spent observing nests varies from project to project, but in the most critically endangered species that are being intensively managed, there may be 24-hour guarding and monitoring. Some nests have been monitored using both video systems and 24-hour human nest minders (e.g. kakapo; see Elliot *et al.*, 2001, and Table 2.2), and field teams live in the field where they have easy access to the breeding birds. Large teams of volunteers have often been involved, necessitating clear guidelines for procedures if nests show signs of failing, as well as on-hand expert supervision and advice. A great advantage of intensive nest guarding is that it may reveal hitherto unknown problems. For example, the first releases of Mauritius kestrels in lowland Mauritius showed that the young kestrels spent a lot of time on the ground playing, where they were very susceptible to both mongoose and cat predation and close monitoring of pink pigeon nests confirmed that monkeys were serious predators. A subsequent nest guarding and trapping programme around some nest sites and most release sites minimized losses to these predators (Jones *et al.*, 1991, 1992, 1994; Swinnerton, 2001).

Egg and brood manipulations

Egg and brood manipulations have also been widely used in our work in New Zealand and Mauritius (Tables 2.2 and 2.3). The most common approach has been to induce double clutching. Many species of birds will lay a replacement clutch if the first nesting attempt fails. Some species can lay several clutches in a season. This ability to re-lay can be exploited by removing the first clutch, and sometimes successive clutches, for artificial incubation and rearing (or fostering). A focal pair can be left with a clutch for them to incubate and to rear the young naturally, or to replace with a clutch of dummy eggs that can later be exchanged with fostered young.

This technique was extensively used in the recovery of the Mauritius kestrel (Cade & Jones, 1994; Jones *et al.*, 1991, 1994; Nicoll *et al.*, 2006; Butler *et al.*, 2009; see Table 2.3). After the first few years of experimental manipulations on Mauritius kestrels a policy for the harvesting of eggs from wild pairs was developed. No pair was made to lay more than one extra clutch in a season and the extending of clutches by sequentially removing eggs was limited to captive birds. Harvested eggs of many species of wild birds have a better hatchability if they have received some natural incubation (Burnham, 1983), yet the birds recycle more readily if the eggs are harvested soon after the first clutch has been laid. In Mauritius kestrels the eggs were harvested about five to seven days after the completion of the clutch. Following the removal of the clutch Mauritius kestrels would usually move their nest site so alternative nest boxes were provided. First-time breeders were left with their first clutch and not encouraged to lay additional eggs. It was considered important that young birds succeeded in rearing young if they were to become good breeders. In those cases where eggs were harvested from young birds (or their eggs were infertile), the eggs were replaced with dummy eggs and subsequently with fostered young (Cade & Jones, 1994; Jones *et al.*, 1991, 1994). The tendency for replacement clutches to be laid in alternative nest sites has also been used as a management approach to move Mauritius kestrels from cavities that were accessible to predators to predator-proof nest boxes.

Double-clutching was also trialled in the echo parakeet recovery programme, but with mixed success, and has not been incorporated as a regular management technique (Jones & Duffy, 1993; Jones *et al.*, 1998; see Table 2.3). A more successful strategy has been downsizing of broods. Wild echo parakeets usually lay clutches of two to four eggs but generally if unmanaged only rear one, sometimes two, young. Brood reduction is usually due to insufficient natural food availability and these methods can be reactive or proactive responses to the situation whereby most species of birds hatch more young than they rear. Reactive downsizing of broods usually follows detection of a chick doing poorly. Rescued chicks are likely to be underweight and dehydrated and without management intervention survival is unlikely. Reactive downsizing of broods has also been implemented in kakapo (Elliot *et al.*, 2001; see Table 2.2), Mauritius kestrel and pink pigeon (Table 2.3).

Downsizing echo parakeet broods proactively, before any of the chicks became compromised by a lack of food, was the manipulation of choice and the harvested young can be fostered to other pairs with small broods or brought into captivity for hand-rearing. Proactive egg manipulations have also been used extensively in the management of other species such as takahe. Takahe often lay two fertile eggs, but chicks raised alone have a higher survival rate than multiple siblings (Lee, 2001). Fertile eggs are, whenever possible, moved between nests to achieve one fertile egg per pair (Maxwell, 2001) and fertile eggs have also been brought into captivity and young reared to one year old before being released into the wild (with much higher survival) (Maxwell, 2001). The artificial incubation of eggs and rearing of young helped establish a captive population and provided management options such as translocation of birds to offshore island refuges. The replacement of infertile eggs with fertile eggs is also very effective and has been used successfully with takahe, echo parakeets and pink pigeons (Maxwell, 2001; Jones *et al.*, 1998; see Table 2.3). Egg replacement is also a useful technique for introducing captive-produced eggs into a wild population.

Egg removal and fostering

The sequential removal of eggs soon after they are laid, and not allowing the bird to complete a clutch, can greatly increase the number of eggs that the bird lays (Jones, 2004). A bird that normally lays a four-egg clutch may lay two or three times this number in succession. This technique works only on birds that have an indeterminate clutch size and is most often used on captive birds where the laying of eggs can be carefully monitored. The few cases where this has been tried in the wild show that there is a great deal of difference in the way species, and individuals, respond to sequential egg removal. Therefore, this technique has limited application to wild birds and can only realistically be used with accessible pairs that will tolerate disturbance at such a sensitive stage. It is usually more effective to remove the whole clutch of eggs to encourage laying of replacement clutches (see above).

The sources of young used in fostering attempts have been captive bred or derived from wild pairs from clutch and brood manipulations. Fostering has been widely practised by aviculturists on a range of species from many different orders. Work on wild birds has been limited, and the most detailed studies have been with birds of prey. Similarly to moving eggs, species that are *K* selected are usually better candidates for fostering manipulations than *r* selected species, where losses of young can be high. It is also important to know the rearing history of the pair when fostering young, especially if a large brood is to be fostered or the pair is to be given young that deviate in age and size from their own. The more experienced and successful the pair at rearing young, the more liberties that can be taken. Some pairs are poor at rearing and can never be trusted with their own or fostered young.

The young to be fostered should not yet have fully formed their species identity or developed their fear reactions because once formed this may cause rejection of attending adults. In birds that produce altricial young this is usually during the second half of the nestling period (although there is a lot of interspecies variation) and is a particular problem if the birds have been hand reared prior to fostering. Fostering attempts with species that produce precocial young are usually done as eggs since the young form their species identity (or their filial attachments) soon after hatching (Hess, 1973).

Breeding pairs that fail in their breeding attempt in the first 48 hours after the eggs have hatched will sometimes stay in the nest cavity and incubate/brood an empty nest scrape and some birds will readily accept fostered eggs or fostered young. This strategy of fostering young has worked with wild Mauritius kestrels and echo parakeets. In some species where there is asynchronous hatching (raptors, parrots) the smaller young have poorer survival. The swapping of young between broods, so that all of the young are about the same size, may enhance the survival of the compromised young and result in increased brood size at fledging.

Fostering was an important tool in the restoration of the Mauritius kestrel and was found to be the easiest of the release techniques (Cade & Jones, 1994; Jones *et al.*, 1991, 1994; see Table 2.3). Replacement fostering was also attempted with pink pigeons as a potential way of releasing captive bred young. It is not as labour intensive as the releases of older birds and allows the young bird to learn social and survival skills from its foster parents (Jones *et al.*, 1991; see Table 2.3).

The most important work on fostering, in terms of being able to inform us about the application and value of this and cross-fostering in a wild situation was with the Chatham Island black robin. Here intra- and intergeneric cross-fostering was tried, first to the Chatham Island warblers, *Gerygone albofrontata*, and then to Chatham Island tits, *Petroica macrocephala chathamensis*. While Chatham Island warbler foster parents successfully incubated eggs and fed young chicks, they were unable to raise Chatham Island black robin nestlings to fledging. In contrast, the Chatham Island tits readily raised fostered Chatham Island black robins to independence but they were imprinted to the tits and tended to reject their own species. Mal-imprinting was solved by reverse fostering Chatham Island black robin chicks back into conspecific nests when they were 14–15 days old. As with the Mauritius kestrel (above), this management technique increased productivity and is heralded as a major factor in restoring Chatham Island black robin numbers (Butler & Merton, 1992).

Head starting

Head starting, where the species is reared in captivity from wild harvested eggs and/or young and then released to the wild when they are older and more able to cope with predators, is a technique developed on chelonians to boost juvenile survival (Jones, 2002), but has been used successfully with kiwi *Apteryx* spp. – perhaps the most intensive use of this technique on birds. This work started in 1994 with a series of studies on the brown kiwi *A. mantelli* and *A. rowi* and has become popularly known as 'Operation Nest Egg' (Colbourne *et al.*, 2005). Kiwi eggs are harvested from the wild, hatched and reared to about six months of age in captivity or a predator-free environment such as an island or fenced area, and are then released back to the wild. At this age kiwi are able to fend for themselves without parental care and fend off stoats, the main predator of juvenile kiwi. Colbourne *et al.* (2005) reported that of 286 eggs removed from the wild 199 (69.6%) hatched and survival to six months in captivity was 81%. This is considerably better than the 11% observed in wild unmanaged birds. Annual survival of released birds was about 74%. Across studied kiwi populations about 21% of eggs collected for head starting result in wild breeding adults three to four years later (Colborne *et al.*, 2005). Again, this is considerably better than the less than 5% of chicks that survive to adulthood without management (Colborne *et al.*, 2005; Robertson *et al.*, 2010).

Supplementary feeding

Understanding the role food plays in influencing the numbers of an endangered bird species may be central to its long-term management. Food shortages impact upon different populations in different ways and there are likely to be seasonal variations to food availability. Supplementary feeding studies have shown a range of effects upon wild bird populations, although not all populations respond in the same way. Some of the main advantages include:

• Increasing the percentage of birds breeding.
• Improving productivity by inducing earlier laying and increased clutch size.
• Enhancing breeding success, by increasing the number of young fledged per breeding attempt.
• Improving adult and juvenile survival.
• Facilitating monitoring: feeding stations can be within trap aviaries/cages so the birds can be easily caught for ringing, disease screening, weighing, etc.

These impacts have all been demonstrated in ecological studies and clearly show the importance of food supply in influencing bird numbers (Newton, 1998). Not surprisingly, supplementary feeding has been an important component in bird recovery projects, often implemented alongside other management (Edmunds *et al.*, 2008). Some species may be easy and straightforward to provide with additional food, such as the pink pigeon, which readily comes to feeding stations for grain (Edmunds *et al.*, 2008) (Table 2.3). Other species are more problematic to feed. For example, Bessinger & Bucher (1992) suggested supplementary feeding as a possible way of increasing parrot productivity but did not know of any field studies that demonstrated this. This was in part because some parrot species do not readily recognize and take supplemental food. Since then supplementary feeding programmes have been initiated for the kakapo and echo parakeet with positive results (Tables 2.2 and 2.3).

For kakapo, hoppers containing food are placed in the home ranges of females and each female is individually managed. Feeding started on Little Barrier Island in 1989 and breeding occurred in the three subsequent summers. Prior to this there was no evidence of breeding during the seven years that the birds had been on the island (Powlesland *et al.*, 1992). While food supplementation sometimes promotes breeding this is not always so with kakapo (Elliot *et al.*, 2001). It seems food supplementation allows females to reach a nutritional threshold that enhances breeding but other environmental triggers are also needed (Elliot *et al.*, 2001). Furthermore, excessive supplementation may lead to sex biases in offspring with females in good condition producing more male chicks (Clout *et al.*, 2002). Feeding is now carefully managed such that female kakapo have access to food supplements until they reach the nutritional threshold required for breeding, but then only have restricted access to feeders (to maintain their interest in food dispensers) until after mating when food is again provided *ad libitum* during egg production, incubation and nestling rearing. This refined management strategy achieves maximal productivity while also preventing an overproduction of male offspring (Robertson *et al.*, 2006; Elliot *et al.*, 2006).

With the echo parakeet, wild birds were reticent about taking supplemental food and learnt to do so from released captive-reared birds that were used to taking the formulated diet from food hoppers. Pairs that took supplementary food subsequently reared larger broods than those pairs that did not. Similar feeding also boosted the productivity of pink pigeons (Jones, 2008).

Captive management

The necessity for captive breeding occurs only when it is not possible to manage free-living populations and ideally is only a transitory step in species recovery. The Mauritius kestrel project, for example, used captive-bred birds as an important management strategy early during the recovery process. However, of 333 birds released between 1984 and 1994, only a third of these were captive-bred and most were eggs removed as first clutches from free-living birds (as a manipulation to encourage double-clutching; see above). Similarly, in the recovery of the echo parakeet and Mauritius fody most of the released birds were derived from clutch and brood manipulations of wild birds rather than from those bred in captivity. Birds reared in captivity from egg and brood manipulations are typically only in captivity for their early development. Such minimal time in captivity avoids many of the problems of adaptation to captivity (both genetic and social) (Snyder *et al.*, 1996). Birds in captive breeding projects may be in captivity for one to several generations before being reintroduced to the wild.

Captive breeding and studies on captive birds have, however, been a major part of the recovery efforts on Mauritius. A captive breeding project offers the opportunity to collect detailed data on the life history and behaviour that would be very difficult to collect from free-living birds. These data can then be used to supplement or help interpret studies on the species in the wild. Captive breeding facilities that are established near wild populations have the advantage that the movements of birds and eggs from the wild to captivity and vice versa can be achieved quickly and easily. There is also the possibility of moving skilled personnel from the captive-breeding facility into the field for the application of aviculture techniques to wild birds, and the reverse. Information on feeding biology, nest site selection and social structure of wild birds has been used to fine-tune captive management, and the intensive management techniques used on free-living birds, such as clutch and brood manipulations, were taken directly from aviculture.

The use of model species for practice and in-cross fostering

On Mauritius we have developed skills and techniques for managing endangered birds by working with common but closely related species, usually with a similar ecology (Table 2.3). We call these common species 'model species'. The most important role of model species has been to develop captive management and breeding techniques, including developing clutch and brood manipulations. Another important role of model species is to allow training of staff in the required management skills before they apply them to endangered species. Work on the echo parakeet was preceded by captive studies on the rose-ringed parakeet *Psittacula krameri*. Wood pigeons *Columba palumbus* and Barbary doves *Streptopelia risoria* were models for the pink pigeon, and common kestrels *Falco tinnunculus* were models for the Mauritius kestrel. Before working on the captive rearing and release of the olive white-eye, a series of trial studies was conducted on the common grey white-eye *Zosterops borbonica*, allowing staff to learn how to care for small insectivorous passerines. The grey white-eyes successfully bred in captivity, and wild harvested eggs were artificially hatched and the young hand-reared. The captive colony of grey white-eyes was released to develop release techniques and they bred at liberty. Subsequently, wild harvested eggs of the critically endangered olive white-eyes were artificially incubated and the young hand-reared and later released on the predator-free island of Ile aux Aigrettes, using the techniques developed on the grey white-eyes (Cristinacce *et al.*, 2006).

Providing high-quality nest sites

In some cases nest sites are an important limiting factor. This is particularly true of species that nest in cavities or on cliffs. In addition to the limited availability of nesting sites, the safety and quality of those sites can also be problematic. Many species increase their breeding populations and productivity as a result of the enhancement of existing nesting sites and/or after new artificial ones have been created (Newton, 1998). The provision of artificial nest sites, where there is a shortage, serves to potentially increase the population in several ways:

- Increase the number of pairs breeding.
- Increase the productivity per nesting attempt by optimizing the features of the nest site so it is the appropriate size and has suitable insulation and protection from the elements.
- Decrease the rate of predation upon the nest by incorporating predator exclusion features or by locating them in safe situations.
- Minimize interspecific competition upon nest sites by designing the nest site to favour the focal species over other competing species.
- Nest sites can be designed to more easily facilitate monitoring, research and management by the incorporation of inspection hatches and cameras.

Where the number of nest sites are limited, a large proportion of hole or cliff nesting birds may attempt to nest in inappropriate locations and unsuitable cavities. For example, in the absence of high-quality nest cavities echo parakeets and Mauritius kestrels have tried to nest in cavities that were prone to predation, flooding or overheating in the sun. Consequently, it has been a policy to improve nest cavities that were considered suboptimal and to provide nest boxes (Jones *et al.*, 1991; Jones & Duffy, 1993; see Table 2.3). Nest boxes are also commonly used by managed New Zealand bird species (e.g. Chatham Island black robin, saddleback and hihi; see Taylor *et al.*, 2005, and Table 2.2). In both the kakapo and echo parakeet, managed nest cavities have an entrance door built into the side of the cavity so that field workers can easily access the eggs and young.

Disease control and management

Disease has emerged, in recent decades, as a major limiting factor among some managed bird populations. Diseases that were previously overlooked may become more obvious because of the more intimate association managers have with their birds, and in some cases management may increase the incidence of disease. Disease transmission at supplemental feeding sites may be common or bacterial and viral disease may be spread from pair to pair by the human managers. The possibility of enhancing disease transmission due to management activities has to be weighed against the benefits of management.

It is necessary to have an idea of the diseases that may be present and how these could impact upon the birds being managed. At the start of any intensive conservation initiative it is important to screen the birds thoroughly to develop an idea of what diseases may be present. While populations are at a low level important diseases may be overlooked or considered as incidental causes of mortality. As populations grow the impact of disease upon the population becomes clearer. Disease control and management can be important in maintaining viable populations in the wild and captivity (see Ewen *et al.*, this volume, Chapter 9).

Disease has proven to be a major limiting factor in both the pink pigeon and echo parakeet while in the Mauritius fody and Mauritius kestrel disease has been encountered (e.g. avian pox and trichominiasis in both species) but is not currently considered a major cause of mortality. The major disease affecting the pink pigeons is trichomoniasis, a common disease of pigeons. This disease is caused by a flagellate protozoan *Trichomonas gallinae*, which infects about 50% of adults (Bunbury *et al.*, 2008). It lives in the mucus membranes and may cause caseous lesions in the throat, mouth and sinuses, causing death by restricting feeding and breathing. The disease is believed to have been introduced to Mauritius with exotic doves, which carry trichomoniasis but are rarely affected, and is thought to be spread in water since the flagellates may survive in warm water (Bunbury, 2006; Bunbury *et al.*, 2007). Adult pink pigeons are resistant to being affected by the parasite but the squabs are highly susceptible and trichomoniasis kills large numbers (Bunbury, 2006; Bunbury *et al.*, 2008). There have been attempts to control exotic doves around pink pigeon supplementary feeding sites, where they may spread the disease by contaminating drinking sources and possibly by contaminating grain with saliva droplets that are then consumed by the pink pigeons. Control of exotic doves was of little success since they are very common and other birds flew in from adjacent areas. It has proved more effective to exclude exotic doves from the feeding stations by altering the design of feeding hoppers. Medicating affected squabs has been successful in improving survival to fledging but these birds usually succumb to trichomoniasis infections post-fledging (Swinnerton *et al.*, 2005). Currently in our most infected subpopulations adults infected with the parasite are being treated to kill the parasite and prevent transmission to the young, and early results suggest that productivity is improved.

The second important disease that has been encountered on Mauritius is psitticine beak and feather disease (PBFD), a circo virus that affects the echo parakeet. This disease is usually spread among the young, which are most susceptible. PBFD is also found in the exotic rose-ringed parakeet and it may have been the original vector. A young echo parakeet with signs of PBFD was found in the wild population in the 1995–1996 breeding season. The level of infection was low until an outbreak of the disease in the release cohort in 2004–2005. Most of the birds became infected and it is believed that this disease was spread among them when an infected bird was inadvertently included in the cohort. The birds were caught and affected birds euthanized. Subsequently, the disease has appeared widely in the wild population. The birds have been screened using an enzyme linked immunosorbent assay (ELISA) test of serum extracted from whole blood. This test detects the presence of antibodies to the virus, which indicates a previous or current infection. Out of 223 birds screened for the disease, between 2005 and 2009, using this test, 79 (35%) proved positive. The disease kills a significant proportion of the young annually, but some affected birds do recover and despite the disease the population continues to grow (Raisin, 2010).

In response to the disease badly affected birds are euthanized. Since it is likely that field workers may spread the virus between nest sites, clutch and brood manipulations are no longer practised. When nests are visited for monitoring nest success strict quarantine procedures are followed and clothes and equipment are sterilized between visits.

Disease is commonly encountered at nest sites and ectoparasites may build up in the nest substrate. Tropical nest flies are a problem on Mauritius in some years, feeding on the blood of chicks, causing high rates of mortality in young echo parakeets, but may also affect Mauritius kestrels, pink pigeons and Mauritius fodies. This problem has been solved by treating the nest substrate with pesticides, originally Carberyl and more recently Frontline. Ectoparasitic nest mites were problematic in some nests of Chatham Island black robin and trials using pesticide treatment resulted in one chick dying in the early years of the recovery program (Ballance & Merton, 2007). Despite this early setback, management of nest mites is now commonly used in some endangered New Zealand species (e.g. hihi; see Ewen et al., 2009).

Most recently, an emergent disease, erysipelas, resulted in the death of three juvenile kakapo shortly after translocation between offshore islands (Gartrell et al., 2005). A vaccination programme has subsequently commenced to provide some protection to kakapo against the bacterium that causes this disease (Gartrell *et al.*, 2005). Concerns have also been raised that disease expression could be stress related concurrent with the increasing levels of 'hands-on' manipulation of individual kakapo.

How long does it take to save a species?

This is a fundamental question. To date most recovery programmes have had unrealistically short time frames. Viewed in terms of population increase it may take many years after the start of a conservation project for a species to show recovery. During the early stages of a recovery programme, populations may persist for generations without any marked increase or, indeed, they may continue to decline, e.g. kakapo (Powlesland *et al.*, 2006). The period from the start of recovery action to the start of substantial population increase and recovery takes a minimum of about ten times the age of first breeding. For example, from the initiation of recovery actions to the start of population recovery took 14 years for the Mauritius kestrel, 18 years for the pink pigeon, 22 years for the echo parakeet and 30 years for the Mauritius fody. It may take 20 or more generations after the start of a recovery programme before we can expect to see full population recovery. A summary of the starting points, time frames and current population estimates of species recovery programmes we have been involved in are presented in Tables 2.2 and 2.3.

This long period before the population starts to increase is typically when work focuses upon understanding the species and its problems. In addition time is required to develop the team needed to restore the population and to develop the political and institutional will to implement intensive management techniques.

A question of great importance is how can we reduce the length of time before the start of population recovery? As we become better at evaluating populations, and the causes of their decline, then this period will be reducible to a time period approaching the theoretical minimum, which is dependent upon the generation time of the species. Experience in Mauritius has shown that one can bring in expertise from other projects, such as from New Zealand, which can accelerate progress. However, it is still important to be able to provide logistic and institutional support and to build local capacity to continue the work. There is a limit on the amount of intensive conservation work that an organization can conduct. Even if the staff, infrastructure and all the techniques are available to restore a population we suspect that even for the most fecund species about a decade is needed before population recovery is witnessed. For the Chatham Island black robin, after several decades of decline the species started an upward trajectory in numbers nine years after the start of conservation action (Butler & Merton, 1992). For K-selected, slow-breeding, long-lived species such as cranes, large raptors and large parrots, we are talking about decades before we are likely to see recovery. The kakapo is a good example of this where there has been a long lead-up to the intensive management techniques that are used today. There were unsuccessful early attempts at captive management and breeding during 1961–1965 and translocation to islands was suggested in the early 1970s, with the first of the modern translocations implemented in 1974 to Maud Island with concurrent enrichment planting of suitable food-plants. However, the birds did not breed (Butler, 1989). The recovery work intensified with the finding of a population on Stewart Island in 1977, but these birds were in severe decline due to cat predation (see the summary references in Powlesland *et al.*, 2006). Subsequently, birds were transferred to predator-free islands in 1980–1991. These were slow to start breeding successfully and there was no recruitment into the adult population until 1991. The population reached 100 birds in 2009, nearly 50 years after the early management attempts in the early 1960s and over a century after the conservation efforts of Richard Henry (Hill & Hill, 1987).

Long-term minimum management

Long-term management is increasingly becoming important in bird conservation. It is clear that where habitats are compromised many species are going to survive only if they are managed for the foreseeable future. In New Zealand and Mauritius it is important to maintain island refugia free of exotic predators through continual guarding and rapid response to any re-invasion. In addition, some species may require nest boxes at certain sites (e.g. Mauritius kestrel, echo parakeet, saddleback and hihi) and some may also require supplementary feeding (e.g. pink pigeon, echo parakeet, kakapo and hihi). Where predators cannot be removed other management actions are required. For example, vulnerable eggs might be removed to captive facilities for hatching and rearing to a point where independent young are safer from exotic predators (e.g. takahe, kiwi spp. and black stilt) or temporary trapping and poisoning of exotic predators around active nest sites (e.g. kokako) might be implemented.

In all cases these species will only be able to survive independent of management once habitat is restored, invasive plants controlled or eradicated, and the numbers of exotic mammalian predators and exotic birds are reduced. With current resources and knowledge these goals are difficult if not impossible to achieve.

It is important that techniques for long-term management are developed so that populations can be maintained with a minimum of interventions and these can be reduced or withdrawn if conditions improve. It has been a commonly held belief by zoo biologists that in the coming decades increasing numbers of species will exist only in captivity as wild populations are extirpated. We believe that the work that has been done in New Zealand and Mauritius illustrates that it is now becoming increasingly likely many species will be managed as free-living populations, still under many of the natural pressures of natural selection, and fulfilling their ecological roles, rather than languishing in captivity for decades or centuries.

An essential way to ensure the feasibility of long-term management is to gain public support. The increase in public involvement for conservation in New Zealand over the last 20 years has facilitated great successes in habitat restoration and endangered species management (Parker, 2008; Galbraith & Hayson, 1995). Increasingly restoration projects are being devised and implemented by community conservation groups with support from traditional government conservation agencies. For many, translocation of extirpated species is their ultimate goal (Parker, 2008). This greatly enhances opportunities for extensive ecological restoration as governments and private organizations have limitations on staff and funding to attempt conservation independently. Furthermore, increased public ownership and participation in conservation provides a catalyst facilitating increased investment of public resources and political will into conservation. Perhaps the most exciting opportunity to prevent further extinctions, and to recover more species and restore ecosystems, will be through greater engagement of people and the organizations they are part of (Parker, 2008).

Concluding remarks and our opinion for the future

The unprecedented growth of the human population and its dramatically expanding 'footprint' is placing an ever-increasing pressure on remaining wild places and wildlife. This same pressure will make the social, political and biological challenges faced by future conservation practitioners increasingly daunting. The issue of habitat retention and restoration must intensify. The proliferation of small and isolated populations requiring regular and repeated transfer of individuals between them to maintain genetic fitness seems inevitable. Such focus on single species management raises questions about the relative merit of species conservation versus habitat conservation. Conservation biologists have long debated the relative merits of each. It obviously makes sense to manage habitats and communities rather than individual species – provided this is achievable. There are exciting advances in our abilities to restore habitats, including removing suites of pest species and reintroducing animal communities. We also believe that the arguments between single species or habitat management are redundant since in all the species that we have been involved with the species work grew and drove habitat protection and ecological restoration. A more fruitful discussion should be focused on appropriate goal setting so that single species management fits within management for community and ecosystem restoration.

We also believe that many future challenges dwell in sound population and genetic management. How do we ensure that the newly established populations that appear successful will persist? This is going to be particularly true of small populations. We tend to see everything in terms of a human lifespan. . .or portions of a lifespan. . .yet genetic diversity may be eroded over generations and this may not express itself for decades (see Chapters 11 to 13, this volume).

Future challenges to birdlife in general and to island endemics in particular seem formidable and intervention will be an ongoing necessity for many. For instance, global climate change will affect bird populations in at least two ways. First, ecosystems that birds depend upon will move or change more quickly as the climate alters. Second, movement of species and habitats will have profound implications for site-based nature conservation – such as on islands or 'mainland islands'. In island nations such as New Zealand and Mauritius the problem is compounded by the fact that endemic birds tend to be poor dispersers (flightless or with limited powers of flight) and the islands very isolated. Many rare species are now effectively confined to offshore or mainland islands – unable to move without human help. Already, for some species survival within their natural range is no longer an option and, in extreme cases, assisted colonization (introductions of species beyond their historical range; see Seddon et al., this volume, Chapter 1) may be a solution.

With an ever-expanding human population, coupled with anticipated impacts of climate change, this situation is set to intensify. Clearly, conservation translocations are going to be crucial. They offer a way forward – a future for threatened life forms – a means by which they might continue to live and evolve in a free-living state.

Finally, the work on New Zealand and Mauritius has been greatly influenced by an exchange of ideas and staff since the early 1980s. This continues. In addition we have frequently relied on international expertise of veterinarians, zoo staff, poultry breeders, aviculturists, population biologists, fertility experts and geneticists. We should also not forget philanthropists, landowners and politicians. People are clearly important and there are a growing number of skill sets contributing to, and taking a stake in, conservation translocations and the management of endangered species. The most exciting developments witnessed in our lifetime are the growth in public support and community involvement in conservation and the development of sound management techniques supported by scientific studies. We hope that this growth in participation and the development of techniques continues and that the number of successful conservation interventions will increase as a result.

References

Armstrong, D.P. (1995) Effects of familiarity on the outcome of translocations, II. A test using New Zealand robins. *Biological Conservation*, 71, 281–288.

Armstrong, D.P. & Craig, J.L. (1995) Effects of familiarity on the outcome of translocations, I. A test using saddlebacks *Philesturnus carunculatus rufusater*. *Biological Conservation*, 71, 133–141.

Armstrong, D.P., Lovegrove, T.G., Allen, D.G. *et al.* (1994) Composition of founder groups for bird translocations: does familiarity matter? In *Reintroduction biology of Australian and New Zealand Fauna*, ed. M. Serena. Surrey Beatty & Sons, Chipping Norton, NSW, Australia.

Atkinson, I.A.E. (1985) The spread of commensal species of *Rattus* to oceanic islands and their effects on island avifaunas. In *Conservation of Island Birds*, ed. P.J. Moors, pp. 35–81. ICBP Technical Publication 3, Cambridge, UK.

Ballance, A. & Merton, D.V. (2007) *Don Merton: the man who saved the black robin*. Reed Publishing (NZ) Ltd, Auckland, New Zealand.

Ballie, J.E.M., Griffiths, J., Turvey, S.T. *et al.* (2010) *Evolution Lost: status and trends of the world's vertebrates*. Zoological Society of London, UK.

Beissinger, S.R. & Bucher, E.H. (1992) Sustainable harvesting of parrots for conservation. In *New World Parrots in Crisis*, eds S.R. Beissinger & N.F.R. Snyder, pp. 73–117. Smithsonian Institution Press, Washington and London.

Bell, B.D. (1978) The Big South Cape Islands rat irruption. In *The Ecology and Control of Rodents in New Zealand Nature Reserves*, eds P.R. Dingwall, I.A.E. Atkinson & C. Hay, pp. 33–45. Department of Lands and Survey Information, New Zealand.

Bell, B.D. & Merton, D.V. (2002) Critically endangered bird populations and their management. In *Conserving Bird Biodiversity*, eds K. Norris & D.J. Pain, pp. 105–138. Cambridge University Press, Cambridge, UK.

Bunbury, N. (2006) *Parasitic disease in the endangered Mauritian pink pigeon Columba mayeri*. Thesis submitted for the degree of Doctor of Philosophy, School of Biological Sciences, University of East Anglia, UK.

Bunbury, N., Jones, C.G., Greenwood, A.G. *et al.* (2007) *Trichomonas gallinae* in Mauritian columbids: implications for an endangered endemic. *Journal of Wildlife Diseases*, 43, 399–407.

Bunbury, N., Jones, C.G., Greenwood, A.G. *et al.* (2008). Epidemiology and conservation implications of *Trichomonas gallinae* infection in the endangered Mauritian pink pigeon. *Biological Conservation*, 141(1), 153–161.

Burnham, W. (1983) Artificial incubation of falcon eggs. *Journal of Wildlife Management*, 47, 158–168.

Butler, D. (1989) *Quest for the Kakapo. The full story of New Zealand's most remarkable bird*. Heinemann Reed, Aukland, New Zealand.

Butler, D. & Merton, D.V. (1992) *The Black Robin: saving the world's most endangered bird*. Oxford University Press, Auckland, New Zealand.

Butler, S.J., Benton, T.G., Nicoll, M.A.C. *et al.* (2009) Indirect population dynamic benefits of altered life-history trade-offs in response to egg harvesting. *The American Naturalist*, 174, 111–121.

Cade, T.J. (1980) The husbandry of falcons for return to the wild. *International Zoo Yearbook*, 20, 23–35.

Cade, T.J. (1986) Propagating diurnal raptors in captivity: a review. *International Zoo Yearbook*, 24/25, 1–20.

Cade, T.J. (1988) Using science and technology to re-establish species lost in nature. In *Biodiversity*, ed. E.O. Wilson, pp. 279–288. National Academic Press, Washington DC, USA.

Cade, T.J. (2000) Progress in translocation of diurnal raptors. In *Raptors at Risk: Proceedings of the 5th World Working Group for Birds of Prey and Owls*, eds R.D. Chancellor & B.U. Meyburg, pp. 343–372. Hancock House, UK.

Cade, T.J. & Burnham, W. (eds) (2003) *Return of the Peregrine: a North American saga of tenacity and teamwork*. The Peregrine Fund, Boise, Idaho, USA.

Cade, T.J. & Jones, C.G. (1994) Progress in restoration of the Mauritius kestrel. *Conservation Biology*, 7(1), 169–175.

Castro, I., Alley, J.C., Empson, R.A. *et al.* (1994) Translocation of hihi or stitchbird *Notiomystis cincta* to Kapiti Island, New Zealand: transfer techniques and comparison of release strategies. In *Reintroduction Biology of Australian and New Zealand Fauna*, ed. M. Serena. Surrey Beatty & Sons, Chipping Norton, NSW, Australia.

Cheke A.S. & Hume, J. (2008) *Lost Land of the Dodo. An ecological history of Mauritius, Reunion and Rodrigues.* T. & A.D. Poyser, London, UK.

Clark, M., Wallace, M. & David, C. (2007) Rearing California condors for release using a modified puppet-rearing technique. In *California Condors in the 21st Century*, eds A. Mee & L.S. Hall, pp. 213–226. Nuttall Ornithological Club, Cambridge, Massachusetts and The American Ornithologists' Union, Washington, DC, USA.

Clout, M.N., Elliot, G.P. & Robertson, B.C. (2002) Effects of supplementary feeding on the offspring sex ratio of kakapo: a dilemma for the conservation of a polygynous parrot. *Biological Conservation*, 107, 13–18.

Colbourne, R., Bassett, S., Billing, T. *et al.* (2005) *The Development of Operation Nest Egg as a Tool in the Conservation Management of Kiwi.* Science for Conservation 259, Department of Conservation, Wellington, New Zealand.

Craig, J., Anderson, S., Clout, M. *et al.* (2000) Conservation issues in New Zealand. *Annual Review of Ecology and Systematics*, 31, 61–78.

Cristinacce, A., Handschuh, M., Switzer, R.A. *et al.* (2009) The release and establishment of Mauritius fodies *Foudia rubra* on Ile aux Aigrettes, Mauritius. *Conservation Evidence*, 6, 1–5.

Cristinacce, A., Ladkoo, A., Kovak, E. *et al.* (2006) The first hand-rearing of Mauritian white-eyes *Zosterops* spp. *Avicultural Magazine*, 112, 150–160.

D'Elia, J. (2010) Evolution of avian conservation breeding with insights for addressing the current extinction crisis. *Journal of Fish and Wildlife Management*, 1(2), 189–210.

Diamond, J. (1989) Overview of recent extinctions. In *Conservation in the 21st Century*, eds D. Western & M.C. Pearl, pp. 37–41. Wildlife Conservation International, New York, USA.

Dingwall, P.R., Atkinson, I.A.E. & Hay, C. (1978) *The Ecology and Control of Rodents in New Zealand Nature Reserves.* Department of Lands and Survey, Information Series 4, Wellington, New Zealand.

Durrell, G. (1990). *The Ark's Anniversary.* Collins, London.

Durrell, G. & Durrell, L.M. (1980) Breeding Mascarene wildlife in captivity. *International Zoo Yearbook*, 20, 112–119.

Eason, D.K. & Willans, M. (2001) Captive rearing: a management tool for the recovery of the endangered takahe. In *The Takahe: Fifty Years of Conservation Management and Research*, eds W.G. Lee & I.G. Jamieson. The University of Otago Press, Dunedin, New Zealand.

Edmunds, K., Bunbury, N., Sawmy, S. *et al.* (2008) Restoring avian island endemics: supplementary food use by the endangered pink pigeon *Columba mayeri*. *Emu*, 108, 74–80.

Elliott, G.P., Eason, D.K., Jansen, P.W. *et al.* (2006) Productivity of kakapo (*Strigops habroptilus*) on offshore island refuges. *Notornis*, 53, 138–142.

Elliot, G.P., Merton, D.V. & Jansen, P.W. (2001) Intensive management of a critically endangered species: the kakapo. *Biological Conservation*, 99, 121–133.

Erickson, R.C. (1980) Propagation studies of endangered wildlife at the Patuxent Wildlife Research Centre. *International Zoo Yearbook*, 20, 40–47.

Ewen, J.G., Acevedo-Whitehouse, K., Alley, M. *et al.* (2011) Empirical consideration of parasites and health in reintroduction. In *Reintroduction Biology: integrating science and management*, eds J.G. Ewen, D.P. Armstrong, K.A. Parker & P.J. Seddon, Chapter 9. Wiley-Blackwell, Oxford, UK.

Ewen, J.G., Thorogood, R., Brekke, P. *et al.* (2009) Maternally invested carotenoids compensate costly ectoparasitism in the hihi. *Proceedings of the National Academy of Sciences USA*, 106, 12 798–12 802.

Galbraith, M.P. & Hayson, C.R. (1995) Tiritiri Matangi Island, New Zealand: public participation in species translocation to an open sanctuary. In *Reintroduction Biology of Australian and New Zealand Fauna*, ed. M. Serena, pp 149–154. Surrey Beatty and Sons, Chipping Norton, NSW, Australia.

Gartrell, B.D., Alley, M.R., Mack, H. *et al.* (2005) Erysipelas in the critically endangered kakapo (*Strigops habroptilus*). *Avian Pathology*, 34, 383–387.

Hess, E.H. (1973) *Imprinting: early experience and the developmental psychobiology of attachment*. Van Nostrand Reinhold Company, London, UK.

Hill, S. & Hill, J. (1987) *Richard Henry and Resolution Island*. John McIndoe, Dunedin.

Jamieson, I.G., Wallis, G.P. & Briskie, J.V. (2006) Inbreeding and endangered species management: is New Zealand out of step with the rest of the world? *Conservation Biology*, 20, 38.

Jones, C.G. (2002) Reptiles and amphibians. In *Handbook of Ecological Restoration*, eds M.R. Perrow & A.J. Davy, pp. 355–375. Cambridge University Press, Cambridge, UK.

Jones, C.G. (2004) Conservation management of endangered birds. In *Bird Ecology and Conservation*, eds W.J. Sutherland, I. Newton & R.E. Green, pp. 269–301. Oxford University Press, Oxford, UK.

Jones, C.G. (2008) Practical conservation on Mauritius and Rodrigues. Steps towards the restoration of devastated ecosystems. In *Lost Land of the Dodo*, eds A.S. Cheke & J. Hume, pp. 226–259. Christopher Helm, London, UK.

Jones, C.G. & Duffy, K. (1993) The conservation management of the echo parakeet *Psittacula eques echo*. *Dodo, Journal of the Wildlife Preservation Trusts*, 29, 126–148.

Jones, C.G. & Swinnerton, K.J. (1997) Conservation status and research for the Mauritius kestrel, pink pigeon and echo parakeet. *Dodo, Journal of the Jersey Wildlife Preservation Trust*, 33, 72–75.

Jones, C.G., Heck, W., Lewis, R.E. *et al.* (1991) A summary of the conservation management of the Mauritius kestrel *Falco punctatus* 1973–1991. *Dodo, Journal of the Jersey Wildlife Preservation Trust*, 27, 81–99.

Jones, C.G., Heck, W., Lewis, R.E. *et al.* (1994) The restoration of the Mauritius kestrel *Falco punctatus* population. *Ibis*, 137, 173–190.

Jones, C.G., Swinnerton, K.J., Hartley, J. *et al.* (1999) The restoration of the free-living populations of the Mauritius kestrel (*Falco punctatus*), pink pigeon (*Columba mayeri*) and echo parakeet (*Psittacula eques*). In *Proceedings of the 7th World Conference on Breeding Endangered Species*, pp. 77–86. Cincinnati, Ohio, USA.

Jones, C.G., Swinnerton, K.J., Taylor, C.J. *et al.* (1992) The release of captive-bred pink pigeons *Columba mayeri* in native forest on Mauritius. A progress report July 1987–June 1992. *Dodo, Journal of the Jersey Wildlife Preservation Trust*, 28, 92–125.

Jones, C.G., Swinnerton, K., Thorsen, M. *et al.* (1998) The biology and conservation of the echo parakeet *Psittacula eques* of Mauritius. In *Proceedings of the 4th International Parrot Convention*, pp. 110–121. Loro Parque, Tenerife.

Kear, J. (1975a) Breeding of endangered wildfowl as an aid to their survival. In *Breeding Endangered Species in Captivity*, ed. R.D. Martin, pp. 49–60. Academic Press, London, UK.

Kear, J. (1975b) Returning the Hawaiian goose to the wild. In *Breeding Endangered Species in Captivity*, ed. R.D. Martin, pp. 115–123. Academic Press, London, UK.

Kear, J. (1986) Captive breeding programmes for waterfowl and flamingos. *International Zoo Yearbook*, 24/25, 21–25.

Kear, J. & Berger, A.J. (1980) *The Hawaiian Goose: an experiment in conservation*. T. & A.D. Poyser, Carlton.

Keedwell, R.J., Maloney, R.F. & Murray, D.P. (2002) Predator control for protecting kaki (*Himantopus novaezelandiae*) – lessons from 20 years of management. *Biological Conservation*, 105, 369–374.

Lack, D. (1954) *The Natural Regulation of Animal Numbers*. Oxford University Press, Oxford, UK.

Lack, D. (1966) *Population Studies in Birds*. Oxford University Press, Oxford, UK.

Lee, W.G. (2001) Fifty years of takahe conservation, research and management: what have we learnt? In *The Takahe: Fifty Years of Conservation Management and Research*, eds W.G. Lee & I.G. Jamieson. The University of Otago Press, Dunedin, New Zealand.

Lee, W.G. & Jamieson, I.G. (eds) (2001) *The Takahe: Fifty Years of Conservation Management and Research*. University of Otago Press, Dunedin, New Zealand.

Lever, C. (2005) *Naturalised Birds of the World*. T. & A.D. Poyser, London, UK.

Long, J.L. (1981) *Introduced Birds of the World*. David and Charles, Newton Abbot, London, UK.

Lovegrove, T.G. (1996) Island releases of saddlebacks *Philesturnus carunculatus* in New Zealand. *Biological Conservation*, 77, 151–157.

Martin, R.D. (ed.) (1975) *Breeding Endangered Species in Captivity*. Academic Press, London, UK.

Maxwell, J.M. (2001) Fiordland takahe: population trends, dynamics and problems. In *The Takahe: Fifty Years of Conservation Management and Research*, eds W.G. Lee & I.G. Jamieson. The University of Otago Press, Dunedin, New Zealand.

Maxwell, J.M. & Jamieson, I.G. (1997). Survival and recruitment of captive-reared and wild-reared takahe in Fiordland, New Zealand. *Conservation Biology*, 2, 28–30.

Merton, D.V. (1965) A brief history of the North Island saddleback and transfer of saddlebacks from Hen Island to Middle Chicken Island January 1964. *Notornis*, 12, 213–222.

Merton, D.V. (1975) The saddleback: its status and conservation. In *Breeding Endangered Species in Captivity*, ed. R.D. Martin, pp 61–74. Academic Press, London, UK.

Merton, D.V., Reed, C. & Crouchley, D. (1999) Recovery strategies and techniques for three free-living, critically-endangered New Zealand birds: kakapo (*Strigops habroptilus*), black stilt (*Himantopus novaezelandiae*) and takahe (*Porphyrio mantelli*). In *Linking Zoo and Field Research to Advance Conservation, The 7th World Conference on Breeding Endangered Species*, pp. 151–162. Cincinnati Zoo, Cincinnati, Ohio, USA.

Newton, I. (1998) *Population Limitation in Birds*. Academic Press, London, UK.

Nicoll, M.A.C., Jones, C.G. & Norris, K. (2004) Comparisons of survival rates of captive-reared and wild-bred Mauritius kestrels (*Falco punctatus*) in a re-introduced population. *Biological Conservation*, 118, 539–548.

Nicoll, M.A.C., Jones, C.G. & Norris, K. (2006) The impact of harvesting on a formerly endangered tropical bird: insights from life-history theory. *Journal of Applied Ecology*, 43, 567–575.

Olney, P.J.S., Mace, G.M. & Feistner, A.T.C. (1994) *Creative Conservation. Interactive management of wild and captive animals*. Chapman & Hall, London, UK.

Parker, K.A. (2008) Translocations: providing outcomes for wildlife, resource managers, scientists, and the human community. *Restoration Ecology*, 16(2), 204–209.

Powlesland, R.G., Lloyd, B.D., Best, H.A. *et al.* (1992) Breeding biology of the kakapo on Stewart Island, New Zealand. *Biological Conservation*, 69, 97–106.

Powlesland, R.G., Merton, D.V. & Cockrem, J.F. (2006) A parrot apart: the natural history of the kakapo (*Strigops habroptilus*) and the context of its conservation management. *Notornis*, 53, 3–26.

Raisin, C. (2010) *Conservation genetics of the Mauritius parakeet*. Thesis submitted for Doctor of Philosophy in Biodiversity Management, Durrell Institute of Conservation and Ecology, University of Kent, Canterbury, Kent, UK.

Robertson, B.C., Elliot, G.P., Eason, D.K. *et al.* (2006) Sex allocation theory aids species conservation. *Biology Letters*, 2, 229–231.

Robertson, H.A., Colbourne, R.M., Miller, G.P.J. *et al.* (2010) Experimental management of brown kiwi *Apteryx mantelli* in central Northland, New Zealand. *Bird Conservation International*, doi: 10.1071/S0 959 270 910 000 444.

Sarrazin, F. & Legendre, S. (2000) Demographic approach to releasing adults versus young in reintroductions. *Conservation Biology*, 14, 488–500.

Saunders, A.J. (2000) A review of Department of Conservation mainland restoration projects and recommendations for further action. Department of Conservation, Wellington, New Zealand.

Seddon, P.J., Strauss, W.M. & Innes, J. (2011) Animal translocations: what are they and why do we do them? In *Reintroduction Biology: integrating science and management*, eds J.G. Ewen, D.P. Armstrong, K.A. Parker & P.J. Seddon, Chapter 1. Wiley-Blackwell, Oxford, UK.

Sherrod, S.K., Heinrich, W.R., Burnham, W.A. *et al.* (1981) *Hacking: a method for releasing Peregrine falcons and other birds of prey*. The Peregrine Fund, Boise, Idaho, USA.

Snyder, N.F.R., Derrickson, S.R., Beissinger, S.R. *et al.* (1996) Limitations of captive breeding in endangered species recovery. *Conservation Biology*, 10, 338–348.

Swinnerton, K.J. (2001) *The ecology and conservation of the pink pigeon Columba mayeri in Mauritius*. Thesis submitted for Doctor of Philosophy in Biodiversity Management, Durrell Institute of Conservation and Ecology, University of Kent, Canterbury, Kent, UK.

Swinnerton, K.J., Greenwood, A.G., Chapman, R.E. *et al.* (2005) The incidence of the parasitic disease trichomoniasis and its treatment in reintroduced and wild pink pigeons *Columba mayeri*. *Ibis*, 147, 772–782.

Taylor, S., Castro, I. & Griffiths, R. (2005) Hihi/stitchbird (*Notiomystis cincta*) recovery plan 2004–09. Threatened Species Recovery Plan 54, Department of Conservation, Wellington, New Zealand.

Temple, S.A. (1978) *Endangered Birds. Management techniques for preserving threatened species*. University of Wisconsin Press, Madison, Wisconsin, USA.

Veitch, R. (1985) Methods of eradicating feral cats from offshore islands in New Zealand. In *Conservation of Island Birds*, ed. P.J. Moors, pp. 125–141. ICBP Technical Publication 3. Cambridge, UK.

Wallace, M. (2000) Retaining natural behaviour in captivity for re-introduction programmes. In *Behaviour and Conservation*, eds I.M. Gosling & W.J. Sutherland, pp. 300–314. Cambridge University Press, Cambridge, UK.

Wallace, M., Clark, M., Vargus, J. *et al.* (2007) Release of puppet reared California condors in Baja California, Mexico: evaluation of a modified rearing technique. In *California Condors in the 21st Century*, eds A. Mee & L.S. Hall, pp. 227–242. Nuttall Ornithological Club, Cambridge, Massachusetts and The American Ornithologists' Union, Washington, DC, USA.

Williams, G.R. (1977) Marooning – a technique for saving threatened species from extinction. *International Zoo Yearbook*, 17, 102–106.

$$3$$

Selecting Suitable Habitats for Reintroductions: Variation, Change and the Role of Species Distribution Modelling

Patrick E. Osborne[1] and Philip J. Seddon[2]

[1]School of Civil Engineering and the Environment,
University of Southampton, United Kingdom
[2]Department of Zoology, University of Otago, New Zealand

'. . . we have stressed the difficulties but offered few solutions. There is no universal panacea because we are essentially balancing unknowns, but uncertainty in outcome may be reduced by being more objective in our choice of release sites, while acknowledging that how we practically do this will differ between species and situations. We may summarize much of the foregoing arguments into a simple maxim: sites should not be selected for reintroductions only on the basis that the species used to be there or that the site looks right.'

Page 87

Introduction

Reviews of the outcomes of reintroduction projects have generally reported low rates of success (e.g. 26% (Fischer & Lindenmayer, 2000), 11% (Beck et al., 1994) and only 7% for bird projects (Cade & Temple, 1995)). Among a

Reintroduction Biology: Integrating Science and Management. First Edition.
Edited by John G. Ewen, Doug P. Armstrong, Kevin A. Parker and Philip J. Seddon.
© 2012 Blackwell Publishing Ltd. Published 2012 by Blackwell Publishing Ltd.

long list of reasons cited for failure (Griffith *et al.*, 1989; Nesbitt & Carpenter, 1993; Powell & Cuthbert, 1993; McLain *et al.*, 1999; Letty *et al.*, 2000; Mathews & Macdonald, 2001; Duncan *et al.*, 2003) is the condition of the habitat in the release area (Wolf *et al.*, 1996, Armstrong *et al.*, 2002, Cook *et al.*, 2010). Indeed, habitat quality may be the main reason for success or failure in many reintroduction projects, although hard data are difficult to come by. Given this background, any improvements that can be made to the way habitats are assessed prior to release are to be welcomed.

When identifying suitable sites for reintroduction, it would seem obvious that 'species should never be released blindly without extensive evaluation of habitat quality' (Lomolino & Channell, 1998). While the IUCN/SSC Guidelines for Reintroductions (IUCN, 1998) make this point, it is not necessarily clear how this should best be done. It is almost 50 years since Charles Elton, a father of modern animal ecology, wrote that 'definition of habitat, or rather lack of it, is one of [our] chief blind spots' (Elton, 1966). Little has changed: definitions of habitat remain confused and their different implications for conservation poorly appreciated (Hall *et al.*, 1997, Gaillard, 2010). If we are to understand what needs to be assessed in order to deem a site suitable for a reintroduction, we need to clear this fog and identify what habitats are and how they affect species. Without being overly critical of reintroduction projects to date, we would argue that this has not always been done adequately. In very broad terms, habitat assessments for reintroductions have usually been done by looking for places with conditions that seem to match where the species is (or has been) known to exist. This may involve quantitative modelling, but most commonly is done intuitively. If we regard this as inadequate, what can provide an adequate framework to help guide future projects?

Southwood's seminal paper on habitat (Southwood, 1977) introduced the term 'favourableness', a property of a habitat that allows species persistence or increased survival and reproduction. In this context, habitat is a species-specific complex of interacting physical and biotic components, including other species (Armstrong & Seddon, 2008; e.g. a predator is part of the habitat of a prey species), that favours persistence. The characteristics of habitats may be expressed along two composite axes as durational stability (i.e. spatial heterogeneity against time) and resource level and constancy (i.e. the temporal heterogeneity of the same space) (Southwood, 1977). In other words, habitat must be viewed in the context of spatial and temporal heterogeneity. This has three practical implications: first, that the characteristics of suitable habitat vary spatially and, through local evolution, so might species adaptations to habitat; second, that habitat has a crucial landscape component that must be assessed alongside the suitability of a habitat patch per se; and third, that environments change over time. There are (at least) eight consequences for reintroductions:

1. Historical locations of a species' presence may not indicate a present-day suitable habitat.
2. Present-day locations of a species' presence may not indicate a currently suitable habitat.
3. Present-day locations where a species is absent may not indicate an unsuitable habitat.
4. Present-day locations of a species' presence may not indicate a future suitable habitat.
5. Not all suitable habitat patches will be colonized because the landscape components may be missing.
6. A habitat's suitability and its characteristics vary across the species range.
7. Individuals from across a species range may not all be equally suited to the chosen release site.

Following from these, that:

8. A suitable habitat may need to be engineered (restored or created) to aid colonization and then managed to maintain its perceived value. The complexity of doing this is often not appreciated.

We argue that these points are rarely considered adequately in current reintroduction projects and there may be insufficient encouragement to do so. The impetus for many projects comes from the bottom up (and, indeed, successful projects often have a 'champion') whereas the framework is set at the policy level. For instance, at the international level, Article 8(f) of the Convention on Biological Diversity requires Contracting Parties to rehabilitate and restore degraded ecosystems and promote the recovery of threatened species through the development and implementation of plans or other management strategies. Under European legislation, the EC Habitats Directive 92/43/EEC obliges Member States to consider the feasibility of restoring species that have become locally extinct. At the national level, legislation such as Britain's Countryside and Rights of Way Act 2000, Section 74, requires government departments to have regard for biodiversity and to take positive steps to further the conservation of listed species and habitats, which is widely interpreted to include translocations. Yet the legislation is silent on the approach to be taken.

To date the main guiding document for reintroductions worldwide has been the IUCN/SSC Guidelines for Reintroductions as approved by the Council of IUCN in May 1995 (IUCN, 1998). Although this document does not have a basis in law, there is a growing expectation that proponents of reintroduction projects will adhere to the guidelines and national governments are at liberty to reject applications that fail to provide the details suggested by them. These guidelines appropriately emphasize the need to ensure that the 'habitat and landscape requirements of the species are satisfied and likely to be sustained for the foreseeable future' (IUCN, 1998), but necessarily are not prescriptive about *how* habitat suitability should be assessed nor what constitutes an adequate test of suitability. In an attempt to move towards a more detailed framework, we summarize the ecological basis for our eight consequences for reintroductions, and then provide examples of good practice in habitat assessment and how more could be done.

Consequence 1. Historical locations may not indicate suitable habitat

Historical studies have shown that species ranges are dynamic, expanding in some regions while contracting in others over time (Hengeveld, 1990). The drivers of these changes are usually environmental, such as climate, land use and human disturbance, as opposed to evolutionary change in the species preferences. The expectation should therefore be of change. When a species is locally extinct, it clearly cannot be seen to be responding to these drivers, but we should not be fooled into thinking that they are not acting. Indeed, it has been argued that the effects (particularly of human activities) are so strong that they render historical patterns of occurrence and density virtually irrelevant to site selection for release (Channell & Lomolino, 2000), although this may be putting it too forcibly. It may therefore be misguided to assume that historical sites offer a suitable habitat and the guideline to reintroduce within the former range may often be inappropriate (Seddon, 2010). The longer the time between local extinction and the planned release, the greater the chance will be that the habitat will no longer be suitable, and the greater the need to evaluate current habitat suitability regardless of historical occupancy.

For example, historical records of the koa finches, an extinct endemic honeycreeper of the genus *Rhodocanthis*, suggest a distribution confined to upland refugia in the Hawaiian islands, but an assessment of palaeohabitats indicates that koa finches were once widespread in lowland areas from which they were extirpated by human disturbance in prehistorical periods (James & Price, 2008). Should restoration of the group have been possible, historical records would have led to the selection of release sites in a less productive upland habitat that became vulnerable to forest harvesting and would currently be inadequate to support population persistence.

It is not only that historical locations may no longer indicate a suitable habitat but also that analysing them against contemporary data sources may give misleading impressions of habitat preferences. While there is much to applaud in the analysis of Metzger *et al.* (2007), which attempts to model suitable areas for the reintroduction of black rhinoceros *Diceros bicornis* in East Africa, their use of animal survey data from 1969 to 1972 against satellite imagery from 2000 is questionable, unless environmental change over the 30-year gap can be assumed to be negligible. This illustrates one of the difficulties with any modelling approach to assessing habitat suitability (which we discuss in much more detail below): the need for good quality species and habitat data from the same time period. How large the gap between the collection of habitat and species data can be depends on environmental stability and there is no fixed or known answer, but we instinctively feel 30 years is a long time given the current rate of climate change.

Consequence 2. Present-day locations may not indicate a currently suitable habitat

We must think in terms of the spatial and temporal components of habitats to understand why present-day locations of a species may not indicate a suitable habitat. Taking the temporal component first, time lags between negative events and their consequences are inevitable in natural systems (Lindenmayer *et al.*, 2008). The event itself may be chronic and only manifest after a build-up over a prolonged period (e.g. pesticide poisoning) or the species' response may be very slow. For example, population viability analysis suggests that great bustards *Otis tarda* could persist for a few decades after conditions become unsuitable for breeding because of their longevity (Osborne, 2005). With a drip-feed of very low recruitment, a population of long-lived animals could continue to occupy a habitat long after it is has become too unfavourable to support new population establishment and long-term persistence.

In the spatial dimension, a critical issue is whether the habitat conditions at a location render it a population sink rather than a source (Pulliam, 1988). Habitat patches do not occur in isolation but exchange individuals depending on demographic processes. Where habitat conditions are excellent, a species may thrive and produce a net excess of offspring that are forced to move elsewhere (the 'source'). These individuals may then occupy habitat patches with conditions that lead to reduced survival (the 'sink'). So long as the source exports an excess of individuals, both source and sink habitats will be occupied, giving the misleading impression of suitability in the absence of any evaluation of habitat-specific population persistence. A reintroduction into a sink habitat will clearly fail as conditions are inadequate for reproduction, although sink habitat patches may form a crucial part of a metapopulation structure.

We often make the mistake of assuming that individuals have perfect knowledge of habitat quality and choose accordingly, yet this is often not the case. Take the 'ecological trap', where a habitat might seem suitable to a species yet not offer the best conditions for maximising fitness (Schlaepfer et al., 2002). Ecological traps can arise where human actions have produced a mismatch between the cues used in habitat selection and the true conditions offered by the habitat. A good example would be many agricultural crops that are structurally good grasslands yet are almost devoid of invertebrates due to widespread use of pesticides (Fuller et al., 1995). A version of the ecological trap often occurs in reintroduction projects when founders reject the habitat near the release site and rapidly move long distances away before settling (see also Le Gouar et al., this volume, Chapter 5). This could be a consequence of Natal Habitat Preference Induction, whereby individuals preferentially seek out similar stimuli to those at the natal site rather than assessing habitat quality per se (Stamps & Swaisgood, 2007). The outcome can be that founders fail to recognize a quality habitat at the release site and preferentially settle where conditions are less good. This post-release dispersal may also be a consequence of the stress and disorientation of the release process, rather than mediated by active habitat selection. The consequence is therefore not that founder animals necessarily specifically select inappropriate habitat but that undirected dispersal immediately after release takes them out of carefully selected release sites where habitat restoration has taken place and into less suitable areas. For example, 32% of 460 black stilts (kaki, *Himantopus novaezelandiae*) settled away from their release sites, with 15% of birds ending up in unmanageable areas not subject to control of introduced predators (van Heezik *et al.*, 2009).

We do not know how often species make direct assessments of a habitat, but evidence for the use of indirect cues in habitat selection is common, many of which have the potential to produce a mismatch between species locations and quality. Conspecific presence or breeding success may influence habitat selection, with conspecific attraction or avoidance potentially leading to aggregations in suboptimal habitats (Mihoub *et al.*, 2009) through selection of poor areas or avoidance of high-quality areas (Railsbeck *et al.*, 2003), or through forcing subordinate individuals into poor habitats (Serrano & Tella, 2007).

The situation is further confused by simultaneous spatial and temporal variation. Many species show seasonal movements between habitats and a single habitat alone is not sufficient to sustain a viable population (e.g. little bustards *Tetrax tetrax* in Spain (Suárez-Seoane *et al.*, 2008) and white storks *Ciconia boyciana* in Japan (Naito & Ikeda, 2007)). This complicates assessments of habitat suitability because multiple habitats are required in the right proportions, with the right spatial pattern and with the right phenology.

Consequence 3. Present-day locations where a species is absent may not indicate an unsuitable habitat

In trying to understand what makes an area suitable for a species, we often contrast it with another area where the species is absent, based on the assumption that 'absence' and 'unsuitable' are interchangeable. This is a dangerous assumption and stems from our failure to appreciate the principles of niche theory and how niches defined in environmental space map on to geographical space.

Put simply, the fundamental niche represents the sets of environmental conditions under which the populations of a species thrive. In its Grinnellian form (Soberon, 2007), and for birds at least, this normally means a range of temperatures, levels of precipitation and net primary production (Pigot *et al.*, 2010). When we translate this niche space into geographical space, however, we cannot assume that a species will occupy *all* locations that meet these basic needs. The reasons for this may be either passive or active. Taking passive examples first, many European trees are not in equilibrium with climate because their ranges are still limited by post-glacial dispersal

(Svenning & Skov, 2004). Similar issues arise when trying to learn from one reintroduction project to guide another: for example, the European population of the Eurasian lynx, *Lynx lynx*, is still expanding, so absence does not equate with lack of suitability (Schadt *et al.*, 2002a). Similarly, because the restoration of breeding populations of white storks *Ciconia ciconia* in Sweden is recent and populations are very small and clustered around release sites, unoccupied sites are not necessarily unsuitable (Olsson & Rogers, 2009). In contrast to passive explanations for limited occupancy of niche space, active reasons consider deliberate avoidance and rejection. One example is the 'perceptual trap', where stimuli for habitat selection are negative rather than positive, leading to a habitat of high quality being *perceived* as low quality and so avoided (Patten & Kelly, 2010). In a study in New Mexico, female lesser prairie-chickens, *Tympanuchus pallidicinctus*, were found to perceive a tebuthiuron-treated habitat to be of low quality even though reproductive data suggest it is not (Patten & Kelly, 2010). Similarly, female greater prairie-chickens, *T. cupidio*, tend to avoid nesting in areas burned the previous season as part of agricultural management, despite there being no reduction in nesting success in burned versus unburned areas (Patten *et al.*, 2007). Of course, other (unmeasured) factors could be involved and we might argue that highly evolved species are better at assessing suitable habitats than are researchers.

When we consider the Eltonian niche (Soberon, 2007), which adds biotic interactions, we find that active rejection of suitable habitats occurs due to the presence of the same or other species. For example, the distribution of great bustards, *Otis tarda*, in Spain is patchy because of a meta-population structure and conspecific attraction (Osborne *et al.*, 2007). The presence of predators may also turn well-resourced habitat patches into no-go areas, creating so-called 'landscapes of fear' (Willems & Hill, 2009). Indeed, many of the arguments why occupied habitat is not always good, expressed in Consequence 2, clearly apply here, although in reverse. The result is that the realized niche is smaller than the fundamental niche and the portion of geographical space that is occupied may only be a fraction of the 'green' habitat that seems suitable. (Again, we must remember that the term habitat encompasses not just vegetation or land use but biotic interactions as well). While these principles apply to all species, they are particularly pertinent for rare or endangered species where populations may be below carrying capacity because of human influence or historical factors. The result is that it is often difficult to discriminate unsuitable habitat from habitat that is merely unoccupied (Davis *et al.*, 2007).

Consequence 4. Present-day locations may not indicate a future suitable habitat

Once we accept that species ranges are dynamic (Hengeveld, 1990; Sexton *et al.*, 2009) (see above) and have shifted over time, it becomes obvious that assessments of present-day habitat conditions provide no guarantee that the habitat will remain suitable in the future (Seddon, 2010). Given the long-term nature of reintroduction projects, it is a curious omission that the IUCN guidelines (IUCN, 1998) did not require an assessment of likely environmental change driven by climate, land use and human disturbance. (This is, however, one focus of a review and revision of the Reintroduction Guidelines by the IUCN/SSC RSG/ISSG Task Force on Moving Plants and Animals for Conservation Purposes (Stanley Price, 2010)). In our view, conservation should always manage for change and the common failure to do this only fuels the institutional obsession with crisis management as opposed to anticipating change (Lindenmeyer *et al.*, 2008). In a modern world faced with unprecedented global climate change, consequent land use changes and huge demographic shifts, the assumption that a habitat will remain suitable is simply untenable. Of course, this leaves us with some very tough decisions: if we suspect that an area will be lost in the future due to sea-level rise, do we simply abandon it now for conservation purposes? The answer must surely lie in how much we need to set priorities and the level of resources available to conservation. In an ideal world, all sites would be conserved, but this is unrealistic. As we argue later, one of the benefits of using species distribution modelling within reintroductions is that it allows us to make objective, informed choices, considering past, present and future conditions.

Consequence 5. Landscape components may be missing

Conservation based on suitable habitat patches alone is very limited because it ignores the flows of biota, water and nutrients across the landscape as well as interactions among different landscape elements (Lindenmeyer *et al.*, 2008). Patch occupancy depends not only on the quality of a patch per se but also the spatial arrangement of the landscape elements and the dispersal abilities of the animal being released (Fernandez *et al.*, 2003; Gaston, 2009; van Langevelde & Wynhoff, 2009; Vogeli *et al.*, 2010). It is not always true that high connectivity is desirable because connected habitats may allow species to leave before the population is large enough and will equally allow predators and diseases in. Issues of source–sink habitats and seasonal habitat use, as mentioned above, clearly have crucial landscape elements. It is therefore not enough to recognize that certain habitat patches are suitable; their spatial arrangement and how they will be used by the reintroduced species both seasonally and as the population grows must be considered. This requires detailed knowledge of the species resource requirements, typical home ranges size and movements. Failure to do this before the project begins is likely to compromise success. The practical issue is that assessing whole landscapes or regions for suitability is a demanding task and for all but the most sedentary species with small home ranges, modelling approaches are required.

Consequence 6. Habitat suitability and habitat characteristics vary across a species range

The issue of how habitat varies across species ranges is one of several key questions that challenge macroecology (Fortin *et al.*, 2005) and has a close relationship with both abundance and fitness. It is also a key argument in the debate over what limits a species range, which has 'often proven frustratingly difficult to explain' (Gaston, 2009). A recent global analysis across all bird species has shown, however, that range shape is limited by temperature, precipitation and net primary production, in decreasing order of importance (Pigot *et al.*, 2010). Furthermore, the role of continental boundaries is small compared with the force of these environmental gradients (Pigot *et al.*, 2010). Therefore, broad-scale patterns of bird species occurrence are climate-limited. The mechanisms behind climatic range limitation in birds are not well documented but are probably only rarely direct effects (e.g. limits of thermal tolerance) as opposed to indirect effects mediated through vegetation and food supply.

As a general rule, habitats at the periphery of a species range are more diverse, more fragmented and less favourable than those at the range core (Lomolino & Channell, 1998; Channell & Lomolino, 2000). Consequently, both population density (Channell & Lomolino, 2000) and reproductive success or survival (Carrascal & Seoane, 2009) are often lower at the periphery than the core, with the latter acting as 'sources' and peripheral areas as 'sinks' (Pulliam, 1988). If a gradient in favourability is envisaged from the core to the periphery, this idea is known as the 'abundant-centre' model (Sagarin & Gaines, 2002), but it may be more accurate to think in terms of a 'rare-periphery' model (Gaston, 2009) where the margin is characterized as being poor, on average, compared with other parts of the range. Success can be just as high at the margins as at the core, but is confined to a few favourable habitat patches in a largely unfavourable matrix (Gaston, 2009; Sexton *et al.*, 2009). Whichever model we follow, what we observe is geographical variation in the constraints affecting a species across its range (Vallecillo *et al.*, 2010) with consequent effects on abundance and fitness.

Of course, it is not quite that simple. For one thing, species' ranges often have multiple cores and interior barriers (Fortin *et al.*, 2005) with environmental gradients within the range as well as towards the periphery (Sexton *et al.*, 2009). Indeed, support for the abundant-centre model is surprisingly difficult to find (Sagarin & Gaines, 2002). Furthermore, there are theoretical reasons why high-density areas may exhibit poor breeding success if settlement follows the Ideal Pre-Emptive Distribution (van Horne, 1983; Pulliam & Danielson, 1991), although this does not seem common, at least in northern hemisphere birds (Bock & Jones, 2004). Interestingly, density may more often be negatively related to reproductive success in areas disturbed by humans, perhaps because birds fail to recognize ecological traps or opportunities in landscapes that differ from those in which they evolved (Bock & Jones, 2004). It may therefore not be a coincidence that studies of range contraction have shown persistence in the periphery rather than the core (Lomolino & Channell, 1995) because it is usually the core that is heavily affected by human influence (Channell & Lomolino, 2000). Core areas may now be so heavily modified by humans that they represent the least important part of the range in terms of habitat and species persistence, and it is incorrect simply to assume that 'peripheral areas are the domain of zombies' (Lomolino & Channell, 1998). For example, before the arrival of humans in New Zealand, takahe (*Porphyrio hochstetteri*), a large endemic flightless rail, was most abundant in lowland forest, but due to habitat modification by humans the relict population of takahe became restricted to higher altitude montane forest that was relatively undisturbed by humans (Bunin & Jamieson, 1995). Active restoration of lowland sites is therefore required before takahe reintroduction to core areas of its former mainland distribution range is possible, but successful translocations to lowland offshore islands have taken place since the early 1990s (Jamieson & Wilson, 2003).

There may also be incidences where the habitats occupied by a species vary greatly across geographical space and many edge-of-range habitats differ markedly from those towards the core (Gaston, 2009). Thus, for example, sand lizards *Lacerta agilis* in Great Britain are restricted to lowland heathlands whereas in France they occur in hedgerows, fields, woodland margins, parks, gardens, roadsides and mountainous habitats. At a finer spatial scale, bluethroats *Luscinia svecica* in northern Spain may occupy dry mattoral with trees over 3 m tall or, in nearby valleys, the scrubby margins of agricultural land or mountain slopes. The issue here is probably that the human eye is misled into seeing the vegetation type and associated landscape as a causal factor determining occurrence whereas, in fact, habitat structure might be more important (Steffens *et al.*, 2005). There is also the issue of researchers using whatever environmental variables are available to model habitat use rather than those that the species may use, particularly within Geographic Information Systems (GIS). This mismatch between human perceptions of the world and the needs of rare species bedevils conservation management (Manning *et al.*, 2004; Lindenmeyer *et al.*, 2008).

So where does this leave us? In an ideal world, we would always assess habitat suitability in terms of its demographic *consequences* (Armstrong, 2005), but this is costly at best and practically impossible for rare species. As a next best, we should use *direct* correlates of survival and reproduction, such as invertebrate abundance for insectivorous birds, because physiology is likely to dictate broadly similar nutritional and energetic needs irrespective of the location in the range. Even this is dangerous, though, because responses of demographic parameters to environmental gradients are usually non-linear and often feature critical thresholds (Lindenmeyer *et al.*, 2008; Gaston, 2009). Therefore only slight errors in habitat characterisation could separate reintroduction success from failure. Once we move to the next level, using *indirect* correlates such as vegetation type and extent, we are tempting fate, although this is often all we can practically manage. We can protect ourselves to some extent when using indirect correlates by studying habitat use across large geographical ranges. This ensures that our knowledge of habitat use is not dominated by local conditions and aids transferability to other situations. Such large-scale studies are impossible in the field and require modelling of often rather coarse-grained habitat details (e.g. captured through satellite as spectral indices, such as the Normalised Difference Vegetation Index). Examples of this approach are discussed in more detail below.

Consequence 7. Individuals from across the range may not all be equally suited to the chosen release site

As habitats vary across a species range and so too do population density and vital rates, these forces may act as drivers of local intraspecies divergence, causing spatial structuring in the phenotype and/or the genotype (Blondel et al., 2006). In other words, species become locally adapted to environmental conditions. Put simply, our concern here is that if the habitat experienced by the donor population does not match the habitat selected for the release, the individuals could be maladapted and success would be compromised. Whereas phenotypic differences at the individual level could be overcome relatively quickly (e.g. through the inherent plasticity that allows acclimation to local temperatures), genetic differences would not. Reintroduced populations often retain a long-term genetic legacy from the source population, characterized by reduced variability, intensified by genetic drift if the founder group is small (Latch & Rhodes, 2005). Getting the genetic characteristics of the founder group right is therefore a good insurance policy if not crucial (see Chapters 11 to 13, this volume).

The spatial structure of adaptive variation is poorly understood, particularly at the genetic level (Sexton et al., 2009), although evidence is growing and general principles have been established. For example, at range margins and other sink habitats where favourability and environmental predictability may be lower, the ability to disperse would be selected for (Holt, 2003). Also, where ranges have expanded, it may be argued that individuals at the expanded margin will be those with better dispersal abilities, a feature that would be repeatedly selected for over generations, leading to runaway selection (Gaston, 2009). Translocating a donor population from such marginal habitats to a small patch elsewhere could lead to higher mortality due to movements into adjacent habitats caused by hard-wired migratory tendencies. There is some evidence through hand-rearing experiments of birds that tendencies towards being sedentary (and therefore the opposite, being prone to move) are indeed genetically determined (Partecke and Gwinner, 2007). Other adaptive differences operate at a more subtle level. For example, evidence that song variation determined by habitat could play a role in ecological speciation (Ripmeester et al., 2010) warns us that individuals moved across habitats may be maladapted to their new local conditions in ways that would reduce breeding success.

The practical consequence when considering habitat selection for reintroductions is always to ensure that the donor population occupies as similar a habitat as possible to that at the chosen release site. The ideal would be to have both a range of candidate release sites and a range of candidate donor populations, and to choose the best match through multivariate comparisons of habitat characteristics. Practicalities will often dictate otherwise, but this ideal is a good starting point.

Consequence 8. Suitable habitat may need to be engineered and managed

One of key lessons of landscape conservation is that we should manage both the species and the ecosystem (Lindenmeyer *et al.*, 2008). Limiting reintroductions to habitats and landscapes that are already suitable is naive, not only because it ignores the likelihood of environmental change but also because it understates the degree to which most habitats are already modified. Few, if any, areas have escaped human influence. In New Zealand, for example, predation by introduced mammalian pests is responsible for ongoing declines and limitation of forest birds (Innes *et al.*, 2010) and extinctions. Mainland restoration of forest birds requires a combination of predator control and translocation, with the greatest success coming from the creation of fenced, predator-free 'mainland islands' (Saunders & Norton, 2001). Even many of New Zealand's pest-free offshore islands, once regarded as 'touchstones of primaeval New Zealand' are recognized as having been substantially modified since human settlement (Bellingham *et al.*, 2010). Indeed, naturalness may not be a desirable state because the landscape we crave (like the look of, feel comfortable in, derive pleasure from) may never have really existed. In addition, the dichotomy between natural (= untouched by humans) and modified landscapes carries a strong cultural bias and underpins a largely Western-derived preservationist model. Many indigenous peoples do not see themselves as separate from their environment, but rather consider humans as being embedded as a functional element, a world-view that focuses on wise use and respect for, rather than isolation and preservation of, nature (Alcorn, 1993). Regardless of the cultural perspective, however, we are biased in what we wish to conserve and where. If you remain in doubt, try fund raising to preserve the polio virus.

A reintroduction is a highly artificial intervention and no more so because habitats have been engineered to increase success. Environmental change and the dynamic nature of species ranges mean that habitat management is almost an inevitable part of species conservation, and increasingly there is awareness that for some species and in some systems some form of management intervention may be required in perpetuity.

Using habitat suitability modelling for reintroductions

Practitioners, in particular, may find the preceding sections rather negative as we have stressed the difficulties but offered few solutions. There is no universal panacea because we are essentially balancing unknowns, but uncertainty in outcome may be reduced by being more objective in our choice of release sites, while acknowledging that how we practically do this will differ between species and situations. We may summarize much of the foregoing arguments into a simple maxim: sites should not be selected for reintroductions only on the basis that the species used to be there or that the site looks right. So what do we need in addition? While detailed autecological studies (e.g. focusing on food supply) may provide essential information on the likely suitability of a proposed release site at the present time, time and financial constraints mean that only modelling can put that knowledge into the landscape context, project it into the future and allow us to weigh up alternative sites from a pool of candidates, chosen objectively. At the very least, modelling should be able to eliminate sites where the uncertainty surrounding their present and future suitability is high and allow us to identify networks of more suitable patches that might allow for population expansion. Caution is needed here, however, since the eight consequences previously discussed are also pitfalls for the unwary in modelling distributions.

The armoury of tools for predicting or generalizing species distributions has developed under the banner of species distribution (or habitat) modelling (Box 3.1). Although species distribution modelling has its roots in the 1990s (Osborne & Tigar, 1992; Buckland & Elston, 1993; Franklin, 1995), its refinement and use exploded from around 2000 onwards due to computer hardware and software developments, with applications to reintroductions coming a year or so later. Our goal here is not to present a manual on how to model species distributions (because that would be a book in itself), but some explanation is required to highlight the issues and some of this is inevitably technical. Most approaches to modelling distributions use presence locations to define suitable environments (because it is the easiest approach) and the crucial first question to ask is whether we have enough locations to build a worthwhile model in this way. The answer depends on the species rarity, its range, which variables we wish to use to define 'suitability' and at what spatial scale.

Box 3.1 **Species distribution modelling**

Species distribution modelling encompasses a variety of approaches for predicting the range of a species over a given area. Models may be empirical or process-based and dynamic or static. Process-based models try to mirror the mechanisms that lead to distribution patterns (e.g. Kearney & Porter, 2009) whereas empirical models are essentially correlative, using the observed relationships between occurrence and environmental (habitat) features at sample locations to predict where else a species should occur. Process-based models require extensive details on how the environment limits a species distribution that are often not available (e.g. physiological limits) and, in practice, most distribution models are empirical, although some are process-empirical hybrids (Sutherst, 2003), which may be a good approach (Morin & Thuiller, 2009). Dynamic models are responsive to differences over time (Osborne & Suarez-Seoane, 2007; Bartel & Sexton, 2009) whereas static models capture only a snapshot view at a particular point.

The literature on empirical species distribution modelling is extensive (Guisan & Zimmermann, 2000; Scott *et al.*, 2002; Guisan *et al.*, 2002, 2006; Lehmann *et al.*, 2002; Moisen *et al.*, 2006; Franklin, 2010). There are many static, correlative distribution modelling algorithms (e.g. Elith *et al.*, 2006), but they essentially fall into two camps. If both presence and absence data are available (e.g. through structured surveys) it is best to employ group discrimination approaches. These contrast the characteristics of sites with known presence against those with known absence. In cases where the locational data lack verified absences, alternative approaches are needed that use presence locations alone (i.e. profile techniques) (Guisan & Zimmermann, 2000) or model presence against the 'background' (Hirzel *et al.*, 2002; Phillips *et al.*, 2006). In all cases, empirical species distribution modelling requires three ingredients: species locations, environmental data layers (often satellite and interpolated ground-station data) and an algorithm to link the two. (Note that empirical species distribution modelling rarely employs data on interacting species although this could be done as in forest gap models; see Bugmann, 2001). The species locations are used to interrogate the environmental layers within a GIS to derive an approximation to the realized niche within the geographic area being studied and within the resource axis considered (Phillips *et al.*, 2006). Once the criteria for identifying the niche have been established (e.g. as some form of regression model) the GIS is used to display all other locations that match the criteria, often giving a probability of membership. This process generalizes the known species locations to define the entire range within the studied geographic extent.

For species that are locally but not globally rare, we will often have enough locations to characterize currently occupied sites. Examples of reintroduced species in this category would be red kites, *Milvus milvus*, and great bustards, *Otis tarda*, in the UK. To avoid the pitfalls of Consequence 2, these sites should ideally be ones where vital rates are known to be high or at least not suspected as being low. Bearing in mind Consequences 6 and 7, the locations used should also be within the local range of the chosen source subpopulation, avoiding geographically isolated subpopulations that may display alternative requirements. If we can be sure that historically documented locations within the former range have not changed in the environmental variables we wish to model, these may also be included. For example, very recent extinctions may be due to known human activities (e.g. hunting) and absence may not be due to an unsuitable climate that we may wish to model (see Consequence 3). Having got this far, the next stage is to consider whether we have verified absence data (e.g. from systematic field surveys) and whether we know enough to judge these as true absence points rather than places where the species was missed due to low detectability. Armed with quality presence locations and verified true absences, distribution modelling may proceed using the most powerful algorithms that contrast presence and absence, such as boosted regression trees and generalized additive models (Hastie *et al.*, 2009). If verified absence data are lacking, modelling may be done instead using the maximum entropy algorithm (Phillips *et al.*, 2006) or Ecological Niche Factor Analysis (Hirzel *et al.*, 2002).

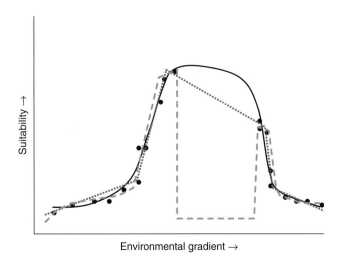

Figure 3.1 **Hypothetical example of how different modelling approaches might fit a response curve to data indicating a species use of an environmental gradient. Black dots are sampled data points; solid line, interpolated curve; dotted line, interpolated piecewise linear model; dashed line, flexible splines but poorly interpolated.**

In the algorithms mentioned above, the key requirement is that sufficient presence locations are used to characterize fully each environmental axis being used in the modelling. For example, if a species is temperature-limited (either directly or via habitat needs or food supply) enough data points are needed to characterize a graph linking suitability to temperature (the so-called 'response curve'; see Figure 3.1). Gaps in the sampling along the curve can be interpolated during the modelling process, but how this is done differs between algorithms and its reliability depends on the spread of the data points. Indeed, interpolation may not always be desirable.

It is often useful to conceptualize species distribution modelling as a nested hierarchy of models built at different spatial resolutions, corresponding to the Grinnellian niche and the Eltonian niche as we move from coarse to fine resolutions. In tackling global ranges, valuable models may be built at resolutions as coarse as 50 km × 50 km or 10 km × 10 km using variables such as temperature, precipitation and net primary production (see Pigot *et al.*, 2010). Land cover classes may also be used at 10 km × 10 km and vegetation classes based on estimates of green biomass at 1 km × 1 km. Habitat classes and measures of human disturbance may be employed below 1 km × 1 km resolution. The key message is that each variable has an appropriate scale and vice versa: measures such as temperature often define a large spatial cell in which the species can survive but play little role in defining which 100 m × 100 m patches within the cell are occupied. (There are clear exceptions too, e.g. plants on south-facing slopes.) In planning a translocation, our pragmatic goal should be to model down to the resolution where choice of sites is usefully reduced by eliminating those that are unsuitable, but not beyond the point where uncertainty becomes unmanageable due to the inclusion of questionable variables.

At the other end of the spectrum are species that are so rare or geographically restricted that we either have very few presence locations to characterize suitable habitat or the range of conditions occupied is so narrow that correlative modelling becomes pointless. In these cases, species distribution modelling is best achieved using physiological models (e.g. Kearney & Porter, 2004, 2009). An excellent introduction to what can be achieved even with very rare species is provided by Porter *et al.*'s (2006) study on the Po'ouli *Melamprosops phaeosoma*, which formerly inhabited the northeast slopes of Haleakala Volcano on the island of Maui, Hawaii, but is now thought to be extinct. The authors show how limited existing data may be used as input to microclimate and endotherm models, particularly by reference to closely related species, and then integrated with remote sensing data to predict likely diet, distribution and levels of environmental resources necessary for survival, growth and reproduction. More recent developments have sought to combine elements of both physiological and correlative mapping where data allow (Kearney & Porter, 2009). Theoretically, a key difference is that distributions (and therefore release sites) identified using physiological data approximate to conditions within the fundamental niche whereas correlative methods reveal portions of the realized niche.

Note that in both correlative modelling and situations where physiology is used, it is often possible to predict potential distributions (and hence reintroduction sites) based only on climatic variables and their proxies (e.g. terrain). This is useful in the context of Consequence 8 because it can reveal locations that could become suitable through habitat management even though the current habitat is not favoured by the species.

Having discussed some general guidance on species distribution modelling for reintroductions, we turn to examples from the literature. Where possible, we illustrate the approach using examples from the reintroduction literature, but readers will find far more explanation and other examples by accessing the citations in Box 3.1.

Perhaps the simplest use of species distribution modelling is to identify a suitable habitat for a release or for range expansion following release. One of the most straightforward ways in which this has been done is not through statistical analysis at all, but by defining a series of rules for patch occupancy based on knowledge of the species, and then applying those rules within a GIS to find all patches meeting those rules. This approach has been applied in reintroduction projects on the Eurasian lynx, *Lynx lynx* (Schadt *et al.*, 2002b), the panther, *Puma concolor* (Thatcher *et al.*, 2006), and the burbot, *Lota lota* (Worthington *et al.*, in review). The assumption is that we know what the rules of habitat selection are (fashionably called expert knowledge), including how to weight different characteristics of the environment to define suitability. Where we doubt our ability to do this or when more objectivity is desirable, the alternative is to define the rules objectively by training a model on patches or locations with known presence and/or absence using statistical procedures (the correlative approach in Box 3.1). Early examples used logistic regression (i.e. a generalized linear model with a logit link and binomial error structure) (Schadt *et al.*, 2002a; Fernandez *et al.*, 2003; Carroll *et al.*, 2003), but the statistical assumptions of generalized linear models are now often thought to be too restrictive and inflexible, and newer studies often feature more complex approaches (Klar *et al.*, 2008). A good example of the use of species distribution modelling within avian reintroductions is work on the bearded vulture *Gypaetus barbatus* in Europe (Box 3.2). This case study illustrates the value of using models for 'adaptive management', i.e. re-evaluating habitat selection as further data are gathered following an initial release (Cook *et al.*, 2010), although caution must always be exercised in extrapolating from a few released individuals.

Box 3.2 **Bearded vulture case studies**

Bearded vultures *Gypaetus barbatus* were extirpated from the European Alps between the end of the 19th and beginning of the 20th centuries. Between 1986 and 2003, 121 captive-bred individuals were reintroduced at four locations and their subsequent dispersal followed. As the population spread, Alexandre Hirzel and co-workers (2004) studied habitat use in the Alps of Valais (SW Switzerland), which itself had no reintroductions but is adjacent to release sites in France. Where range use is limited by dispersal, it is unwise to equate absence with lack of suitability (see also Consequence 3 above) so modelling employed a 'presence-only' technique called Ecological Niche Factor Analysis (ENFA) (Hirzel *et al.*, 2002). ENFA is a data reduction and ordination algorithm, rather like principal components analysis except that it operates on a species location and a global environmental dataset simultaneously, looking for differences between the two. What Hirzel and colleagues found was that during the prospecting phase, mostly carried out by immature birds, food supply was the most important variable explaining habitat use. In the later settling phase, however, the presence of limestone substrates became more important than food supply, the birds tending to avoid areas of the Alps dominated by silicate substrates. The logical conclusion was that future reintroductions would be more likely to succeed if releases focused on the large limestone massifs.

In a complementary study in the Spanish Pyrenees, Margalida *et al.* (2008) looked at variation in nest-cliff selection by bearded vultures during long-term population recovery. They assessed whether a previously published habitat selection model (Donazar *et al.*, 1993) would correctly predict nest-cliff use over the following decade when the population was larger (93 territories in 2002 compared with 53 in 1991). Results showed this to be the case and, interestingly, the new pairs occupied cliffs among the existing territories rather than going elsewhere, forcing territory shrinkage and reduced breeding success (Carrete *et al.*, 2006). Future conservation efforts should therefore spread the species elsewhere through promotion of dispersal and reintroductions.

Going beyond simple applications of species distribution modelling, attempts have been made to integrate predictions of patch suitability with population models that assess viability at potential reintroduction sites (Carroll *et al.*, 2003). Such integrated studies have the potential to model habitat suitability, the probability of occupancy through dispersal and population viability as founder populations grow and spread. Others studies have identified both potential release sites and areas where habitat restoration would relatively easily improve suitability (Olsson & Rogers, 2009). This approach satisfies our Consequence 8, which calls for serious consideration of habitat manipulation in reintroductions. Again, we can envisage here a role for adaptive management where we learn about the success of model predictions and how well restoration meets the desired goals as a reintroduction project progresses.

Most species distribution modelling studies are snapshots, assessing habitat suitability at only a single point in time (Bartel & Sexton, 2009), largely because multitemporal studies are very data-hungry. However, satellite data are beginning to have a long enough history to make this easier and work has been done using data from the Advanced Very High Resolution Radiometer (AVHRR) satellite from 1983 to 1999 (Osborne & Suarez-Seoane, 2007) and Landsat data from 1985 to 2007 (Bartel & Sexton, 2009). Such studies allow identification of potential release sites that are defined as suitable in most years, overcoming the possibly erroneous conclusions that would be drawn from a snapshot survey taken by chance in a good or poor year. Of course, use of satellite data restricts the types of variables modelled and their spatial resolution, and this needs to be balanced against the benefits of an output with perhaps less temporal uncertainty in its predictions.

Probably one of the most important uses for species distribution modelling is, however, in forecasting habitat suitability in the future (Box 3.3). Much effort has recently gone into building species distribution models based on climate data and then forecasting the impacts of climate change by inserting future climatic variables from the Intergovernmental Panel on Climate Change (IPCC) projections. The approach can be refined still further by considering not just the direct effects of climatic variables but also consequent land use changes (Wisz et al., 2008) and dispersal ability. There is a good deal of sense in these approaches and they have been underused in reintroductions to date, but there is a caveat: future projections are subject to uncertainty, which should be factored into decision making rather than be ignored. Forecasts of the effects of climate change on birds are affected by choice of the modelling algorithm, general climate model, greenhouse gas emissions scenario and set of predictor variables (Synes & Osborne, 2011). Indeed, even the crucial basic question of whether climate is a limiting factor for the species is often overlooked.

Box 3.3 Forecasting future suitable climate space

As a generalization, we may assume that the distributions of birds are limited hierarchically by climate, land cover type, habitat and then patch characteristics, moving from global to local scales. A valid first step in modelling suitable areas for reintroductions is therefore to consider climate space and how this might change in the future. As part of the long-running series of studies on great bustards *Otis tarda*, Patrick Osborne and student Nick Synes modelled where suitable climate space lies in Europe now and where it would be by 2050 and 2080 (Synes & Osborne, 2011). Their study considered multiple ways of expressing temperature and precipitation along with three global climate models and two greenhouse gas emissions scenarios, so capturing some of the uncertainty in climate change projections.

Models of present-day climate space showed strong agreement with the present-day distribution, particularly in Spain and Hungary (Figure 3.2). Small areas of France were also deemed climatically suitable (Consequence 3) and are within the historical range, but may lack appropriate habitat at more local scales. Large tracts of England, France and Germany which were historically suitable were emphatically rejected by all the models (Consequence 1). When projected into the future, there was a noticeable increase in uncertainty and the geographic locations of strong agreement changed (Figure 3.3). Both the southern half of Spain and much of Hungary were no longer considered suitable (Consequence 4). Clearly these areas would not be a good bet for population re-inforcement. Looking towards potential reintroduction sites, parts of northern France and SE England were predicted to have suitable climatic conditions by 2050. Perhaps crucially, however, the locations in England excluded the release site of the current reintroduction project on Salisbury Plain (Osborne, 2005), which suggests that a free-ranging, self-sustaining population might be difficult to establish there. On climatic grounds, and with the knowledge that climate change is forcing northern hemisphere species ranges northwards, it might be better to focus on areas that are north of major population centres, currently unoccupied but predicted to be suitable both now and in the

future. Small areas of France and Poland might meet these requirements, contingent, of course, on detailed field-scale studies finding
suitable habitat and food supplies, or areas where they could be created
(Consequence 8).

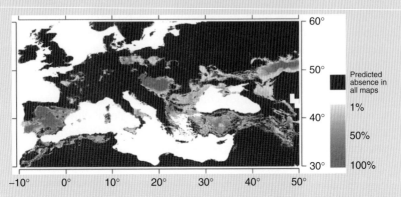

Figure 3.2 Percentage agreement in models of the present day climate
space for great bustards. (Reproduced from Synes & Osborne, 2011, with
permission from Wiley-Blackwell).

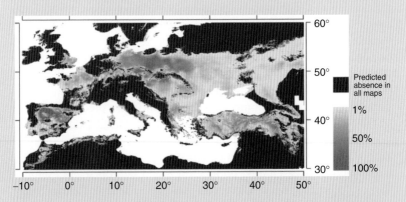

Figure 3.3 Percentage agreement in models of suitable climate space for
great bustards in 2050. (Reproduced from Synes & Osborne, 2011, with
permission from Wiley-Blackwell).

Conclusion

Assessment of habitat quality is an essential part of the feasibility study for any reintroduction, yet guidance on what needs to be taken into account is lacking. By returning to robust attempts to define habitat and its role in ecology, we have shown that at least eight consequences of ecological theory must be considered, all of which affect how we should assess a site's suitability for a proposed release. Many of these considerations require a combination of detailed field knowledge and the tools of species distribution or habitat modelling. Indeed, without predictive tools, conservation will fail to rise to the challenges of environmental change. We recognize, however, that the application of species distribution modelling to reintroductions is relatively new and there is little evidence to judge its success. Models do not and cannot make decisions but they can provide evidence to guide the setting of realistic objectives and build a realistic framework for action. We need a more honest, open approach when attempting reintroductions that acknowledge our lack of ecological knowledge. Provided there is no detriment in carrying out a trial release, detailed monitoring of a few individuals can yield data on a species' ecology that historical literature and analogue studies in neighbouring regions cannot reveal. Proper evaluation, however, should never be optional and all reintroductions should be carried out within an experimental framework, as has been advocated for landscape management (Lindenmeyer *et al.*, 2008). If we adopt this approach, reintroductions will contribute to the theoretical basis of ecology as well as achieving the immediate aims of conserving a target species.

References

Alcorn, J.B. (1993) Indigenous peoples and conservation. *Conservation Biology*, 7, 424–426.

Armstrong, D.P. (2005) Integrating the metapopulation and habitat paradigms for understanding broad-scale declines of species. *Conservation Biology*, 19, 1402–1410.

Armstrong, D.P. & Seddon, P.J. (2008) Directions in reintroduction biology. *Trends in Ecology and Evolution*, 23, 20–25.

Armstrong, D.P., Davidson, R.S., Dimond, W.J. *et al.* (2002) Population dynamics of reintroduced forest birds on New Zealand islands. *Journal of Biogeography*, 29, 609–621.

Bartel, R.A. & Sexton, J.O. (2009) Monitoring habitat dynamics for rare and endangered species using satellite images and niche-based models. *Ecography*, 32, 888–896.

Beck, B.B., Papaport, L.G., Stanley Price, M.R. *et al.* (1994) Reintroduction of captive-born animals. In *Creative Conservation: interactive management of wild and captive animals*, eds P.J.S. Olney, G.M. Mace & A.T.C. Feistner, pp. 265–284. Chapman & Hall, London, UK.

Bellingham, P. J., Towns, D.R., Cameron, E.K. *et al.* (2010) New Zealand island restoration: seabirds, predators, and the importance of history. *New Zealand Journal of Ecology*, 34, 115–136.

Blondel, J., Thomas, D.W., Charmantier, A. *et al.* (2006) A thirty-year study of phenotypic and genetic variation of blue tits in Mediterranean habitat mosaics. *Bioscience*, 56, 661–673.

Bock, C.E. & Jones, Z.F. (2004) Avian habitat evaluation: should counting birds count? *Frontiers in Ecology and the Environment*, 2, 403–410.

Buckland, S.T. & Elston, D.A. (1993) Empirical-models for the spatial-distribution of wildlife. *Journal of Applied Ecology*, 30, 478–495.

Bugmann, H. (2001) A review of forest gap models. *Climatic Change*, 51, 259–305.

Bunin, J.S. & Jamieson, I.G. (1995) New approaches towards a better understanding of the decline of takahe (*Porphyrio mantelli*) in New Zealand. *Conservation Biology*, 9, 100–106.

Cade, T.J. & Temple, S.A. (1995) Management of threatened bird species: an evaluation of the hands on approach. *Ibis*, 137, 161–172.

Carrascal, L.M. & Seoane, J. (2009) Linking density, productivity and trends of an endangered species: the Bonelli's eagle in Spain. *Acta Oecologica – International Journal of Ecology*, 35, 341–348.

Carrete, M., Donazar, J.A. & Margalida, A. (2006) Density-dependent productivity depression in Pyrenean bearded vultures: implications for conservation. *Ecological Applications*, 16, 1674–1682.

Carroll, C., Phillips, M.K., Schumaker, N.H. *et al.* (2003) Impacts of landscape change on wolf restoration success: planning a reintroduction program based on static and dynamic spatial models. *Conservation Biology*, 17, 536–548.

Channell, R. & Lomolino, M.V. (2000) Dynamic biogeography and conservation of endangered species. *Nature*, 403, 84–86.

Cook, C.N., Morgan, D.G. & Marshall, D.J. (2010) Reevaluating suitable habitat for reintroductions: lessons learnt from the eastern barred bandicoot recovery program. *Animal Conservation*, 13, 184–195.

Davis, F.W., Seo, C. & Zielinski, W. J. (2007) Regional variation in home-range-scale habitat models for fisher (*Martes pennanti*) in California. *Ecological Applications*, 17, 2195–2213.

Donazar, J.A., Hiraldo, F. & Bustamante, J. (1993) Factors influencing nest-site selection, breeding density and breeding success in the bearded vulture (*Gypaetus barbatus*). *Journal of Applied Ecology*, 30, 504–514.

Duncan, R.P., Blackburn, T.M. & Sol, D. (2003) The ecology of bird introductions. *Annual Review of Ecology and Systematics*, 34, 71–98.

Elith, J., Graham, C.H., Anderson, R.P. *et al.* (2006) Novel methods improve prediction of species' distributions from occurrence data. *Ecography*, 29, 129–151.

Elton, C. (1966) *The Pattern of Animal Communities*. Methuen, London.

Fernandez, N., Delibes, M., Palomares, F. *et al.* (2003) Identifying breeding habitat for the Iberian lynx: inferences from a fine-scale spatial analysis. *Ecological Applications*, 13, 1310–1324.

Fischer, J. & Lindenmayer, D.B. (2000) An assessment of the published results of animal relocations. *Biological Conservation*, 96, 1–11.

Fortin, M.J., Keitt, T.H., Maurer, B.A. *et al.* (2005) Species' geographic ranges and distributional limits: pattern analysis and statistical issues. *Oikos*, 108, 7–17.

Franklin, J. (1995) Predictive vegetation mapping: geographic modelling of biospatial patterns in relation to environmental gradients. *Progress in Physical Geography*, 19, 474–499.

Franklin, J. (2010) *Mapping Species Distributions: Spatial Inference and Prediction*. Cambridge University Press, Cambridge, UK.

Fuller, R.J., Gregory, R.D., Gibbons, D.W. *et al.* (1995) Population declines and range contractions among lowland farmland birds in Britain. *Conservation Biology*, 9, 1425–1441.

Gaillard, J.M., Hebblewhite, M., Loison, A. *et al.* (2010) Habitat-performance relationships: finding the right metric at a given spatial scale. *Philosophical Transactions of the Royal Society B – Biological Sciences*, 365, 2255–2265.

Gaston, K.J. (2009) Geographic range limits: achieving synthesis. *Proceedings of the Royal Society B – Biological Sciences*, 276, 1395–1406.

Griffith, B., Scott, J.M., Carpenter, J.W. *et al.* (1989) Translocations as a species conservation tool: status and strategy. *Science*, 245, 477–480.

Guisan, A. & Zimmermann, N.E. (2000) Predictive habitat distribution models in ecology. *Ecological Modelling*, 135, 147–186.

Guisan, A., Edwards, T.C. & Hastle, T. (2002) Generalised linear and generalised additive models in studies of species distributions: setting the scene. *Ecological Modelling*, 157, 89–100.

Guisan, A., Lehman, A., Ferrier, S. *et al.* (2006) Guest editorial: making better biogeographical predictions of species' distributions. *Journal of Applied Ecology*, 43, 386–392.

Hall, L.S., Krausman, P.R. & Morrison, M.L. (1997) The habitat concept and a plea for standard terminology. *Wildlife Society Bulletin*, 25, 173–182.

Hastie, T.J., Tibshirani, R. & Friedman, J. (2009) *The Elements of Statistical Learning: data mining, inference, and prediction.* Springer-Verlag, New York.

Hengeveld, R. (1990) *Dynamic Biogeography.* Cambridge University Press, Cambridge, UK.

Hirzel, A.H., Posse, B., Oggier, P.-A. *et al.* (2004) Ecological requirements of reintroduced species and the implications for release policy: the case of the bearded vulture. *Journal of Applied Ecology*, 41, 1103–1116.

Hirzel, A.H., Hausser, J. Chesse, D. *et al.* (2002) Ecological-niche factor analysis: how to compute habitat-suitability maps without absence data? *Ecology*, 83, 2027–2036.

Holt, R.D. (2003) On the evolutionary ecology of species' ranges. *Evolutionary Ecology Research*, 5, 159–178.

Innes, J., Kelly, D., Overton, J.McC. *et al.* (2010) Predation and other factors currently limiting New Zealand forest birds. *New Zealand Journal of Ecology*, 34, 86–114.

IUCN (1998) *Guidelines for Re-introductions.* IUCN, Gland, Switzerland.

James, H.F. & Price J.P. (2008) Integration of palaeontological, historical, and geographical data on the extinction of koa-finches. *Diversity and Distributions*, 14, 441–451.

Jamieson, I.G. & Wilson, G.C. (2003) Immediate and long-term effects of translocations on breeding success in takahe *Porphyrio hochstteri. Bird Conservation International*, 13, 299–306.

Kearney, M. & Porter, W. (2004) Mapping the fundamental nice: physiology, climate, and the distribution of a nocturnal lizard. *Ecology*, 85, 3119–3131.

Kearney, M. & Porter, W. (2009) Mechanistic niche modelling: combining physiological and spatial data to predict species' ranges. *Ecology Letters*, 12, 334–350.

Klar, N., Fernandez, N., Kramer-Schadt, S. *et al.* (2008) Habitat selection models for European wildcat conservation. *Biological Conservation*, 141, 308–319.

Latch, E.K. & Rhodes, O.E. (2005) The effects of gene flow and population isolation on the genetic structure of reintroduced wild turkey populations: are genetic signatures of source populations retained? *Conservation Genetics*, 6, 981–997.

Le Gouar, P., Mihoub, J.-B. & Sarrazin, F. (2011) Dispersal and habitat selection: behavioural and spatial constraints for animal translocations. In *Reintroduction Biology: integrating science and management*, eds J.G. Ewen, D.P. Armstrong, K.A. Parker & P.J. Seddon, Chapter 5. Wiley-Blackwell, Oxford, UK.

Lehmann, A., Overton, J.McC. & Austin, M.P. (2002) Regression models for spatial prediction: their role for biodiversity and conservation. *Biodiversity and Conservation* 11, 2085–2092.

Letty, J., Marchandeau, S., Clobert, J. *et al.* (2000) Improving translocation success: an experimental study of antistress treatment and release method for wild rabbits. *Animal Conservation*, 3, 211–219.

Lindenmayer, D., Hobbs, R.J., Montague-Drake, R. *et al.* (2008) A checklist for ecological management of landscapes for conservation. *Ecology Letters*, 11, 78–91.

Lomolino, M.V. & Channell, R. (1995) Splendid isolation: patterns of range collapse in endangered mammals. *Journal of Mammalogy*, 76, 335–347.

Lomolino, M.V. & Channell, R. (1998) Range collapse, re-introductions, and biogeographic guidelines for conservation. *Conservation Biology*, 12, 481–484.

Manning, A.D., Lindenmayer, D.B. & Nix, H.A. (2004) Continua and Umwelt: novel perspectives on viewing landscapes. *Oikos*, 104, 621–628.

Margalida, A., Donazar, J.A., Bustamante, J. *et al.* (2008) Application of a predictive model to detect long-term changes in nest-site selection in the bearded vulture *Gypaetus barbatus*: conservation in relation to territory shrinkage. *Ibis*, 150, 242–249.

Mathews, F. & Macdonald, D.W. (2001) The sustainability of the common crane (*Grus grus*) flock breeding in Norfolk: insights from simulation modelling. *Biological Conservation*, 100, 323–333.

McLain, D.K., Moulton, M.P. & Sanderson, J.G. (1999) Sexual selection and extinction: the fate of plumage-dimorphic and plumage-monomorphic birds introduced onto islands. *Evolutionary Ecology Research*, 1, 549–565.

Metzger, K.L., Sinclair, A.R.E., Campbell, K.L.I. *et al.* (2007) Using historical data to establish baselines for conservation: the black rhinoceros (*Diceros bicornis*) of the Serengeti as a case study. *Biological Conservation*, 139, 358–374.

Mihoub, J.B., Le Gouar, P. & Sarrazin, F. (2009) Breeding habitat selection behaviors in heterogeneous environments: implications for modeling reintroduction. *Oikos*, 118, 663–674.

Moisen, G.G., Edwards, T.C & Osborne, P.E. (2006) Further advances in predicting species distributions. *Ecological Modelling*, 199, 129–131.

Morin, X. & Thuiller, W. (2009) Comparing niche- and process-based models to reduce prediction uncertainty in species range shifts under climate change. *Ecology*, 90, 1301–1313.

Naito, K. & Ikeda, H. (2007) Habitat restoration for the reintroduction of Oriental white storks. *Global Environmental Research*, 11, 217–221.

Nesbitt, S.A. & Carpenter, J.W. (1993) Survival and movements of greater sandhill cranes experimentally released in Florida. *Journal of Wildlife Management*, 57, 673–679.

Olsson, O. & Rogers, D.J. (2009) Predicting the distribution of a suitable habitat for the white stork in Southern Sweden: identifying priority areas for reintroduction and habitat restoration. *Animal Conservation*, 12, 62–70.

Osborne, P.E. (2005) Key issues in assessing the feasibility of reintroducing the great bustard *Otis tarda* L. to Britain. *Oryx*, 39, 22–29.

Osborne, P.E. & Suarez-Seoane, S. (2007) Identifying core areas in a species' range using temporal suitability analysis: an example using little bustards *Tetrax tetrax* L. in Spain. *Biodiversity and Conservation*, 16, 3505–3518.

Osborne, P.E. & Tigar, B.J. (1992) Interpreting bird atlas data using logistic-models – an example from Lesotho, southern Africa. *Journal of Applied Ecology*, 29, 55–62.

Osborne, P.E., Suarez-Seoane, S. & Alonso, J.C. (2007) Behavioural mechanisms that undermine species envelope models: the causes of patchiness in the distribution of great bustards *Otis tarda* in Spain. *Ecography*, 30, 819–828.

Partecke, J. & Gwinner, E. (2007) Increased sedentariness in European blackbirds following urbanization: a consequence of local adaptation? *Ecology*, 88, 882–890.

Patten, M.A. & Kelly, J.F. (2010) Habitat selection and the perceptual trap. *Ecological Applications* [online], doi: 10.1890/09-2370.1.

Patten, M.A., Shochat, E., Wolfe, D.H. *et al.* (2007) Lekking and nesting response of the greater prairie-chicken to burning of tallgrass prairie. In *Proceedings of the 23rd Tall Timbers Fire Ecology Conference: fire in grassland and shrubland ecosystems*, eds R.E. Masters & K.E.M. Galley, pp. 149–155. Tall Timbers Research Station, Tallahassee, Florida, USA.

Phillips, S.J., Anderson, R.P. & Schapire, R.E. (2006) Maximum entropy modelling of species geographic distributions. *Ecological Modelling*, 190, 231–259.

Pigot, A.L., Owens, I.P.F. & Orme, C.D.L. (2010) The environmental limits to geographic range expansion in birds. *Ecology Letters*, 13, 705–715.

Porter, W.P., Vakharia, N., Klousie, W.D. *et al.* (2006) Po'ouli landscape bioinformatics models predict energetics, behavior, diets, and distribution on Maui. *Integrative and Comparative Biology*, 46 (6), 1143–1158.

Powell, A.N. & Cuthbert, F.J. (1993) Augmenting small populations of plovers – an assessment of cross-fostering and captive-rearing. *Conservation Biology*, 7, 160–168.

Pulliam, H.R. (1988) Sources, sinks, and population regulation. *American Naturalist* 132, 652–661

Pulliam, H.R. & Danielson, B.J. (1991) Sources, sinks, and habitat selection – a land-scape perspective on population-dynamics. *American Naturalist*, 137, S50–S66.

Railsbeck, S.F., Stauffer, H.B. & Harvey, B.C. (2003) What can habitat preference models tell us? Tests using a virtual trout population. *Ecology*, 13, 1580–1594.

Ripmeester, E.A.P., Mulder, M. & Slabbekoorn, H. (2010) Habitat-dependent acoustic divergence affects playback response in urban and forest populations of the European blackbird. *Behavioral Ecology*, 21, 876–883.

Sagarin, R.D. & Gaines, S.D. (2002) The 'abundant centre' distribution: to what extent is it a biogeographical rule? *Ecology Letters*, 5, 137–147.

Saunders, A. & Norton, D.A. (2001) Ecological restoration at Mainland Islands in New Zealand. *Biological Conservation*, 99, 109–119.

Schadt, S., Revilla, E., Wiegand, T. *et al.* (2002a) Assessing the suitability of central European landscapes for the reintroduction of Eurasian lynx. *Journal of Applied Ecology*, 39, 189–203.

Schadt, S., Knauer, F., Kaczensky, P. *et al.* (2002b) Rule-based assessment of suitable habitat and patch connectivity for the Eurasian lynx. *Ecological Applications*, 12, 1469–1483.

Schlaepfer, M.A., Runge, M.C. & Sherman, P.W. (2002) Ecological and evolutionary traps. *Trends in Ecology and Evolution*, 17, 474–480.

Scott, J.M., Heglund, P.J. & Morrison, M.L. (eds) (2002) *Predicting Species Occurrences: Issues of Accuracy and Scale.* Island Press, Covelo, California.

Seddon, P.J. (2010) From reintroduction to assisted colonization: moving along the conservation translocation spectrum. *Restoration Ecology*, 18, 796–802.

Serrano, D. & Tella, J.L. (2007). The role of despotism and heritability in determining settlement patterns in the colonial lesser kestrel. *American Naturalist*, 169, E53–E67.

Sexton, J.P., McIntyre, P.J., Angert, A.L. *et al.* (2009) Evolution and ecology of species range limits. *Annual Review of Ecology Evolution and Systematics*, 40, 415–436.

Soberon, J. (2007) Grinnellian and Eltonian niches and geographic distributions of species. *Ecology Letters*, 10, 1115–1123.

Southwood, T.R.E. (1977) Habitat, templet for ecological strategies – Presidential Address to British Ecological Society, 5 January 1977. *Journal of Animal Ecology*, 46, 337–365.

Stamps, J.A. & Swaisgood, R.R. (2007) Someplace like home: experience, habitat selection and conservation biology. *Applied Animal Behaviour Science*, 102, 392–409.

Stanley Price, M.R. (2010) Assisted colonization: move ahead with models. *Science*, 330, 1317.

Steffens, K.E., Seddon, P.J., Mathieu, R. *et al.* (2005) Habitat selection by South Island saddlebacks and Stewart Island robins reintroduced to Ulva Island. *New Zealand Journal of Ecology*, 29 (2), 221–229.

Suárez-Seoane, S., de la Morena, E.L.G., Prieto, M.B.M. *et al.* (2008) Maximum entropy niche-based modelling of seasonal changes in little bustard (*Tetrax tetrax*) distribution. *Ecological Modelling*, 219, 17–29.

Sutherst, R.W. (2003) Prediction of species geographical ranges. *Journal of Biogeography*, 30, 805–816.

Svenning, J.C. & Skov, F. (2004) Limited filling of the potential range in European tree species. *Ecology Letters*, 7, 565–573.

Synes, N.W. & Osborne, P.E. (2011) Choice of predictor variables as a source of uncertainty in continental-scale species distribution modelling under climate change. *Global Ecology and Biogeography*, Article first published online 17 February 2011, doi: 10.1111/j.1466-8238.2010.00635.x.

Thatcher, C.A., van Manen, F.T. & Clark, J.D. (2006) Identifying suitable sites for Florida panther reintroduction. *Journal of Wildlife Management*, 70, 752–763.

Vallecillo, S., Brotons, L. & Osborne, P.E. (2010) Geographical variation in ecological constraints on species distributions along a gradient of spatial aggregation. *Acta Oecologica – International Journal of Ecology*, 36, 666–674.

van Heezik, Y. Maloney, R.F. & Seddon, P.J. (2009) Movements of translocated captive-bred and released critically endangered kaki (black stilts) *Himantopus novaezelandiae* and the value of long-term post-release monitoring. *Oryx*, 43, 639–647.

van Horne, B. (1983) Density as a misleading indicator of habitat quality. *Journal of Wildlife Management*, 47, 893–901.

van Langevelde, F. & Wynhoff, I. (2009) What limits the spread of two congeneric butterfly species after their reintroduction: quality or spatial arrangement of habitat? *Animal Conservation*, 12, 540–548.

Vogeli, M., Serrano, D., Pacios, F. *et al.* (2010) The relative importance of patch habitat quality and landscape attributes on a declining steppe-bird metapopulation. *Biological Conservation*, 143, 1057–1067.

Willems, E.P. & Hill, R.A. (2009) Predator-specific landscapes of fear and resource distribution: effects on spatial range use. *Ecology*, 90, 546–555.

Wisz, M., Dendoncker, N., Madsen, J. *et al.* (2008) Modelling pink-footed goose (*Anser brachyrhynchus*) wintering distributions for the year 2050: potential effects of land-use change in Europe. *Diversity and Distributions*, 14, 721–731.

Wolf, C.M., Griffith, B., Reed, C. *et al.* (1996) Avian and mammalian translocations: update and reanalysis of 1987 survey data. *Conservation Biology*, 10, 1142–1154.

Worthington, T.A. *et al.* (in review) A spatial analytical GIS and RHS approach to selecting reintroduction sites for the reestablishment of the burbot in English rivers. *Freshwater Biology*.

4

The Theory and Practice of Catching, Holding, Moving and Releasing Animals

Kevin A. Parker[1], Molly J. Dickens[2], Rohan H. Clarke[3] and Tim G. Lovegrove[4]

[1]Massey University, Auckland, New Zealand
[2]University of Liège, Liège, Belgium
[3]Monash University, Melbourne, Australia
[4]Auckland Council, Auckland, New Zealand

'. . .translocation has no evolutionary precedent. While the physiological system has adapted to respond to specific, expectable and relatively quick stressors, it is not adapted to handle the multiple, continuous, unpredictable and novel nature of the stressors facing translocated individuals.'

Page 112

Introduction

The translocation process is essentially a forced dispersal event with no evolutionary precedent. Mortality immediately post-release is a critical determinant in the successful establishment of a new population (Armstrong & Seddon, 2008; see Box 6.1 in Armstrong & Reynolds, this volume, Chapter 6). Therefore, although few reintroduction plans include success criteria at the individual level (Jule *et al.*, 2008), the condition of animals immediately post-release is likely to be an important factor for a successful translocation outcome (Dickens *et al.*, 2010). Translocation is an inherently and unavoidably stressful

Reintroduction Biology: Integrating Science and Management. First Edition.
Edited by John G. Ewen, Doug P. Armstrong, Kevin A. Parker and Philip J. Seddon.
© 2012 Blackwell Publishing Ltd. Published 2012 by Blackwell Publishing Ltd.

procedure (Teixeira *et al.*, 2007; Dickens *et al.*, 2009, 2010) and it is likely that this stress is a significant factor in the mortality and dispersal that is frequently observed soon after the initial release (Armstrong & Seddon, 2008; Pinter-Wollman *et al.*, 2009; Tavecchia *et al.*, 2009; Swaisgood, 2010; Terhune *et al.*, 2010). There are also ethical considerations for the translocated animals (Bekoff, 2010); indeed, from a strict animal welfare perspective the translocation process might be considered unacceptable (Swaisgood, 2010). Despite these issues, translocation is an increasingly popular conservation tool (Griffith *et al.*, 1989; Wolf *et al.*, 1996; Fischer & Lindenmayer, 2000; Seddon *et al.* 2007; Armstrong & Seddon, 2008). It is being used in recovery programmes for everything from plants (Maschinski & Haskins, 2011), butterflies (Harris, 2008) and freshwater mussels (Wilson *et al.*, 2011) to reptiles (Germano & Bishop, 2008), birds (Parker & Laurence, 2008), marsupials (Hardman & Moro, 2006) and elephants (Pinter-Wollman *et al.*, 2009). Furthermore, the use of translocation seems likely to increase as it is applied in increasingly diverse ways, particularly as a tool for adaptation to anthropogenic climate change (Hunter Jr, 2007; McLachlan *et al.*, 2007; Hoegh-Guldberg *et al.*, 2008; Richardson *et al.*, 2009; Seddon *et al.*, 2009) and other human-mediated impacts (Seddon, 2010). Therefore, it is critical that translocation practice is scrutinized, researched and refined at all levels of the process.

There are many factors that can influence the outcome of a translocation but the objectives across projects are generally consistent: (1) high post-release survival; (2) settlement in the release area (Le Gouar *et al.*, this volume, Chapter 5) and (3) persistence through successful breeding, recruitment and population increase (IUCN, 1987; IUCN, 1998; Seddon 1999; Gosling & Sutherland, 2000; Teixeira *et al.*, 2007; Armstrong & Seddon, 2008). Here we discuss the practicalities of planning a translocation, including catching, holding, moving and releasing animals. We define this as the translocation process, the initial means by which a population is established or re-enforced, including management intended to increase post-release survival during the acclimation period following release (Armstrong & Reynolds, this volume, Chapter 6; Hamilton *et al.*, 2010). This is clearly a critical phase of the translocation process as it is the implicit means by which animals establish at a new site. New populations are comprised of individual animals that are subject to a succession of novel procedures during the translocation process, followed by an acclimation period (cf. Hamilton *et al.*, 2010) in a novel environment. This is an important point because animals display considerable individual variation in their ability to cope with novel and stressful situations (Wilson *et al.*, 1994;

Sih *et al.*, 2004; Smith & Blumstein, 2008). Therefore, the specific aspects of the carefully planned procedures aimed at establishing a successful population might impose a direct cost on the released individuals; for example, extended quarantine and handling procedures (Calvete *et al.*, 2005) or the fitting of monitoring devices (Godfrey *et al.*, 2001; Wilson *et al.*, 2011) might induce negative stress-related outcomes. We acknowledge that a similar challenge exists for plant translocations, although the specific issues often differ (Maschinski & Haskins, 2011). Here, our discussion is restricted to animal translocations.

Given the great diversity of species being translocated (see references above and Soorae, 2010) it is impossible to offer overarching guidelines for methodologies to guide the translocation process, even for a particular group, e.g. mammals, reptiles, birds, invertebrates or fish. However, there are themes common to successful projects: (1) they are carefully planned and carried out by multidisciplinary teams including managers, scientists and animal husbandry experts (e.g. see Jones & Merton, this volume, Chapter 2; Bristol *et al.*, 2005; Robertson *et al.*, 2006; Reynolds *et al.*, 2008); (2) they consider stress, either explicitly or implicitly, which is often cited as a significant reason for translocation failure (Letty *et al.*, 2000; Teixeira *et al.*, 2007; Chipman *et al.*, 2008; Dickens *et al.*, 2009, 2010) and (3) their translocation planning and methodology is underpinned by an intimate biological and ecological knowledge of the translocated species coupled with appropriate husbandry and release techniques. Here, we discuss how each of these factors contributes to translocation success.

First, we discuss the role of research and science in devising and improving translocation processes with a specific focus on the benefits of a multidisciplinary approach. We then provide a theoretical framework for understanding stress and the implications of stress in devising translocation processes. Finally we discuss how a fundamental understanding of biology can inform capture, holding, moving and release of animals with a range of examples from the great diversity of species being translocated. We consider the transport, handling, release and post-release management of both captive and wild animals because of the differing challenges associated with each.

The role of science and research in devising translocation processes

Science has contributed a great deal to reintroduction biology, particularly, as demonstrated in this book, in the fields of genetics (Jamieson & Lacy, this volume, Chapter 13), health management (Ewen *et al.*, this volume, Chapter 9), habitat selection (Osborne & Seddon, this volume, Chapter 3) and population modelling (Armstrong & Reynolds, this volume, Chapter 6). There has been some excellent research published on some aspects of the translocation process (Lovegrove & Veitch, 1994; Groombridge *et al.*, 2004a; Taylor & Jamieson, 2007; Dickens *et al.*, 2009; Adams *et al.*, 2010) but publications on the actual practicalities (catching, holding, moving, releasing) are relatively rare. This is interesting because the reintroduction biology literature is usually characterized as being dominated by case studies and post hoc descriptive analyses (Seddon *et al.*, 2007). However, these studies are usually restricted to descriptions of outcome, the demographic characteristics of the release group and comparisons of delayed (soft) versus immediate (hard) release strategies. It is a rare scientific paper that provides sufficient detail to guide the translocation process for any particular species and rarer still to find comparisons of different translocation processes. There are two explanations for this. First, the origins of reintroduction biology are in applied conservation management and many of the past and present pioneers of reintroduction biology do not have formal scientific training. Some have documented their translocation experiences (e.g. Jones & Merton, this volume, Chapter 2) but many do not have the institutional time or support to conduct and publish scientific studies. Second, scientists are generally evaluated on the scientific robustness of their studies (e.g. sample size and experimental design) (Armstrong & McCarthy, 2007) and are increasingly under pressure to publish in the 'best' (i.e. high impact) journals. However, translocations, and conservation biology in general, often result in sparse, messy data where robust conclusions from a single study are difficult to achieve, and publication in peer-reviewed journals can be challenging (Armstrong & McCarthy, 2007). As a consequence, the most appropriate methods for capturing, holding, moving and releasing animals have largely remained a matter of expert opinion rather than research-based practice, despite the clear desire of many scientists to be involved. This may not have hindered reintroduction biology because many conservation managers, particularly those with experience in captive situations, have a finely honed intuitive knowledge of the best way to work with a particular species and in many cases might perceive little need for formal evidence-based approaches. This is common throughout conservation management where management decisions are often based on experience rather than formal research (Pullin *et al.*, 2004).

However, there is room to improve post-release survival rates and to minimize the deleterious post-release consequences of the translocation process

(Box 6.1 in Armstrong & Reynolds, this volume, Chapter 6). Science, particularly the hypothetico-deductive method, offers a means to increase our knowledge beyond that inferred through observational means (e.g. induction and retroduction) (Romesburg, 1981), and translocations offer important opportunities for testing and refining methodologies. The adaptive management approach advocated by McCarthy *et al.* (this volume, Chapter 8), in particular, offers a valuable means to use and model existing data (e.g. through meta-analyses), set strategic monitoring to obtain new data, test and update models, and optimize management. It is widely acknowledged in the literature that improving communication between managers and scientists is essential for successful conservation (Romesburg, 1981; Pullin *et al.*, 2004; Fazey *et al.*, 2005; Milner-Gulland *et al.*, 2009; Burbidge *et al.*, 2011). Our collective experience, and that of others (e.g. see Jones & Merton, this volume, Chapter 2; Bristol *et al.*, 2005; Robertson *et al.*, 2006; Reynolds *et al.*, 2008), suggests that successful, well-planned translocation projects typically consist of a mix of conservation managers, captive rearing experts, veterinarians and scientists. Thus, a multidisciplinary approach gives the best chance of success (Burbidge *et al.*, 2011) by providing diverse skills and expertise for successful translocation outcomes including leadership, management and logistical expertise, biological, ecological, captive husbandry and veterinary expertise, technical field abilities, research skills, innovativeness and political acumen. This also provides a means to utilize knowledge from other scientific disciplines, for example physiological (see the discussion on physiological stress below), animal behaviour and animal welfare research. The needs of any particular translocation project should dictate that this occurs (e.g. see Jones & Merton, this volume, Chapter 2), and Burbidge *et al.* (2011) provide a useful model for linking management and science that is particularly relevant and appropriate for conservation translocations. The multidisciplinary approach requires strong leadership to translate planning into action, particularly when dealing with critically endangered species (Groombridge *et al.*, 2004b), and we note that decisive action has saved several species from extinction (Jones & Merton, this volume, Chapter 2).

Stress and translocations

Stress is an easy target for blame for failed translocations. Stress may be unavoidable during the translocation process, but by understanding what stress is, what an animal perceives as stressful and what the long-term consequences of stress are on physiology and behaviour, we can improve our approach to planning the translocation process to ensure the lowest possible impact on individual animals and hence on the translocated population (Dickens *et al.*, 2010). We discuss the concepts of acute and chronic stress with reference to the translocation process, and then address the question 'what is a stressor?'

The acute stress response

On the physiological and behavioural level, when faced with a single, relatively quick but potentially life-threatening situation, an animal will mount an *acute stress response*. This physiological and behavioural response is considered a *beneficial* reaction to the noxious stimulus of the stressor. Since this response has been well conserved across species, we assume the adaptive importance of having such a response in the wild (Romero, 2004). Box 4.1 summarizes the physiological response mechanisms that are triggered during the acute stress response.

Box 4.1 The acute stress response

One can consider the adaptive goal of the acute stress response to be *immediate survival*. The response acts to defer temporarily normal daily activity in favour of physiologically and behaviourally coping with the imminent threat (Wingfield *et al.*, 1998). Two main physiological systems aid in this effort: the fast-acting cardiovascular (fight-or-flight) response and the slower glucocorticoid (GC) response. Within seconds upon perceiving the stressor, the fight-or-flight response is initiated, mediated by the catecholamines: epinephrine and norepinephrine (adrenaline and noradrenaline) (Figure 4.1). The role of these mediators is to aid the animal in escaping the threat; therefore, this fast response facilitates such behavioral shifts as increasing vigilance, as well as physiological changes such as increasing heart rate, blood flow and energy mobilization to fuel the muscles and brain for a quick escape or fight.

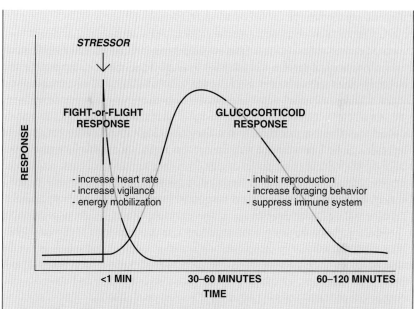

Figure 4.1   Theoretical representation of both the rapid fight-or-flight response and the slower glucocorticoid response as they pertain to the introduction of a single acute stressor. The immediate fight-or-flight response mediates rapid effects such as changes in heart rate, vigilance and energy mobilization. The glucocorticoid response is slower, with the hormone taking several minutes to rise in the bloodstream and 15–60 minutes to reach peak secretion. This hormonal response acts temporarily to suppress physiological and behavioural systems that are unnecessary to immediate survival. The hormone is cleared from the system within hours due to a strong negative feedback response on its secretion.

The GC response, the increase in glucocorticoid hormones (cortisol or corticosterone depending on the vertebrate species) in the blood, mediates the survival of that initial threat by temporarily suspending aspects of behaviour and physiology that are not essential to immediate survival, such as reproduction and immune system building (Sapolsky *et al.*, 2000). Peak elevation in glucocorticoid concentrations occurs between 15 and 60 min after the stressor (again, this depends on the vertebrate species) and, once the stressor ceases, concentrations of the hormone will return to baseline within hours.

Chronic stress: beneficial becomes detrimental

As stated earlier, translocation has no evolutionary precedent. Although an animal's physiological system has adapted to respond to specific, expectable and relatively quick stressors, it is not adapted to handle the multiple, continuous, unpredictable and novel nature of the stressors that face translocated individuals. As a result, the scenario presented with translocation pushes the system beyond the adaptive norm such that the physiological regulators throughout the body can be knocked out of balance; this imbalance may place unnatural strain on the system (Romero *et al.*, 2009). When this shift occurs, referenced here as 'chronic stress', the temporary suspension of non-essential systems (beneficial during the acute stress response) becomes long-term suppression (detrimental to individual fitness); for example, short-term suspension of reproduction becomes long-term reproductive suppression (McEwen, 1998). Since exposing animals to stress during translocation procedures is generally unavoidable, understanding the consequences of chronic stress (Box 4.2) and how to decrease overall exposure to individual stressors should become a vital component of translocation planning (Dickens *et al.*, 2010).

Box 4.2 **Chronic stress and translocations**

Stress becomes problematic and potentially pathological when it is pushed beyond its adaptive acute response role into a state of chronic stress. We expect that translocation inevitably results in some degree of chronic stress because the translocation procedure itself represents a scenario that stress physiologists often employ to induce chronic stress: multiple, consecutive and/or continuous, unpredictable stressors that cause an animal to mount multiple or sustained stress responses. As demonstrated experimentally, employing a series of unpredictable stressors to a wild-caught animal in the lab for multiple weeks (see Rich & Romero, 2005) has the same physiological effects as an experimental translocation (see Dickens *et al.*, 2009).

From the physiology literature, we know that chronic stress can result in immune and reproductive suppression (McEwen, 1998), disruption of metabolism (Dallman & Bhatnagar, 2001) and changes in the fight-or-flight response (Dickens *et al.*, 2009). However, for the purposes of translocation planning, the pathological consequences of stress is best visualized as a measure of *vulnerability to environmental factors*. For example, the chronic stress of translocation may lead to immune system suppression, but, in the end, it is the microbes or viruses in the wild that kill the translocated individual. As another example, translocation stress may alter how an animal responds to a predator in the wild, but it is the physical act of predation that results in the death of that individual. Figure 4.2 demonstrates how chronic stress can alter physiology and behaviour, and how these alterations make an animal more *vulnerable* to outcomes that contribute to translocation failure, such as reproductive failure, disease, starvation, predation and long-range dispersal following release.

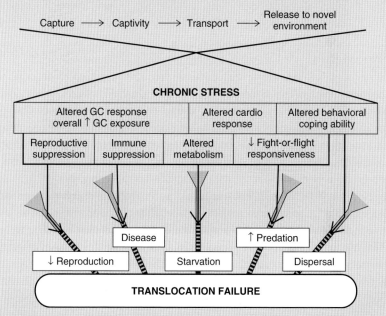

Figure 4.2   **Figure adapted from Dickens *et al.* (2010), demonstrating the negative effects of chronic stress and how they relate to causes of translocation failure. The figure shows that alterations in physiology and behaviour due to the chronic stress of translocation increase the vulnerability to external sources at the release site such that both the internal and external factors contribute to pathology that can result in translocation failure.**

Reducing the vulnerability of the individuals and translocated population can therefore be approached from both sides. First, it is important
to take into account that the *number* of stress responses mounted can
increase the degree to which a state of chronic stress is realized. Therefore, it is important to consider what is stressful to an animal and,
specifically, the species being translocated (see text). Second, it should
be recognized that released individuals are not as capable of handling
the environmental challenges of the release site as they would have been
prior to translocation. Therefore, carefully considering aspects of the
release (the predation pressure, food availability, disease prevalence,
etc.) is also key to helping this more vulnerable population survive the
initial stages following release into the wild (for a more thorough review
on this subject see Dickens *et al.*, 2010).

In conservation, the term 'distress' is sometimes preferred to describe the
negative impacts of stress on the individual animal (Linklater *et al.*, 2010).
Linklater & Gedir (2011) argue that use of this terminology over that put
forth in the physiological literature (acute stress versus chronic stress versus
allostasis) is more applicable to conservation research because it affords
measurable consequences of stress with direct links to individual survival and
reproduction. They also state that it provides '*what to measure and when to act*'.
However, we view the concept of distress as a complimentary measurement
tool best applied to observe and document the effects of stress at the level
of the individual animal; for example, a captive animal that is losing weight
and failing to breed can be classified as 'in distress'. For this reason, measures
of distress may be most suitable to translocation at the endpoint – diagnosis
and treatment of individuals rather than prevention. We believe that the
concept of distress is less useful at the planning stages of a translocation,
although, of course, studies where distress has been identified in translocated
animals will be informative. In the planning stages, we must consider the
long-term *potential* effects of stress for the overall population of translocated
animals. To this end, we take a more theoretical approach for the concept of
stress such that here the general effects of stress on physiology and behaviour
can be applied to the wide range of translocated species. In addition, we
continue to use the term 'chronic stress' to describe the potential effects

that occur when the stress response system is pushed beyond its adaptive homeostatic boundaries.

What is a 'stressor'?

How do we know what an animal perceives as a stressor? Also, since the stress response is not all or nothing (animals can modulate how they respond to certain stressful scenarios (Nephew *et al.*, 2003)), how do we know how severe a stressor is perceived to be? It is impossible to fully comprehend how each species and each individual perceives and responds to each stressful situation. However, three general but overlapping scenarios are known to initiate the stress response system (especially the glucocorticoid (GC) system) in both laboratory and wild animals. These scenarios include lack of control, unpredictability and novelty:

1. *Lack of control.* Animals that have the capacity to behaviourally cope with stressors may not respond to the stressor by activation of the GC response. The link between the capacity to behaviourally cope versus initiation of the GC response appears to be the perception of control. In the laboratory this has been shown by testing pairs of rats exposed to the same electric shock stress. If the rat can behaviourally control the duration of the shock (via lever press), there is a significantly smaller increase in glucocorticoid concentrations in response to that stress, as compared to the rat that receives the same shock but has no control over the stressor (Weiss, 1968). An example from the wild is the ability to escape. Comparing bird species challenged by a direct nest predation threat, the species with an escape route (open-cup nesters) have little or no glucocorticoid response to this stressor compared to species with limited escape options (cavity nesters) (Butler *et al.*, 2009). Open-cup nesters have the option of escape and thus have control over the situation, whereas cavity nesters that are unable to escape have limited control. Translocated prey species such as rabbits may fare better after release when they are provided with food (control of nutritional demands) and shelter (behavioural control of predation threat), possibly an effect of removing the potential for these factors to be physiological stressors (Cabezas & Moreno, 2007).

2. *Unpredictability.* Animals have the capacity to acclimate to stressors, essentially to *learn* to differentiate between potentially life-threatening and benign situations. The predictability of a stressor is a critical part of this

acclimation and the perception of its severity (Levine & Ursin, 1991). For example, the glucocorticoid response to handling is significantly reduced in laboratory rodents when they are handled by the same person each day for consecutive days, but the response returns or is even greater if the handler changes (Dobrakovova *et al.*, 1993). In birds captured from the wild and brought into captivity, feeding on an irregular, unpredictable schedule results in higher baseline stress hormone concentrations than those in birds fed at the same time each day (Reneerkens *et al.*, 2002).

3. *Novelty.* Exposing an animal to an unfamiliar environment is such a potent psychological stressor that biomedical research will often use a novel cage as a way to study the effects of severe acute stress (Marin *et al.*, 2007). In an experimental translocation of chukar (*Alectoris chukar*) where only the final release point differed between groups (release to the home site versus release at a new site with identical habitat to that of the home site), the group released at the novel site experienced the greatest stress-related physiological disruption, including significant weight loss (Dickens *et al.*, 2009). Perhaps the severity of the stress of novelty is due to its ties to the two factors discussed above: a novel environment is the epitome of unpredictability, as well as being a difficult environment in which to sense control. However, herein lies a major problem for the translocation process: every additional move, from transport cage to captivity to the release site, is a novel environment. Therefore, the best way to reduce the severity of this type of stress is to minimize novelty wherever possible. This contrasts with captive rearing programmes where novelty is used to enhance survival skills prior to the translocation process, for example through learning foraging and predator avoidance behaviours.

How severe a stressor is perceived to be varies among individuals, sexes, and species, and animals may detect sources of stress not detectable by humans (Swaisgood, 2010). Therefore, the best approach is to understand thoroughly the species being translocated and to consider how best to tailor protocols to decrease the amount of stress experienced throughout the translocation procedure.

Husbandry, biology and translocation processes

The taxonomic variety of animals selected for translocation projects is matched by the variety of techniques and approaches for conducting translocations. The fundamental criteria for success is an understanding of the basic ecology and biology of the target species (Swaisgood, 2010) and a tailoring of techniques to minimize the various potential stressors and provide released animals with cues that will enhance settlement at the release site. Ideally, this information will be gleaned from studies of wild conspecifics: what do they feed on; where do they den, roost, nest; what are their mating systems; how do they respond to conspecifics; are there seasonal variations in behaviour that might complicate or assist in tailoring translocation strategies? In the absence of detailed wild studies, captive institutions can provide useful experience and information, although we caution that wild behaviours can be altered by captivity.

The translocation of burrow-nesting petrels (*Procellariidae*) provides a good example of how an understanding of biological characteristics is being used to devise successful translocation processes (Miskelly *et al.*, 2010). Seabird chicks are removed from burrows in the late stages of nestling development, before they have imprinted on the natal area (Miskelly *et al.*, 2010). They are translocated to a new site, housed in artificial burrows and fed an artificial diet (Miskelly *et al.*, 2010). Upon fledging, they imprint on the new 'natal' site. When the translocated chicks mature they return to the breeding colony from which they fledged, supporting the subsequent establishment of new breeding colonies (Priddel & Carlile, 2001; Miskelly *et al.*, 2010). In subsequent years acoustic cues may also be broadcast from speakers at the translocation site to promote further the return of individuals (Miskelly *et al.*, 2010), although the effectiveness of this is unclear.

Where there is uncertainty about the best methods for a given species, aspects of translocation processes should be tested and compared prior to an actual translocation, ideally on the target species and even the target population. Such trials would ideally be designed as controlled experiments allowing assessment of individual procedures, the cumulative stress effects of each procedure (Dickens *et al.*, 2009) and, through catch and release at the same site, control for the impacts of a novel release site relative to the translocation procedure itself (Dickens *et al.*, 2009). However, as Swaisgood (2010) stated, small sample sizes, endangered species and limited experimental controls are common features of translocation projects and it is often simply not feasible to conduct trial translocations. A common approach to this problem is to use surrogate species; Groombridge *et al.* (2004a) trialled several transportation techniques over varying distances using the Maui creeper (*Paroreomyza montana*) to devise translocation techniques for the critically endangered po'ouli (*Melamprosops phaecsoma*). Surrogate species have been used with great success in several reintroduction programmes (see Jones & Merton, this volume, Chapter 2). However, as with the observed differences between wild and captive-bred individuals (see below), practitioners should be aware of possible differences between even closely related species. The alternative approach is simply to start by translocating relatively few individuals dependent on species characteristics, e.g. solitary versus social species, and then release more individuals if the responses of the initial individuals to translocation are satisfactory. It may also be sensible to start with individuals perceived to have lower value, for example males in many situations.

We now discuss the different phases of the translocation process, beginning by examining differences between captive- and wild-sourced animals, because an appreciation of these differences is essential when selecting the most appropriate methods to use (Jones & Merton, this volume, Chapter 2).

Captive- versus wild-sourced individuals for translocation

In some species population declines have been so extensive and dramatic that there has been little choice but to capture an entire extant wild population and bring it into captivity; for example, captive rearing has played a crucial role in saving the black-footed ferret (*Mustela nigripes*) (Biggins *et al.*, 1999), the California condor (*Gymnogyps californianus*) (Synder & Synder, 1989) and the Mauritius kestrel (*Falco punctatus*) (Jones & Merton, this volume, Chapter 2) from imminent extinction. However, captive rearing is often viewed as a last resort (Synder *et al.*, 1996). This is largely because of the well-documented impacts of long-term captivity on behaviour, morphology, physiology and genetics (Frankham *et al.*, 1986; Allendorf, 1993; Synder *et al.*, 1996; McPhee, 2004; Hakånsson & Jensen, 2005, 2008) and the often fundamental differences in the behaviour of captive and wild animals. In particular, while captive-reared animals will invariably be stressed by the novelty of release to the wild, wild-caught animals will generally be more stressed by the novelties associated with capture, holding and transport. Subsequently, comparative studies suggest that post-release survival is generally lower in captive-reared animals relative to that of wild-sourced translocated animals or free-living conspecifics (Griffith *et al.*, 1989; Fischer & Lindenmayer, 2000). However, there is great variation in the impact of captivity on different species. Jule *et al.* (2008) state that wild carnivores do significantly better than captive-reared carnivores but note that captivity is especially hard on canids and slightly less on ursids and mustelids (see McCarthy *et al.*, this volume, Chapter 8, for a further analysis of the same data set that accounts for the case-specific confounds that usually make comparative analyses difficult to interpret). Several studies have also directly assessed the differences between captive and wild stock by comparing post-release survival of individuals of the same species translocated in similar circumstances. Captive-reared northern water snakes (*Nerodia sipedon sipedon*) show reduced survival relative to wild snakes (Roe *et al.*, 2010) as do galliformes (Sokos *et al.*, 2008), Aplomado falcons (*Falco femoralis septentrionalis*) (Brown *et al.*, 2006), otters (*Lutra lutra*) (Sjoasen, 1996), Vancouver Island marmots (*Marmota vancouverensis*) (Aaltonen *et al.*, 2009) and griffon vultures (*Gyps fulvus fulvus*) (Sarrazin *et al.*, 1996). In contrast, Santos *et al.* (2009) found no difference in captive-bred versus wild lacertid lizards (*Psammodromus algirus*) and Tweed *et al.* (2003) attained an impressive 100% survival to 56 days for captive-reared puaiohi, a Hawaiian bird (*Myadestes palmeri*). We also note that in the examples cited above, while captive-reared individuals fared worse than their wild counterparts, success could be improved by varying release strategies. For example, in Vancouver Island marmots individuals released at two years of age survived better than those released at one year (Aaltonen *et al.*, 2009), griffon vultures released as immatures had greater breeding success than those released at two years of age (Sarrazin *et al.*, 1996) and otters released soon after separation from their mothers survived better than those held for longer periods in captivity (Sjoasen, 1996).

Jule *et al.* (2008) note that the generally poor success of captive-reared animals has to be traded off against the ability and cost of rearing animals for release relative to sourcing animals from wild populations – particularly those of endangered species. For example, Cristinacce *et al.* (2008) used captive rearing to produce large numbers of Mauritius fodies (*Foudia rubra*) for release relatively quickly with subsequent successful population establishment.

Captive rearing protocols should be carefully designed to reduce stresses associated with the post-release period and generally to improve the ability of the animals to acclimate to the new environment (Roe *et al.*, 2010; Swaisgood, 2010). Methods for acclimating captive-reared animals to the wild involve pre-release training or conditioning based on situations faced in the wild. For instance, Biggins *et al.* (1999) showed that pre-release experience in captive-reared black-footed ferrets affected survival, with animals reared in large outdoor pens surviving at a higher rate than those reared in small cages. Zidon *et al.* (2009) noted that differences in rearing environments affected post-release survival of Persian fallow deer (*Dama mesopotamica*) and McLean *et al.* (1996) taught captive-reared rufous hare-wallabies (*Lagorchestes hirsutus*) to recognize predators prior to release, although the effects of that training on survival are unknown. However, van Heezik *et al.* (1999) showed increased post-release survival of captive-reared houbara bustards (*Chlamydotis [undulate] macqueenii*) subject to pre-release training through exposure to a live red fox (*Vulpes vulpes*). Hand rearing with puppets and complex cross-fostering protocols have also been used to prevent imprinting in captivity and the wild (Jones & Merton, this volume, Chapter 2). Complex migratory behaviours provide a particular biological challenge for captive rearing programmes. However, the learning of migration routes by translocated individuals, particularly key foraging and breeding sites, has been achieved using wild conspecifics as instructors (e.g. loggerhead shrikes (*Lanius ludovicianus*) (Nichols *et al.*, 2010) and orange-bellied parrots (*Neophema chrysogaster*) (C. Tzaros, personal communication)) and in the case of whooping cranes (*Grus americana*) by training released individuals to follow microlight aircraft (Ellis *et al.*, 2003).

These approaches can be challenging, with Swaisgood (2010) noting that welfare in captivity is often incompatible with conservation goals (e.g. by exposing captive animals to stressful situations during predator training) but that short-term poor welfare in captivity might pay dividends in post-release welfare, i.e. reduction of post-release stress.

While captive animals may generally pose greater challenges than wild-caught animals, it is important to remember that the relative stresses faced by captive and wild animals will depend on the stage of the translocation process. We now discuss these phases of the translocation process in turn.

Capture

Humans have derived a virtually infinite variety of methods for catching animals. The method of capture used in a translocation is an important factor for ensuring appropriate welfare (Swaisgood, 2010); therefore, various capture options should be thoroughly explored. Discussion with appropriately skilled colleagues will be the usual manner for selecting a capture technique, but a treatise such as Bub's (1995) work on bird-catching methods might also provide effective and efficient alternatives.

For an active wild animal, capture is probably perceived as an attempted predation event, and may therefore elicit a violent stress response appropriate to this life-and-death situation (Dickens *et al.*, 2010). It is therefore important to recognize the trade-off between efficiency and effectiveness when choosing a capture method. An efficient method is the fastest and most economical way to capture an animal. An effective method is the best way to capture an animal in accordance with the desired objective. Existing methods can be modified in a relatively simple manner; for example, Terhune *et al.* (2010) covered traps to minimize stress when catching Northern bobwhite quail (*Colinus virginianus*) for translocation and Casper (2009) recommended covered traps and dawn trap checks when catching nocturnal species such as bettongs (*Bettongia* spp.). For particularly stress-prone animals, transmitters that signal a captured animal or constant monitoring of nets are also appropriate in order to minimize the time spent in traps (Koch *et al.*, 2006; Churchill, 2008; O'Neill *et al.*, 2008). Green (2005) used a novel method for capture that negated any handling of the targeted animals by providing artificial refugia for an invertebrate, the Auckland tree weta (*Hemideina thoracica*), six months prior to translocation. The refugia were readily used, collected with weta inside and translocated to the release site, so the process is assumed to be stress free for the animals. Stress can be avoided also by translocating animals during their dormant period (e.g. by moving snakes between hibernacula; see McMillan, 1995) or at life history strategies, when they are less likely to perceive stress. Examples include the translocation of juvenile seabirds, discussed above, and cross-fostering of eggs, discussed by Jones & Merton (this volume, Chapter 2).

Holding

Captivity reduces an animal's ability to behave in a natural manner and, critically, by placing them in a novel and unpredictable environment their perception of control is sharply reduced. There is also great variation between species and individuals in response to captivity (Mason, 2010). Some readily adapt while close relatives do not, in both vertebrates (Mason, 2010) and invertebrates (Carere *et al.*, 2011). With movement restricted, avoidance behaviours curtailed and density orders of magnitude above natural levels, the risk of serious injury and stress-related problems are increased. Considerable care needs to be taken in deciding how to house animals between capture and translocation. For stress-prone species, such as New Zealand rifleman (*Acanthisitta chloris*) (Leech *et al.*, 2007), capture and immediate transfer to the release site is preferable, whereas for others logistical constraints might make a period of confinement unavoidable. Highly territorial or solitary species such as New Zealand robins (*Petroica* spp.) or fernbirds (*Bowdleria punctata vealeae*) require individual enclosures to prevent conspecific aggression (Lovegrove & Veitch, 1994; Parker, 2002), whereas entire colonies of social species such as Australian black-eared miners can be housed together (Clarke *et al.*, 2002). Ultimately the choice of enclosure design will be a trade-off between biological requirements and practical limitations, but as with captive-reared animals wild conditions should be mimicked where possible (Swaisgood, 2010) by providing suitable denning, roosting, nesting and feeding opportunities. Larger enclosures are usually preferable unless the holding period is very short (Sherwin, 2004; Gebhardt-Henrich & Steiger, 2006) or if natural movement needs to be restricted to reduce stress or injury potential. For example, a small passerine might be held in a bag for a move taking less than an hour, in a small transport box if being held for less than a day or in a larger aviary with conspecifics if being held for several days. The key factor is keeping enclosure *changes* to a minimum, thereby minimizing novelty and unpredictability. In captive European starlings (*Sturnus vulgaris*), enclosure shape seems more important than size; long cages reduced stereotypical behaviour relative to square cages, although size (bigger is better) is still an important factor (Asher *et al.*, 2009). Furthermore, in starlings the effect of density in enclosures is dependent on season (Dickens *et al.*, 2006). Gelling *et al.* (2010) found that reduced density in captive colonies minimized stress in water voles. In dark-eyed juncos (*Junco hyemalis*) the relative abundance of males to females affected behaviour, condition and immune function in captive groups, with significantly more negative effects (reduced mass, pectoral muscle condition and cell mediated immune responses) observed in male biased groups versus female or equal sex ratio groups (Greives *et al.*, 2007).

Replicating a wild diet in captivity will probably facilitate both acclimation to captivity by wild animals and acclimation to the wild by captive-reared animals. This may not always be possible or feasible, but suitable alternatives are usually available. For example, small insectivores will consume commercially reared invertebrates such as mealworms (*Tenebrio molitor*) and waxmoth larvae (*Galleria mellonella*) and frugivores often quickly adapt to commercially available fruit and vegetables. There is also a large industry providing balanced diets for captive husbandry that will be suitable nutritionally for many species, but may not facilitate behavioural acclimation. Behavioural enrichment, particularly in food provision, has been demonstrated to provide significant benefits in captive institutions (Mason *et al.*, 2007; Mason, 2010; Swaisgood, 2010) and is likely to contribute to the successful maintenance of animals during extended periods of captivity for translocation purposes.

Movement

During transportation from capture to release sites, animals are potentially moved from a large enclosure to a smaller transport box and then exposed to an unpredictable sequence of fluctuating temperature and humidity along with unexpected noise, knocks, vibrations and light. Given experimental evidence showing the cumulative impacts of translocation procedures (Dickens *et al.*, 2009), and the effects of novelty and unpredictability on captive animals (see discussion above and Weiss, 1968), it is clear that these effects need to be minimized. Attention to detail, ranging from finding the quietest place on the vehicle that will transport the animals through to guarding the transport boxes from human voices and fluctuating temperatures, will all be beneficial. Transfer boxes and transportation methods should protect the translocated individuals as much as possible, and the duration of transfer trips minimized. For example, Groombridge *et al.* (2004a) found no difference in containment method when moving Maui creepers, but they did find an effect of duration in captivity; birds held longer displayed a higher stress response, as measured by heterophil–lymphocyte white blood cell ratios, than those held for short periods. However, when extended moves are unavoidable various techniques can improve success. For example, individual Laysan teal (*Anas laysanenis*) were provided with intensive veterinary and nutritional support (Reynolds *et al.*, 2008) during transfer by ship, and Gelling (2010) found that reduced density in transportation cages minimized stress in water voles. Letty *et al.* (2000) tried tranquilizing European rabbits during capture and translocation to minimize the effects of stress, but found it had no effect on subsequent survival whereas Ancreaz *et al.* (1996) found long-acting tranquilizers and pre-release transport training beneficial in Arabian oryx (*Oryx leucoryx*) translocations.

Release

Critical questions facing translocation practitioners are those relating to release: how many individuals should be released, what age and sex should the release group be, when should they be released and what, if any, supportive management will be required post-release? It is obvious that there will be some relationship between translocation success and number released, and meta-analyses support this notion (Griffith *et al.*, 1989; Wolf *et al.*, 1996; Fischer & Lindenmayer, 2000; Cassey *et al.*, 2008; Germano & Bishop, 2008).

Part of this is simply increasing the probability of sufficient individuals surviving and breeding to establish the population, but in some cases larger numbers will improve individual survival and reproduction. This is an Allee effect (see Box 4.2, Armstrong & Reynolds, this volume, Chapter 6), and Allee effects are expected in species whose normal behaviour in the wild requires large groups, i.e. social versus solitary species. For example, meta-analysis of 21 black-faced impala (*Aepyceros melampus petersi*) translocations showed higher success when groups of greater than 15 were released, and this was associated with lower vulnerability to predation, as is expected in a species that relies on group anti-predator behaviour (Matson *et al.*, 2004). In the black-eared miner (*Manorina melanotis*), a species with extreme social complexity (Ewen *et al.*, 2001), entire wild-caught colonies (10–30 individuals) were translocated to new sites, including recently fledged young (Clarke *et al.*, 2002). The colonies remained as cohesive units 1–2 months after release and the translocations were subsequently successful. In contrast, in less social species individual survival and reproduction may be unaffected by the number released, so populations can potentially be established with very small numbers, although this may not be advisable for genetic reasons (Jamieson & Lacy, this volume, Chapter 13). For example, translocations of Seychelles magpie-robins (*Copsychus sechellarum*) and Chatham Island black robins (*Petroica traversi*) were successful with just five individuals translocated (Lopez-Scepulcre *et al.*, 2008; Butler & Merton, 1992).

There often appears to be age- and sex-related mortality following release. This indicates that negative stress responses can be related to demography and will influence the optimal composition of release groups, although we acknowledge that this can sometimes only be inferred following initial release. For instance, Pinter-Wollman *et al.* (2009) found more adult male and calf mortality than expected in African elephants (*Loxodonta africana*) following translocation while Imlay *et al.* (2010) recorded high survival of captive-reared migratory loggerhead shrikes, but noted that most deaths were of females.

The timing of a release can also be a significant factor in success. Tavecchia *et al.* (2009) found that late winter releases of crested coots (*Fulica cristata*) had a higher survival than those conducted in summer when droughts, i.e. unfavourable weather conditions, promoted dispersal or in autumn when hunting mortality was high. In New Zealand, small passerine translocations are typically conducted in the autumn when weather conditions are moderate and food abundant, which may reduce post-release acclimation stress.

Animals are often held captive at the release site, a so-called 'soft' or delayed release as opposed to a 'hard' or immediate release. Delayed releases are frequently recommended as the release strategy of choice, although these recommendations are often made without supporting evidence (e.g. see Wanless *et al.*, 2002; Teixeira *et al.*, 2007). Swaisgood (2010) states that despite their intuitive appeal, delayed releases are expensive, not supported by meta-analyses and ultimately offer little benefit. However, some species benefit from this approach (see Jones & Merton, this volume, Chapter 2). For example, delayed releases enhanced the survival of female European rabbits but reduced the survival of male rabbits (Letty *et al.*, 2000), while Rouco (2010) found that rabbits confined for six days had better survival rates than those confined for three days. Ryckman *et al.* (2010) showed that extended conditioning appeared beneficial for elk (*Cervus elaphus*), whereas short-term holding resulted in a greater post-release movement. Captive-reared delayed-release Western burrowing owls (*Athene cunicularia hypugaea*) showed a higher site-affinity, survival and reproductive performance relative to immediate-released birds (Mitchell *et al.*, 2011), but release protocol made no difference when immediate and delayed releases were compared in hare wallabies (Hardman & Moro, 2006), North Island saddleback (*Philesturnus rufaster*) (Lovegrove, 1996) or New Zealand hihi (*Notiomystis cincta*) (Castro *et al.*, 1995).

A delayed release is clearly beneficial to some species, taxa or situations, but may, in other cases, have little effect or even be harmful. Jones & Merton (this volume, Chapter 2) suggested that captive-reared birds should normally be held at the release site with supportive management, whereas wild-caught birds should be released immediately. This makes theoretical sense in terms of stress responses; i.e. a soft release strategy may help acclimate a captive-reared animal to the wild, and therefore reduce stress, whereas an additional holding period for a wild-caught animal adds another novel situation that will probably increase its cumulative stress response and increase the opportunity for traumatic injury. However, in addition to expected differences in response between captive-reared and wild-caught animals, it is important to consider the available data for the translocated species or closely related taxa. Decisions may also be made on an individual level rather than the whole translocation group. For example, Reynolds & Klavitter (2006) used a flexible release strategy based on individual conditions.

Various other methods have been used to increase post-release survival during the acclimation period. Swaisgood (2010) states that post-release support in the form of supplemental food, roost and nest sites are advantageous, and that external stresses, such as a novel environment, can be buffeted by release with familiar conspecifics. Armstrong (1995) and Armstrong & Craig (1995) found no effect of familiarity in translocated North Island saddlebacks (*Philesturnus rufusater*) or North Island robins (*Petroica longipes*), but they are not particularly social species. Experimental pre-release exposure of guppies (*Poecilia reticulata*) to a ubiquitous parasite (*Gyrodactylus turnbulli*) increased post-release survival (Faria *et al.*, 2010), Laysan teal were given a feather trim to reduce post-release dispersal (Reynolds & Klavitter, 2006) and Stoen *et al.* (2009) recommend that white rhinoceroses (*Ceratotherium simum*) should not be released into areas where conspecifics have established territories as a means to prevent unwanted dispersal. Behavioural cues could also potentially be useful in either attracting animals or minimizing dispersal. For example, sound anchoring, the playback of conspecific calls, has been employed to attract and minimize dispersal of both seabirds (Miskelly *et al.*, 2010) and terrestrial birds (Molles *et al.*, 2008; Bradley *et al.*, 2011), although there is no clear evidence yet that it has a useful effect.

As a final comment we note that the release itself is often a celebratory event and many people might want to be involved. This can often be accommodated but the primary concern is the welfare of the translocated animals with photos, filming, speeches and spectators taking second place to ensuring that animals are released in an appropriate manner. Observers must be sited appropriately so that upon release animals have an unimpeded view of suitable habitat and a safe direct dispersal route. There is often a desire to see the released animals up close and in the hand. This can have considerable advocacy benefits as people generally respond very positively to close contact with animals (Mankin *et al.*, 1999; Muth & Jamieson, 2000). However, we suggest that in most circumstances this should be avoided as it almost certainly represents an additional avoidable stressor. We also suggest that it is imperative not to allow ceremony to prolong yet another stressful situation for the animals.

Conclusions and research directions

There is clearly a great variety of species-specific methods and outcomes when translocating animals and it is difficult to offer general guidelines. However, we feel that by using multidisciplinary teams and incorporating a research-based approach, where appropriate, translocation processes can be continually improved. By understanding what stress is and how it might affect translocated individuals we can review our translocation processes to eliminate as many potential stressors as possible. Finally, by having an intimate understanding of our translocated species we can tailor our translocation processes by finding the most appropriate means to catch, hold, move and release animals. Carefully planned and researched capture, holding, movement and release protocols are essential to translocation success and ongoing translocation of a species engenders continued refinement of procedures. Therefore we make a plea, along with others (Sutherland *et al.*, 2004, 2010; Seddon *et al.*, 2007; Armstrong & Seddon, 2008), for ongoing experimentation, communication and collaboration within the reintroduction community through publication and communication of all translocation outcomes and techniques, both positive and negative. Needless repetition of failed techniques is financially and ethically unsustainable. We think that all translocation outcomes should be documented, including not only the technical aspects but also the social and political environments in which translocations do – or do not – succeed. We hope that relationships between managers, scientists and community groups affected or involved in translocation projects continue to form and persist, and we encourage the questioning of everything, including what 'works'. We also note that while science is an important tool, intuitive expert opinion will always play a vital role in successful translocations. However, as long as there is mortality there is room for improvement. Finally, we make that most characteristic call of the reintroduction biology literature and emphasize, yet again, as so many others have before us, the vital importance of post-release monitoring in translocations. It is the only way by which our methods can truly be tested and our management refined.

Acknowledgements

This chapter benefited from comments and discussion with Doug Armstrong, Raewyn Empson, John Ewen, Kalinka Rexer Huber, Don Merton, Luis Ortiz Catedral, Graham Parker, Philip Seddon, Chris Smuts Kennedy and Kirsty Swinnerton.

References

Aaltonen, K., Bryant, A.A., Hostetler, J.A. *et al.* (2009) Reintroducing endangered Vancouver Island marmots: survival and cause-specific mortality rates of captive-born versus wild-born individuals. *Biological Conservation*, 142, 2181–2190.

Adams, N.J., Parker, K.A., Cockrem, J.F. *et al.* (2010) Corticosterone responses and post release survival in translocated North Island saddlebacks (*Philesturnus rufusater*) in New Zealand. *Emu*, 110, 296–301.

Allendorf, F.W. (1993) Delay of adaptation to captive breeding by equalizing family size. *Conservation Biology*, 7, 416–419.

Ancrenaz, M., Ostrowski, S., Anagariyah, S. *et al.* (1996). Long-duration anesthesia in Arabian oryx (*Oryx leucoryx*) using a medetomidine: etorphine combination. *Journal of Zoo and Wildlife Medicine*, 27, 209–216.

Armstrong, D.P. (1995) Effects of familiarity on the outcome of translocations 2. A test using New Zealand robins. *Biological Conservation*, 71, 281–288.

Armstrong, D.P. & Craig, J.L. (1995) Effects of familiarity on the outcome of translocations 1. A test using saddlebacks *Philesturnus carunculatus rufusater*. *Biological Conservation*, 71, 133–141.

Armstrong, D.P. & McCarthy, M.A. (2007) Big decisions and sparse data: adapting scientific publishing to the needs of practical conservation. *Avian Conservation and Ecology*, 2, 14.

Armstrong, D.P. & Reynolds, M.H. (2011) Modelling reintroduced populations: the state of the art and future directions. In *Reintroduction Biology: integrating science and management*, eds J.G. Ewen, D.P. Armstrong, K.A. Parker & P.J. Seddon, Chapter 6. Wiley-Blackwell, Oxford, UK.

Armstrong, D.P., & Seddon, P.J. (2008) Directions in reintroduction biology. *Trends in Ecology and Evolution*, 23, 20–25.

Asher, L., Davies, G.T.O., Bertenshaw, C.E. *et al.* (2009) The effects of cage volume and cage shape on the condition and behaviour of captive European starlings (*Sturnus vulgaris*). *Applied Animal Behaviour Science*, 116, 286–294.

Bekoff, M. (2010) First do no harm. In *New Scientist*, p. 28. Reed Business Information, London.

Biggins, D.E., Vargas, A., Godbey, J.L. *et al.* (1999) Influence of prerelease experience on reintroduced black-footed ferrets (*Mustela nigripes*). *Biological Conservation*, 89, 121–129.

Bradley, D.W., Ninnes, C.E., Valderrama, S.V. *et al.* (2011) Does 'acoustic anchoring' reduce post-translocation dispersal of North Island robins? *Wildlife Research*, 38, 69–76.

Bristol, R., Millett, J. & Shah, N.J. (2005) Best practise handbook for management of a critically endangered species: the Seychelles magpie robin. In Nature Seychelles, Seychelles.

Brown, J.L., Collopy, M.W., Gott, E.J. *et al.* (2006) Wild-reared aplomado falcons survive and recruit at higher rates than hacked falcons in a common environment. *Biological Conservation*, 131, 453–458.

Bub, H. (1995) *Bird Trapping and Bird Banding*. Cornell University Press, Ithaca, New York.

Burbidge, A.H., Maron, M., Clarke, M.F. *et al.* (2011) Linking science and practise in ecological research and management: how can we do it better? *Ecological Management and Restoration*, 12, 54–60.

Butler, D. & Merton, D.V. (1992) *The Black Robin. Saving the world's most endangered bird*. Oxford University Press, Oxford, UK.

Butler, L.K., Bisson, I.A., Hayden, T.J. *et al.* (2009) Adrenocortical responses to offspring-directed threats in two open-nesting birds. *General and Comparative Endocrinology*, 162, 313–318.

Cabezas, S. & Moreno, S. (2007) An experimental study of translocation success and habitat improvement in wild rabbits. *Animal Conservation*, 10, 340–348.

Calvete, C., Angulo, E., Estrada, R. *et al.* (2005) Quarantine length and survival of translocated European wild rabbits. *Journal of Wildlife Management*, 69, 1063–1072.

Carere, C., Wood, J.B. & Mather, J. (2011) Species differences in captivity: where are the invertebrates? *Trends in Ecology and Evolution*, 26, 211.

Casper, R.M. (2009) Guidelines for the instrumentation of wild birds and mammals. *Animal Behaviour*, 78, 1477–1483.

Cassey, P., Blackburn, T.M., Duncan, R.P. *et al.* (2008) Lessons from introductions of exotic species as a possible information source for managing translocations of birds. *Wildlife Research*, 35, 193–201.

Castro, I., Alley, J.C., Empson, R.A. *et al.* (1995) Translocation of hihi or stitchbird *Notiomystis cincta* to Kapiti Island, New Zealand: transfer techniques and comparison of release strategies. In *Reintroduction Biology of Australian and New Zealand Fauna*, ed. M. Serena. Surrey Beatty & Sons, Chipping Norton, NSW, Australia.

Chipman, R., Slate, D., Rupprecht, C. *et al.* (2008) Downside risk of wildlife translocation. *Developments in Biologicals* (IABS Symposia Series), 131, 223–232.

Churchill, S. (2008) *Australian Bats*. Allen and Unwin, Crows Nest, Sydney, Australia.

Clarke, R.H., Boulton, R.L. & Clarke, M.F. (2002) Translocation of the socially complex black-eared miner *Manorina melanotis*: a trial using hard and soft release techniques. *Pacific Conservation Biology*, 8, 223–234.

Cristinacce, A., Ladkoo, A., Switzer, R. *et al.* (2008) Captive breeding and rearing of critically endangered Mauritius fodies *Foudia rubra* for reintroduction. *Zoo Biology*, 27, 255–268.

Dallman, M. F. & Bhatnagar, S. (2001) Chronic stress and energy balance: role of the hypothalamo-pituitary-adrenal axis. In *Handbook of Physiology*, Section 7: *The endocrine system, coping with the environment: neural and endocrine mechanisms*, eds B.S. McEwen & H.M. Goodman, vol. IV. Oxford University Press, New York.

Dickens, M.J., Delehanty, D.J. & Romero, L.M. (2009) Stress and translocation: alterations in the stress physiology of translocated birds. *Proceedings of the Royal Society B*, 276, 2051–2056.

Dickens, M.J., Delehanty, D.J. & Romero, L.M. (2010) Stress: an inevitable component of animal translocation. *Biological Conservation*, 143, 1329–1341.

Dickens, M.J., Nephew, B.C. & Romero, L.M. (2006) Captive European starlings (*Sturnus vulgaris*) in breeding condition show an increased cardiovascular stress response to intruders. *Physiological and Biochemical Zoology*, 79, 937–943.

Dobrakovova, M., Kvetnansky, R., Oprsalova, Z. *et al.* (1993) Specificity of the effect of repeated handling on sympathetic-adrenomedullary and pituitary-adrenocortical activity in rats. *Psychoneuroendocrinology*, 18, 163–174.

Ellis, D.H., Sladen, W.J.L., Lishman, W.A. *et al.* (2003) Motorized migrations: the future or mere fantasy? *BioScience*, 53, 260–264.

Ewen, J.G., Clarke, R.H., Moysey, E. *et al.* (2001) Primary sex ratio bias in an endangered cooperatively breeding bird, the black-eared miner, and its implications for conservation. *Biological Conservation*, 101, 137–145.

Ewen, J.G., Acevedo-Whitehouse, K., Alley, M. *et al.* (2011) Empirical consideration of parasites and health in reintroduction. In *Reintroduction Biology: integrating science and management*, eds J.G. Ewen, D.P. Armstrong, K.A. Parker & P.J. Seddon, Chapter 9. Wiley-Blackwell, Oxford, UK.

Faria, P.J., Van Oosterhaut, C. & Cable, J. (2010) Optimal release strategies for captive-bred animals in reintroduction programs: experimental infections using the guppy as a model organism. *Biological Conservation*, 143, 35–41.

Fazey, I., Fischer, J. & Lindenmyer, D.B. (2005) What do conservation biologists publish? *Biological Conservation*, 124, 63–73.

Fischer, J. & Lindenmayer, D.B. (2000) An assessment of the published results of animal relocations. *Biological Conservation*, 96, 1–11.

Frankham, R., Loebel, D.A., Ryder, O.A. *et al.* (1986) Modeling problems in conservation genetics using captive *Drosophila* populations: rapid genetic adaptation to captivity. *Zoo Biology*, 11, 333–342.

Gebhardt-Henrich, S.G. & Steiger, A. (2006) Effects of aviary and box sizes on body mass and behaviour of domesticated budgerigars (*Melopsittacus undulatus*). *Animal Welfare*, 15, 353–358.

Gelling, M., Montes, I., Moorhouse, T.P. *et al.* (2010) Captive housing during water vole (*Arvicola terrestris*) reintroduction: does short-term social stress impact on animal welfare? *PLoS One*, 5, e9791.

Germano, J.M. & Bishop, P.J. (2008) Suitability of amphibians and reptiles for translocation. *Conservation Biology*, 23, 7–15.

Godfrey, J.D., Bryant, D.M. & Williams, M.J. (2001) Radio-telemetry increases free-living energy costs in the endangered takahe *Porphyrio mantelli*. *Biological Conservation*, 114, 35–38.

Gosling, L.M. & Sutherland, W.J. (2000) *Behaviour and Conservation*. Cambridge University Press, Cambridge, UK.

Green, C. (2005) Using artificial refuges to translocate and establish Auckland tree weta *Hemideina thoracica* on Korapuki Island, New Zealand. *Conservation Evidence*, 2, 94–95.

Greives, T.J., Casto, J.M. & Ketterson, E.D. (2007) Relative abundance of males to females affects behaviour, condition and immune function in a captive population of dark-eyed juncos *Junco hyemalis*. *Journal of Avian Biology*, 38, 255–260.

Griffith, B., Scott, J.M., Carpenter, J.W. *et al.* (1989) Translocation as a species conservation tool: status and strategy. *Science*, 245, 477–480.

Groombridge, J.J., Massey, J.G., Bruch, J.C. *et al.* (2004a) Evaluating stress in a Hawaiian honeycreper, *Paroreomyza montana*, following translocation. *Journal of Field Ornithology*, 75, 183–187.

Groombridge, J.J., Massey, J.G., Bruch, J.C. *et al.* (2004b) An attempt to recover the Po'ouli by translocation and an appraisal of recovery strategy for bird species of extreme rarity. *Biological Conservation*, 118, 365–375.

Hakånsson, J. & Jensen, P. (2005) Behavioural and morphological variation between captive populations of red junglefowl (*Gallus gallus*) – possible implications for conservation. *Biological Conservation*, 122, 431–439.

Hakånsson, J. & Jensen, P. (2008) A longitudinal study of antipredator behaviour in four successive generations of two populations of captive red junglefowl. *Applied Animal Behaviour Science*, 114, 409–418.

Hamilton, L.P., Kelly, P.A., Williams, D.F. *et al.* (2010) Factors associated with survival of reintroduced riparian brush rabbits in California. *Biological Conservation*, 143, 999–1007.

Hardman, B. & Moro, D. (2006) Optimising reintroduction success by delayed dispersal: is the release protocol important for hare-wallabies? *Biological Conservation*, 128, 403–411.

Harris, J.E. (2008) Translocation of the silver-studded blue *Plebejus argus* to Cawston Heath, Norfolk, England. *Conservation Evidence*, 5, 1–5.

Hoegh-Guldberg, O., Huhes, L., Mcintyre, S. *et al.* (2008) Assisted colonization and rapid climate change. *Science*, 321, 345–346.

Hunter Jr, M.L. (2007) Climate change and moving species: furthering the debate on assisted colonization. *Conservation Biology*, 21, 1356–1358.

Imlay, T.I., Crowley, J.F., Argue, A.M. *et al.* (2010) Survival, dispersal and early migration movements of captive-bred juvenile eastern loggerhead shrikes (*Lanius ludovicianus migrans*). *Biological Conservation*, 143, 2578–2582.

IUCN (1987) *The IUCN Position Statement on Translocation of Living Organisms*. IUCN, Gland, Switzerland.

IUCN (1998) *IUCN Guidelines for Re-introductions*. IUCN, Gland, Switzerland and Cambridge, UK.

Jamieson, I.G. & Lacy, R.C. (2011) Managing genetic issues in reintroduction biology. In *Reintroduction Biology: integrating science and management*, eds J.G. Ewen,

D.P. Armstrong, K.A. Parker & P.J. Seddon, Chapter 13. Wiley-Blackwell, Oxford, UK.

Jones, C.G. & Merton, D.V. (2011) A tale of two islands: the rescue and recovery of endemic birds in New Zealand and Mauritius.. In *Reintroduction Biology: integrating science and management*, eds J.G. Ewen, D.P. Armstrong, K.A. Parker & P.J. Seddon, Chapter 2. Wiley-Blackwell, Oxford, UK.

Jule, K.R., Leaver, L.A. & Lea, S.E.G. (2008) The effects of captive experience on reintroduction survival in carnivores: a review and analysis. *Biological Conservation*, 141, 355–363.

Koch, N., Munks, S.A., Tutesch, M. *et al.* (2006) The platypus *Ornithorhynchus anatinus* in headwater streams, and effects of pre-code forest clearfelling, in the South Esk River catchment, Tasmania, Australia. *Australian Zoologist*, 33, 458–473.

Leech, T.J., Craig, E., Beaven, B. *et al.* (2007) Reintroduction of rifleman *Acanthisitta chloris* to Ulva Island, New Zealand: evaluation of techniques and population persistence. *Oryx*, 41, 369–375.

Le Gouar, P., Mihoub, J.-B. & Sarrazin, F. (2011) Dispersal and habitat selection: behavioural and spatial constraints for animal translocations. In *Reintroduction Biology: integrating science and management*, eds J.G. Ewen, D.P. Armstrong, K.A. Parker & P.J. Seddon, Chapter 5. Wiley-Blackwell, Oxford, UK.

Letty, J., Marchandeau, S. Clobert, J. *et al.* (2000) Improving translocation success: an experimental study of anti-stress treatment and release method for wild rabbits. *Animal Conservation*, 3, 211–219.

Levine, S. & Ursin, H. (1991) What is stress? In *Stress: Neurobiology and Neuroendocrinology*, eds M.R. Brown, G.F. Goob & C. Rivier, pp. 3–20. Dekker, New York.

Linklater, W.L. & Gedir, J.V. (2011) Distress unites animal conservation and welfare towards synthesis and collaboration. *Animal Conservation*, 14, 25–27.

Linklater, W.L., Macdonald, E.A., Flamond, J.R.B. *et al.* (2010) Declining and low fecal corticoids are associated with distress, not acclimation to stress, during the translocation of African rhinoceros. *Animal Conservation*, 13, 104–111.

Lopez-Scepulcre, A., Doak, N. Norris, K. *et al.* (2008) Population trends of Seychelles magpie-robins *Copsychus sechellarum* following translocation to Cousin Island, Seychelles. *Conservation Evidence*, 5, 33–37.

Lovegrove, T.G. (1996) Island releases of saddlebacks *Philesturnus carunculatus* in New Zealand. *Biological Conservation*, 77, 151–157.

Lovegrove, T.G. & Veitch, C.R. (1994) Translocating wild forest birds. *Ecological Management*, 2, 23–35.

Mankin, P.C., Warner, R.E. & Anderson, W.L. (1999) Wildlife and the Illinois public: a benchmark study of attitudes and perceptions. *Wildlife Society Bulletin*, 27, 465–472.

Marin, M.T., Cruz, F.C. & Planeta, C.S. (2007) Chronic restraint or variable stresses differently affect the behavior, corticosterone secretion and body weight in rats. *Physiology and Behavior*, 90, 29–35.

Maschinski, J. & Haskins, K. (eds) (2011) *Plant Reintroduction in a Changing Climate: Promises and Perils*. Island Press, Redding, California.

Mason, G.J. (2010) Species differences in responses to captivity: stress, welfare and the comparative method. *Trends in Ecology and Evolution*, 25, 713–721.

Mason, G., Clubb, R. Latham, N. *et al.* (2007) Why and how should we use environmental enrichment to tackle stereotypic behaviour? *Applied Animal Behaviour Science*, 102, 163–188.

Matson, T.K., Goldizen, A.W. & Jarman, P.J. (2004) Factors affecting the success of translocations of the black-faced impala in Namibia. *Biological Conservation*, 116, 359–365.

McCarthy, M.A., Armstrong, D.P. & Runge, M.C. (2011) Adaptive management of reintroduction. In *Reintroduction Biology: integrating science and management*, eds J.G. Ewen, D.P. Armstrong, K.A. Parker & P.J. Seddon, Chapter 8. Wiley-Blackwell, Oxford, UK.

McEwen, B.S. (1998) Seminars in medicine of the Beth Israel Deaconess Medical Center: protective and damaging effects of stress mediators. *New England Journal of Medicine*, 338, 171–179.

McLachlan, J.S., Hellman, J.J. & Schwartz, M.W. (2007) A framework for debate of assisted migration in an era of climate change. *Conservation Biology*, 21, 297–302.

McLean, I.G., Lundie-Jenkins, G. & Jarman, P.J. (1996) Teaching an endangered mammal to recognise predators. *Biological Conservation*, 75, 51–62.

McMillan, S. (1995) Restoration of an extirpated red-sided garter snake *Thamnophis sirtalis sirtalis* population in the interlake region of Manitoba, Canada. *Biological Conservation*, 72, 13–16.

McPhee, M.E. (2004) Generations in captivity increases behavioural variance: considerations for captive breeding and reintroduction programmes. *Biological Conservation*, 115, 71–77.

Milner-Gulland, E.J., Fisher, M., Browne, S. *et al.* (2009) Do we need to develop a more relevant conservation literature? *Oryx*, 44, 1–2.

Miskelly, C.M., Taylor, G.A. Gummer, H. *et al.* (2010) Translocations of eight species of burrow-nesting seabirds (genera Pterodroma, Pelecanoides, Pachyptila and Puffinus: Family Procellariidae). *Biological Conservation*, 142, 1965–1980.

Mitchell, A.M., Wellicome, T.I., Brodie, D. *et al.* (2011) Captive-reared burrowing owls show higher site-affinity, survival, and reproductive performance when reintroduced using a soft-release. *Biological Conservation*, 144, 1382–1391.

Molles, L.E., Calcott, A., Peters, D. *et al.* (2008) 'Acoustic anchoring' and the successful translocation of North Island kokako (*Callaeas cineras wilsoni*) to a New Zealand mainland site within continuous forest. *Notornis*, 55, 57–68.

Muth, R.M. & Jamieson, W.M. (2000) On the density of deer camps and duck blinds: the rise of the animal rights movement and the future of wildlife conservation. *Wildlife Society Bulletin*, 28, 841–851.

Nephew, B.C., Kahn, S.A. & Romero, L.M. (2003) Heart rate and behavior are regulated independently of corticosterone following diverse acute stressors. *General and Comparative Endocrinology*, 133, 173–180.

Nichols, R.K., Steiner, J., Woolaver, L.G. *et al.* (2010) Conservation initiatives for an endangered migratory passerine: field propagation and release. *Oryx*, 44, 171–177.

O'Neill, L., Wilson, P., De Jong, A. *et al.* (2008) Field techniques for handling, anaesthetising and fitting radio-transmitters to Eurasian otters (*Lutra lutra*). *European Journal of Wildlife Research*, 54, 681–687.

Osborne, P.E. & Seddon, P.J. (2011) Selecting suitable habitats for reintroductions: variation, change and the role of species distribution modelling. In *Reintroduction Biology: integrating science and management*, eds J.G. Ewen, D.P. Armstrong, K.A. Parker & P.J. Seddon, Chapter 3. Wiley-Blackwell, Oxford, UK.

Parker, K.A. (2002) *Ecology and Managment of North Island Fernbird (Bowdleria punctata vealeae)*. University of Auckland, Auckland, New Zealand.

Parker, K.A. & Laurence, J. (2008) Translocation of North Island saddleback *Philesturnus rufusater* from Tiritiri Matangi Island to Motuihe Island, New Zealand. *Conservation Evidence*, 5, 47–50.

Pinter-Wollman, N., Isbell, L.A. & Hart, L.A. (2009) Assessing translocation outcome: comparing behavioral and physiological aspects of translocated and resident African elephants (*Loxodonta africana*). *Biological Conservation*, 142, 1116–1124.

Priddel, D. & Carlile, N. (2001) A trial translocation of Gould's Petrel (*Pterodroma leucoptera leucoptera*). *Emu*, 101, 79–88.

Pullin, A.S., Knight, T.M., Stone, D.A. *et al.* (2004) Do conservation managers use scientific evidence to support their decision making? *Biological Conservation*, 119, 245–252.

Reneerkens, J., Piersma, T. & Ramenofsky. M. (2002) An experimental test of the relationship between temporal variability of feeding opportunities and baseline levels of corticosterone in a shorebird. *Journal of Experimental Zoology*, 293, 81–88.

Reynolds, M.H. & Klavitter, J.L. (2006) Translocation of wild Laysan duck *Anas laysanensis* to establish a population at Midway Atoll National Wildlife Refuge, United States and US Pacific Possession. *Conservation Evidence*, 3, 6–8.

Reynolds, M.H., Seavy, N.E., Vekasy, M.S. *et al.* (2008) Translocation and early post-release demography of endangered Laysan teal. *Animal Conservation*, 11, 160–168.

Rich, E.L. & Romero, L.M. (2005) Exposure to chronic stress downregulates corticosterone responses to acute stressors. *American Journal of Physiology – Regulatory Integrative and Comparative Physiology*, 288, R1628–R1636.

Richardson, D.M., Hellman, J.J., McLachlan, J.S. *et al.* (2009) Multidimensional evaluation of managed relocation. *Proceedings of the National Academy of Sciences of the United States of America*, 106, 9721–9724.

Robertson, H.A., Karika, I. & Saul, E.K. (2006) Translocation of Rarotonga Monachs *Pomarea dimidiata* within the southern Cook Islands. *Bird Conservation International*, 16, 197–215.

Roe, J.H., Frank, M.R., Gibson, S.E. *et al.* (2010) No place like home: an experimental comparison of reintroduction strategies using snakes. *Journal of Applied Ecology*, 47, 1253–1261.

Romero, L.M. (2004) Physiological stress in ecology: lessons from biomedical research. *Trends in Ecology and Evolution*, 19, 249–255.

Romero, L.M., Dickens, M.J. & Cyr, N.E. (2009) The reactive scope model – a new model integrating homeostasis, allostasis, and stress. *Hormone Behavior*, 55, 375–389.

Romesburg, H.C. (1981) Wildlife science: gaining reliable knowledge. *Journal of Wildlife Management*, 45, 293–313.

Rouco, C., Ferreras, P., Castro, F. *et al.* (2010) A longer confinement period favours European wild rabbit (*Oryctolagus cuniculus*) survival during soft releases in low cover habitats. *European Journal of Wildlife Research*, 56, 215–219.

Ryckman, M.J., Rosatte, R.C., Mcintosh, T. *et al.* (2010) Postrelease dispersal of reintroduced elk (*Cervus elaphus*) in Ontario, Canada. *Restoration Ecology*, 18, 173–180.

Santos, T., Perez-Tris, J., Carbonell, R. *et al.* (2009) Monitoring the performance of wild-born and introduced lizards in a fragmented landscape: implications for *ex situ* conservation programmes. *Biological Conservation*, 142, 2923–2930.

Sapolsky, R.M., Romero, L.M. & Munck, A.U. (2000) How do glucocorticoids influence stress responses? Integrating permissive, suppressive, stimulatory, and preparative actions. *Endocrinology Review*, 21, 55–89.

Sarrazin, F., Bagnolini, C., Pinna, J.L. *et al.* (1996) Breeding biology during establishment of a reintroduced Griffon vulture *Gyps fulvas* population. *Ibis*, 138, 315–325.

Seddon, P.J. (1999) Persistence without intervention: assessing success in wildlife reintroductions. *Trends in Ecology and Evolution*, 14, 503.

Seddon, P.J. (2010) From reintroduction to assisted colonization: moving along the conservation translocation spectrum. *Restoration Ecology*, 18, 796–802.

Seddon, P.J., Armstrong, D.P. & Maloney, R.F. (2007) Developing the science of reintroduction biology. *Conservation Biology*, 21, 303–312.

Seddon, P.J., Armstrong, D.P., Soorae, P. *et al.* (2009) The risks of assisted colonization. *Conservation Biology*, 23, 788.

Sherwin, C.M. (2004) The motivation of group-housed laboratory mice, *Mus musculus*, for additional space. *Animal Behaviour*, 67, 711–717.

Sih, A., Bell, A. & Chadwick Johnson, J. (2004) Behavioural syndromes: an ecological and evolutionary overview. *Trends in Ecology and Evolution*, 19, 372–378.

Sjoasen, T. (1996) Survivorship of captive-bred and wild-caught reintroduced European otters *Lutra lutra* in Sweden. *Biological Conservation*, 76, 161–165.

Smith, B.R. & Blumstein, D.T. (2008) Fitness consequences of personality: a meta-analysis. *Behavioral Ecology*, 19, 448–457.

Sokos, C.K., Birtsas, P.K. & Tsachalidis, E.P. (2008) The aims of galliforms release and choice of techniques. *Wildlife Biology*, 14, 412–422.

Soorae, P.S. (2010) *Global Re-introduction Perspectives: 2010.* IUCN/SSC Re-introduction Specialist Group and Environment Agency–Abu Dhabi, Abu-Dhabi.

Stoen, O., Pitlagano, M.L. & Moe, S.R. (2009) Same-site multiple releases of translocated white rhinoceroses *Ceratotherium simum* may increase the risk of unwanted dispersal. *Oryx*, 43, 580–585.

Sutherland, W.J., Armstrong, D.P., Butchart, S.H.M. *et al.* (2010) Standards for documenting and monitoring bird reintroduction projects. *Conservation Letters*, 3, 229–235.

Sutherland, W.J., Pullin, A.S., Dolman, P.M. *et al.* (2004) The need for evidence-based conservation. *Trends in Ecology and Evolution*, 19, 305–308.

Swaisgood, R.R. (2010) The conservation-welfare nexus in reintroduction programmes: a role for sensory ecology. *Animal Welfare*, 19, 125–137.

Synder, N.F.R. & Synder, H.A. (1989) Biology and conservation of the California condor. *Current Ornithology*, 6, 175–263.

Synder, N.F.R., Derrickson, S.R., Bessinger, S.R. *et al.* (1996) Limitations of captive breeding in endangered species recovery. *Conservation Biology*, 10, 338–348.

Tavecchia, G., Viedma, C., Martinez-Abrain, A. *et al.* (2009) Maximizing re-introduction success: assessing the immediate cost of release in a threatened waterfowl. *Biological Conservation*, 142, 3005–3012.

Taylor, S.S. & Jamieson, I.G. (2007) Factors affecting the survival of founding individuals in translocated New Zealand saddlebacks *Philesturnus carunculatus*. *Ibis*, 149, 783–791.

Teixeira, C.P., De Azevedo, C.S., Mendl, M. *et al.* (2007) Revisiting translocation and reintroduction programmes: the importance of considering stress. *Animal Behaviour*, 73, 1–13.

Terhune, T.M., Sisson, D.C., Palmer, W.E. *et al.* (2010) Translocation to a fragmented landscape: survival, movement, and site fidelity of Northern bobwhites. *Ecological Applications*, 20, 1040–1052.

Tweed, E.J., Foster, J.T., Woodworth, B.L. *et al.* (2003) Survival, dispersal, and home-range establishment of reintroduced captive-bred puaiohi, *Myadestes palmeri*. *Biological Conservation*, 111, 1–9.

van Heezik, Y., Seddon, P.J. & Maloney, R.F. (1999) Helping reintroduced bustards avoid predation: effective anti-predator training and the predictive value of pre-release behaviour. *Animal Conservation*, 2, 155–163.

Wanless, R.M., Cunningham, J., Hockey, P.A.R. *et al.* (2002) The success of a soft-release reintroduction of the flightless Aldabra rail (*Dryolimnas* [cuvieri] *aldabranus*) on Aldabra Atoll, Seyschelles. *Biological Conservation*, 107, 203–210.

Weiss, J.M. (1968) Effects of coping responses on stress. *Journal of Comparative and Physiological Psychology*, 65, 251–260.

Wilson, C.D., Arnott, G., Reid, N. *et al.* (2011) The pitfall with PIT tags: marking freshwater bivalves for translocation induces short-term behavioural costs. *Animal Behaviour*, 81, 341–346.

Wilson, D.S., Clark, A.B., Coleman, K. *et al.* (1994) Shyness and boldness in humans and other animals. *Trends in Ecology and Evolution*, 9, 442–446.

Wingfield, J.C., Maney, D.L., Breuner, C.W. *et al.* (1998) Ecological bases of hormone–behavior interactions: the emergency life history stage. *American Zoologist*, 38, 191–206.

Wolf, C.M., Griffith, B., Reed, C. *et al.* (1996) Avian and mammalian translocations: update and reanalysis of 1987 survey data. *Conservation Biology*, 10, 1142–1154.

Zidon, R., Saltz, D., Shore, L.S. *et al.* (2009) Behavioural changes, stress, and survival following reintroduction of Persian fallow deer from two breeding facilities. *Conservation Biology*, 23, 1026–1035.

5

Dispersal and Habitat Selection: Behavioural and Spatial Constraints for Animal Translocations

Pascaline Le Gouar[1,2,3], *Jean-Baptiste Mihoub*[2] *and François Sarrazin*[2]

[1]Université Rennes 1, UMR 6553 Ecobio, Paimpont, France
[2]Université Pierre et Marie Curie, UMR 7204 CERSP, Paris, France
[3]NIOO-KNAW, Dutch Centre for Avian Migration and Demography, Wageningen, the Netherlands

'*The increasing number of studies of dispersal in conservation biology (Reed & Dobson, 1993; Clobert et al., 2001; Imms & Hjermann, 2001; Macdonald & Johnson, 2001; Trakhtenbrot et al., 2005) can help identify the individual, environmental and social components that influence post-release movements that could, at least in part, be controlled within translocation programmes.*'

Page 139

Introduction

Analysis of factors influencing the success or failure of translocation programmes is crucial to understand broadly the process of establishment of new populations (Seddon *et al.*, 2007) and to improve future programmes (Griffith *et al.*, 1989; Wolf *et al.*, 1996; Fischer & Lindenmayer, 2000). Translocation failure occurs when individuals disappear from where they are released (see

Reintroduction Biology: Integrating Science and Management. First Edition.
Edited by John G. Ewen, Doug P. Armstrong, Kevin A. Parker and Philip J. Seddon.
© 2012 Blackwell Publishing Ltd. Published 2012 by Blackwell Publishing Ltd.

Box 5.1 for definitions of release site and area). Mortality is the most commonly documented cause of failure and is dependent on habitat quality, including predation, on the characteristics of founders, and on methods of releases. However, post-release dispersal (Box 5.1) also results in the disappearance of individuals after release. Unfortunately, due to inadequate post-release monitoring, the fate of missing individuals is often unknown (only 45% of 336 projects analysed by Wolf *et al.*, 1996, used banding or radio-tagging). Post-release dispersal and the importance of habitat selection behaviours have been neglected in previous reviews of translocation programmes (but see Stamps & Swaisgood, 2007). Nevertheless, the association of post hoc assessment of low-release habitat quality with translocation failure (Wolf *et al.*, 1996) implies a general lack of knowledge of the habitat requirements and habitat selection behaviours of reintroduced species (Stamps & Swaisgood, 2007).

Box 5.1 Definition of post-release dispersal from the release area

Usually, dispersal is defined as: (i) the departure of an individual from its natal or breeding habitat, (ii) the movement itself and (iii) the settlement in a new habitat (Clobert *et al.*, 2001). In the particular context of translocation we defined post-release dispersal as any movement involving the departure of released animals from the release area definitively or temporarily. We thus included cases when movement did not end up in settlement into a new habitat or in breeding. In the literature, the release location may be either named as the release site or the release area without distinction, while we distinguished here the release area – a habitat where animals are hoped to remain in order to found a population – from the release site – a smaller location within the release area where they are actually released. Following our definitions, we considered homing behaviour, i.e. the return of released individuals to the area that they occupied before capture, as post-release dispersal, although some authors considered homing as fidelity to their natal or breeding habitat.

The increasing number of studies of dispersal in conservation biology (Reed & Dobson, 1993; Clobert *et al.*, 2001; Imms & Hjermann, 2001; Macdonald

& Johnson, 2001; Trakhtenbrot *et al.*, 2005) can help identify the individual, environmental and social components that influence post-release movements that could, at least in part, be controlled within translocation programmes. Understanding these factors has been listed as one of the 10 key questions to address in reintroduction biology (Armstrong & Seddon, 2008). It is important to account for post-release dispersal when making population projections for reintroduced populations (Armstrong & Reynolds, this volume, Chapter 6) and also to distinguish between mortality and dispersal in order to adopt appropriate strategies to maximize future establishment success (Tweed *et al.*, 2003). This issue encompasses the monitoring of released individual movements and the comparison of dispersal patterns between successful and unsuccessful programmes (Nichols & Armstrong, this volume, Chapter 7).

Our aim was to quantify the extent to which dispersal has been considered by translocation practitioners during a project's post-release phase through examination of published literature. Then, by focusing on those published studies that investigated dispersal of translocated individuals, we identified the problems associated with dispersal, the factors that influence propensity to disperse and the methods that have been proposed to reduce the deleterious effects of dispersal on translocation success.

Is dispersal and habitat selection behaviour considered in translocation programmes?

To quantify consideration of dispersal in post-release monitoring of wildlife translocations, we reviewed the abstracts of 1007 studies of translocations of birds, mammals, reptiles, amphibians and insects published from 1950 to 2008. Publications from 1950 to early 1996 ($n = 712$) were found in the online free access annotated bibliography of wildlife translocations (Griffith *et al.*, 1998) and publications from 1996 to 2008 ($n = 295$) were located among natural science databases of CSA Illumina using the search terms 'reintroduction', 're-establishment' and 'translocation'. In each case, we noted whether or not post-release dispersal was mentioned in the abstract. This procedure will have under-represented the proportion of papers in which dispersal of translocated individuals was considered, as dispersal may not be mentioned in the abstract. However, we believe that mentioning post-release dispersal in the abstract indicates whether dispersal was considered an issue.

We found that post-release dispersal or habitat selection behaviour of translocated individuals was considered in 282 cases (28%). Because of the potential influence of Griffith *et al.*'s (1989) review of translocations, we compared consideration of dispersal in papers published before and after 1989. Consideration of dispersal significantly increased between the 1950–1989 ($n = 170$, 27.8%) and the 1990–2008 ($n = 112$, 37.9%) periods ($\chi^2 = 32.2$, df $= 1$, $p< 0.0001$). Additionally, we observed a recent increase in the number of studies focusing principally on the dispersal of translocated individuals (e.g. Ostro *et al.*, 2000; Saltz *et al.*, 2000; Letty *et al.*, 2002; Gerber *et al.*, 2003; Moehrenschlager & Macdonald, 2003; Tweed *et al.*, 2003; Gardner & Gustafson, 2004; Dzialak *et al.*, 2005; King & Gurnell, 2005; Wichrowski *et al.*, 2005; Hardman & Moro, 2006).

Only 39 studies assumed post-release dispersal or suboptimal habitat selection as causes of failure in re-establishing populations (18 for birds, 17 for mammals, 2 for reptiles, 2 for invertebrates). For example, reintroduction of 65 red grouse (*Lagopus l. scotocus*) in Ireland failed because '*all individuals left the release site and the most evident factor is spontaneous and extensive dispersal and lack of ability in habitat selection*' (Lance, 1974). In another case 29 thick billed parrots (*Rhynchopsitta pachyrhyncha*) reintroduced in south-eastern Arizona '*left the release area and assumed new location*' (Snyder & Wallace, 1987). In addition, translocation of wolves (*Canis lupus*) in Minnesota from 1975 to 1978 was '*largely unsuccessful at keeping wolves out of livestock production areas*' due to long-distance movements away from the release site by translocated individuals (Fritts *et al.*, 1984). Partial dispersal by reintroduced mammals was also recorded for two species of hare-wallaby (*Lagorchestes hirsutus* and *Lagostrophus fasciatus*) by Hardman & Moro (2006), where '*32% of the reintroduced individuals left the release site in the four weeks following the reintroduction*'. Deleterious post-release dispersal does not occur only in vertebrates; for instance, '*experimental translocation trials of the tiger beetle Cicindela dorsalis dorsalis in Virginia by using adults were unsuccessful because the adults dispersed from the translocation sites within 1–2 weeks*' (Knisley *et al.*, 2005).

Critical problems associated with dispersal in translocation programmes

Post-release dispersal could affect translocation success in various ways and its impacts may differ between the different phases of population foundation, i.e.

establishment, growth and regulation (Sarrazin, 2007). We thus distinguished between the main problems associated with dispersal during the establishment phase and during the growth and regulation phase (the persistence phase in Armstrong & Seddon, 2008).

Establishment is sensitive to dispersal, through unexpectedly high dispersal distances or dispersal rates, particularly in cases with a small number of founders. For instance, some translocation failures have been due to homing to the source habitat (Linnell *et al.*, 1997). Homing behaviour in mammals is usually inversely related to the displacement distance between the release site and the area of origin (Allen & Sargeant, 1993; VanVuren *et al.*, 1997). The consequences of short-term, long-distance dispersal leading individuals outside the release area are often negative. First, individuals who disperse and settle far away from the release area will not contribute demographically or to the future gene pool of the translocated population, which thus faces an enhanced risk of extinction. Second, the consequences of dispersal depend on the habitat matrix in which the animals will travel or settle. If habitats around the release site are of sufficient quality, or if managed (protected/restored) areas around the release site are large enough, it is unlikely there will be negative consequences. In contrast, by settling outside a managed release area, released individuals could suffer from threats that led to the previous extinction of the wild population (van Heezik *et al.*, 2009). For instance, unforeseen dispersal can induce conflicts with human societies and thus affect success of reintroduction programmes (Box 5.2). Moreover, mortality due to dispersal, or increased vulnerability to predation during dispersal, can increase proportionally with distance from the release site (Nappee, 1982; Matthews, 2003; Spinola *et al.*, 2008).

Box 5.2 **Dispersal of translocated individuals and conflicts with human populations**

Translocations and particularly reintroductions remain biased towards animals with a high dispersal ability such as large vertebrate species. The social consequences of dispersal of translocated individuals are twofold.

First, local people who waited for or who may have sometimes initiated the re-establishment of the population may miss the expected cultural or economical benefits and can be strongly disappointed (Lorimer, 2006; Nilsen *et al.*, 2007). This may be the case if ecotourism is set up around the translocation program (Lindsey *et al.*, 2005) or if the translocation aims at restoring an ecological function (Nilsen *et al.*, 2007). As an example, we consider one reintroduction program in France aiming at restoring a population of griffon vulture (*Gyps fulvus*), a cliff nester and scavenger raptor (Figure 5.1). This reintroduction was conducted at Navacelles, Southern France, and 50 vultures were released from 1993 to 1997 at a radius of 40–50 km from a successfully reintroduced breeding population in the Grands Causses region. Despite a rigorous monitoring scheme, the availability of suitable nesting habitat and food supply management, only two successful breeding events have been observed at the release area between 1993 and 1998. In 1998, the Navacelles reintroduction program was declared a failure because there were no griffon vultures left in the release area. Actually, many of the birds released at Navacelles moved and settled into the Grands Causses population, probably because of conspecific attraction (Le Gouar *et al.*, 2008). The social impact of this failure prevents the initiation of expected economic activities (such as bird watching) and the restoration of ecological function for farmers (i.e. the ecological service of livestock carcass removal by vultures) in the Navacelles area.

Second, if translocated individuals disperse from the target release area, and then cross or settle in a different area, they might have to coexist unexpectedly with humans. Unexpected coexistence between released animals and humans can be felt as a violation of the human living space and security, causing or exacerbating social conflicts. This often occurs with translocations of carnivores or large herbivorous mammals. A relevant illustration of this problem is the re-inforcement of brown bears (*Ursus arctos*) in the French Pyrenees. Between 1996 and 2006, eight bears from Slovenia were released in the Pyrenean Mountains. Among them, a 4-year-old male brown bear, named 'Balou', was released on

Figure 5.1   Griffon vultures in the Grands Causses, Southern France (photo: Jean-Baptiste Mihoub).

1 June 2006. This bear quickly moved long distances towards urban areas. Five days after release, it was recorded 61 km away from its release site less than 30 km from Toulouse, the fourth most populated city in France (Figure 5.2; from ONCFS, unpublished data). The re-inforcement of brown bears in this region had already been strongly debated, mainly because farmers and hunters claimed that bears may cause severe injuries to sheep livestock or to game species. After the post-release dispersal of Balou, the situation became even worse due to fears for human safety.

These cases illustrate how post-release dispersal can complicate public acceptance of translocation projects. Conservation managers or scientists may also lose their credibility, and this could be detrimental for future translocations.

Figure 5.2 **Movement of Balou, a reintroduced brown bear in the south of France, during the first month after release (Data from ONCFS, équipe technique Ours, unpublished data).**

Movement intensity, either the propensity to move, distance moved or frequency of movement, can also be high just after release, during an exploratory phase (Rittenhouse *et al.*, 2007; Hester *et al.*, 2008). The time used to explore the new environment is costly and may reduce the time allocated to foraging, reproduction or predator vigilance. This exploratory phase is thus associated with low survival and low reproductive success (Kurzejeski & Root, 1988; Letty *et al.*, 2000; Moehrenschlager & Macdonald, 2003), and thus with reduced probability of population establishment.

Managers may also encounter problems if post-release dispersal is low, although this situation seems rarer. For example, if released individuals stay in close proximity they might attract predators (Toepfer, 1988) through cues such as high concentrations of faecal matter (Banks *et al.*, 2002).

Once the translocated population is established, reduced immigration could alter rescue effects and be problematic for long-term population viability, while reduced emigration could enhance negative density-dependent competition processes through increased competition for resources (Gardner & Gustafson, 2004). In addition, a pattern of low dispersal is a risk for the long-term persistence of reintroduced migratory species (Shandruk, 1986) as migratory movements have been selected to utilize essential resources at the right time and in the right place. The ability to colonize new habitats, i.e. to disperse and select suitable habitat, is crucial for the long-term viability of the translocated population since it limits the extinction risks of the whole population if unpredictable perturbation occurs at the release area or if the habitat is modified because of global change. Moreover, if the translocation aims to restore a metapopulation, low dispersal would be detrimental through preventing genetic and demographic rescue effects (Reed & Levine, 2005; Mihoub *et al.*, in preparation).

Moderate post-release dispersal of translocated individuals is necessary to enhance chances of survival in a changing environment and to promote gene flow among established populations (Trakhtenbrot *et al.*, 2005). Understanding the factors influencing dispersal is crucial in managing the trade-off between site fidelity and adaptive dispersal.

Critical factors for managing post-release dispersal of reintroduced individuals

Individual factors

The sex of individuals can potentially affect post-release dispersal. For example, male-biased dispersal occurs in deer (Persians fallow deer, *Dama dama mesopotamica*, Dolev *et al.*, 2002; and red deer, *Cervus elaphus*, Nussey *et al.*, 2006), whereas dispersal is female-biased in capercaillie (*Tetrao urogallus*, Angelstam & Sandegren, 1982) and river otters (*Lontra canadensis*, Spinola *et al.*, 2008). Sex biases in post-release dispersal could be handled by studying mating systems in natural populations of the target species (Legendre *et al.*, 1999). The age of translocated individuals may also affect dispersal rates (Kress & Nettleship, 1988). Young individuals often disperse further than adults. For example, long-distance dispersal of young ural owls (*Strix uralensis*) was proposed to explain low reintroduction success in the Bavarian National Park in Germany (Scherzinger, 1987). However, the relationship between age and dispersal might be species-specific (Larkin *et al.*, 2004), and for now theory predicting the sex bias in any species is not set up. Therefore, when possible, sex- and age-dependent dispersal should be investigated for every species prior to reintroduction, to enhance fidelity to the released area. Managers should focus on individuals that are more likely to accept a new environment and are less likely to disperse widely in the short term.

Recent studies have also shown that consistent individual differences in personality influence natal dispersal in wild bird populations (Dingemanse *et al.*, 2003). Individuals that quickly explored a novel environment (i.e. risk-prone individuals) later dispersed further than did slow-exploring individuals (i.e. risk-adverse individuals). Releasing bold individuals that are more likely to settle in a novel environment could be risky as they are also more likely to disperse away from the release site and into unsuitable areas (Bremmer-Harrison *et al.*, 2004). However, releasing shy individuals could also lead to reintroduction failure if individuals are unable to forage in new environments or to disperse. Therefore, careful management of behavioural types could enhance post-release success (Watters & Meehan, 2007).

The origin and history of an individual can also influence post-release dispersal patterns. Various studies have investigated the difference between captive-rearing versus wild-rearing or hand-rearing versus natural-rearing methods on post-release movement patterns (Maxwell & Jamieson, 1997; Stoinski *et al.*, 2003; Hellstedt & Kallio, 2005). Contrasting patterns have been reported: either captive- or hand-reared individuals dispersed more than their wild- and natural-reared conspecifics (Jamieson & Wilson, 2003; Van Zant & Wooten, 2003; Mathews *et al.*, 2005) or they showed higher fidelity to their rearing site (Morgart *et al.*, 1987; Toepfer, 1988). Behavioural variance between individuals could be explained by the duration of time the species has spent in captivity, since captivity can relax or exert different selection pressures, which might be maladaptive in a natural environment (Mcphee, 2003). This process may also interact with factors affecting dispersal already mentioned, such as age at release or sex.

Environmental factors

Dispersal from the release area can occur if threats remain in the target habitat (e.g. non-native competitors; see Danielson & Gaines, 1987) or if some resources are missing. Eradication of known threats that decrease habitat quality should be achieved before releases (IUCN, 1998), e.g. by controlling predators (Lenain & Warrington, 2001) or competitors. However, ethical questions may arise from potential conflicting conservation strategies, as illustrated by the recovery management of two endangered bird species in Hawaii. Efforts to restore a Hawaiian crow (*Corvus hawaiiensis*) population through releases of captive-reared birds was compromised by predation from Hawaiian hawks (*Buteo solitarius*), requiring managers to revise their initial recovery plans (Banko *et al.*, 2009).

Stamps & Swaisgood (2007) proposed that a positive experience for individuals in their natal habitat would increase their establishment in release areas that contain similar features to the natal habitat (i.e. 'natal habitat preference induction' hypothesis); in other words, if the release habitat differs from the natal habitat, individuals may disperse to familiar habitats to which they are better adapted. The role of prior habitat experience on fidelity to release areas could be integrated in release site selection and for defining the rearing conditions of potential founder individuals.

Providing food at the release site could reduce dispersal after release (Finlayson & Moseby, 2004). However, the benefit of such management supply should be carefully considered, as it could sometimes interrupt migratory behaviour (Doligez *et al.*, 2004) or lead to settlement in an 'ecological trap' (Frair *et al.*, 2007). Moreover, using fences in the short term can prevent dispersal just after release and force individuals to acclimatize to the release area (Hunter, 1998).

The habitat matrix could affect both the intensity of dispersal and dispersal-related mortality (Calvete & Estrada, 2004; Larkin *et al.*, 2004; Dzialak *et al.*, 2005; Castellon & Sieving, 2006). The availability of higher habitat quality surrounding a release area may favour dispersal (Erickson & McCullough, 1987), but this may have fewer consequences for reintroduction success than if individuals had to cross a hostile matrix that may increase mortality.

Social factors

Conspecific density could also influence the intensity of post-release dispersal and could be managed in some cases. For some species, high conspecific density leads to high rates of dispersal (Dunham, 2000). In such cases, if large numbers of individuals are released then the dispersal rate might be high. However, if density is low, individuals might disperse to search for a mate, or for social species, to join a bigger or a better structured group (Le Gouar et al., 2008). This mechanism can be seen as a behaviourally mediated Allee effect, and release group sizes may need to be carefully considered to avoid such negative effects of density-dependent dispersal (Deredec & Courchamp, 2007).

Familiarity between released individuals, especially adults, may provide an incentive for animals to remain near the release site (Hunter, 1998; Van Zant & Wooten, 2003), but not in all cases (Armstrong, 1995). In contrast, inbreeding could affect dispersal (Szulkin & Sheldon, 2007), as individuals generally avoid related conspecifics. Animals may avoid familiar individuals, as this is often a cue for relatedness; hence familiarity among captive-reared unrelated juveniles may also enhance post-reintroduction dispersal (Mootnick et al., 2005).

For social species, promoting settlement through conspecific attraction could potentially be achieved by using visual models of individuals and/or acoustic playbacks (Kress, 1997). Keeping captive individuals on release sites in cages (see below) might also help to fix the reintroduced individuals on the release area, since effects of conspecific attraction may reduce dispersal distances. However, conspecific cues may provide little or no attraction to stressed animals after release, and several experiments recently conducted with translocated New Zealand birds suggest that playback of conspecific calls is generally ineffective at keeping birds near release sites (David Bradley & Laura Molles, personal communication).

Release method factors

The most commonly documented factor affecting post-release dispersal in translocations was the release method; more specifically, the influence of delayed (or soft) versus immediate (or hard) release methods. Several studies have shown that holding individuals in captivity (e.g. by penning or in aviaries) for acclimatizing to the release site reduced post-release dispersal

(Stanley-Price, 1989; Stout *et al.*, 1989; Bright & Morris, 1994; Fritts *et al.*, 2001; King & Gurnell, 2005; Tuberville *et al.*, 2005). However, several others have shown no effect on dispersal (Short *et al.*, 1992; Castro *et al.*, 1994; Lovegrove, 1996; Campbell & Croft, 2001; Clarke *et al.*, 2002; Hardman & Moro, 2006), and in some cases holding animals in captivity has reduced their post-release survival. A systematic review of the effectiveness of various release methods for any given species is thus needed before selecting the most appropriate release method.

In addition, variation in dispersal behaviours between seasons has been studied in established populations of mammals (*Sigmodon, Microtus* and *Peromyscus*, see Diffendorfer *et al.*, 1995; *Lynx pardinus*, see Ferreras *et al.*, 2004) as well as for translocated birds (*Alectoris chukar*, see Dickens *et al.*, 2009). This behaviour would be worth investigating for future reintroductions by releasing animals when they are less liable to move, e.g. before breeding seasons for birds or after annual prospecting periods (Sarrazin *et al.*, 1996; Le Gouar *et al.*, 2008).

Repeated releases could affect the dispersal of individuals and the establishment of reintroduced populations. For example, Dolev *et al.* (2002) showed that repeated releases of Persian fallow deer into the same area slowed the home range establishment of newly reintroduced males because of the territorial behaviour of residents. For females, however, the presence of earlier released conspecifics acted as a cue for home range establishment (Dolev *et al.*, 2002). Releasing dependent young with their parents could be used to anchor breeders on the release area (Clarke *et al.*, 2002). Releasing adults with offspring with reduced movement ability may indeed reduce post-release dispersal and enhance the settlement of released individuals, such as during the translocation of kaki (*Himantopus novaezelandiae*) in New Zealand (van Heezik *et al.*, 2009).

Integrating dispersal and habitat selection behaviour into the pre-release phase

Mapping the distribution of ecological requirements of reintroduced species may help to reduce errors in choosing a release site (Osborne & Seddon, this volume, Chapter 3). However, suitability maps are often based on environmental cues only (Schadt *et al.*, 2002; Hirzel *et al.*, 2004; Osborne & Seddon, this volume, Chapter 3), i.e. abiotic environmental factors such as a favourable landscape for breeding or feeding, although biotic factors may strongly affect habitat suitability and habitat selection choice (Guisan & Thuillier, 2005). Animals may use social cues, such as conspecifics (Danchin *et al.*, 2001, 2004) or heterospecifics (Mönkkönen *et al.*, 1999; Forsman *et al.*, 2008; Hromada *et al.*, 2008), more than abiotic environmental cues. Thus, flawed perceptions of the criteria used by individuals to select foraging and reproductive habitats may lead managers to select *apparently* suitable sites more often than *actually* suitable sites. For instance, animals with a high degree of sociability could desert release areas to join established populations simply because of their attractiveness. The first empirical example of this phenomenon is the failure of five caribou (*Rangifer tarandus*) reintroduction programmes in eastern North America because all the animals probably joined nearby herds (Bergerud & Mercer, 1989). A second example concerns attempts to establish a population of griffon vultures (*Gyps fulvus*), in which almost all released individuals were quickly attracted to the nearest and biggest colony (see details in Box 5.2; Le Gouar *et al.*, 2008). Finally, for kaki translocated in New Zealand, the larger the size of the population being re-inforced through releases, the smaller the dispersal distances from the release site (van Heezik *et al.*, 2009).

Unusual home-range size and use of suboptimal habitat by released animals have been observed (Zwank *et al.*, 1988; Stoinski *et al.*, 2003; Steffens *et al.*, 2005). Such behaviours are associated with stress in the new environment (Teixeira *et al.*, 2007) or potentially with mal-adaptation arising from captivity (Robert, 2009). Due to recognition of the importance of behavioural studies in conservation (Buccholz, 2007; Caro, 2007), researchers have developed methods to reduce perturbation of behaviour in captivity. In the context of translocation, efforts focus on understanding natural social and mating systems, strategies to reduce stress during manipulation and training of individuals to improve predator avoidance or foraging abilities (Wielebnowski, 1998; Wallace, 2000; Shier & Owings, 2006). However, training individuals to select appropriate habitats has yet to be investigated to restore adaptive habitat selection behaviours. Nevertheless, training strategies might be logistically and financially prohibitive for many species.

A lack of familiarity between individuals or an established social structure is known to enhance dispersal in social species (Lambin *et al.*, 2001, but see Armstrong, 1995, and Armstrong & Craig, 1995). Even if captivity allowed social structure to develop within the pre-release group, this may be disrupted once individuals are released into the new environment. For example, reintro-duction of collared peccaries (*Pecari tajacu*) in Arizona, which were kept for six months in captivity, failed as the group broke up just after releases (Day, 1985). Leadership or decision making could differ between captive and free conditions as individuals might react differently to new conditions. According to Couzin *et al.* (2005), despite inherent differences among individuals, e.g. dominance can explain leadership, there are often other cues such as different motivation preferences that can influence group movements. In a translocation context, newly released individuals are naive or have the same information about their environment, potentially making leadership decisions more difficult to achieve. Similarly, unless the released group is very strongly structured, individuals with different preferences but the same motivations could divide the group, as observed in translocation experiments of house sparrows (*Passer domesticus*) in Ecuador (Pierre-Yves Henry, personal communication).

Conclusion

Unfortunately, the effects of age, sex, origin and release method on post-release dispersal behaviour of reintroduced individuals have been reported mostly a posteriori. Integrating the effects of all these factors systematically and a priori for the composition of the release groups, the choice of animals to release, as well as the identification of release sites that are more likely to inhibit extra-post-release movement, are necessary to avoid dispersal-related project failure. This means a need for more experimental studies assessing the effects of pre-release manipulations and/or habitat on the dispersal of individuals (e.g. Calvete & Estrada, 2004; Dickens *et al.*, 2009). Testing the efficiency of training for habitat selection is also an important issue. Spatially explicit population viability analysis (Box 5.3), as advocated by Seddon *et al.* (2007), could be useful as a complementary tool to experimental testing, which can then act to validate or refine our understanding of post-release dispersal behaviour. In addition, quantitative meta-analyses, constructed around sound theoretical frameworks of the factors affecting dispersal and habitat selection in translocations, should be conducted to reveal consistent patterns according to taxa, habitat, release methods, etc. These meta-analyses require improvements in post-release monitoring (Sarrazin & Barbault, 1996; Sutherland *et al.*, 2010) and detailed international databases of translocations in which to store monitoring data (e.g. the avian reintroductions and translocations database held at Lincoln Park zoo (http://www.lpzoo.org/artd)). The influence of dispersal on reintroduction success also relies on a standard definition of success at different temporal and spatial scales. More and more programs aim to restore meta-populations of endangered species, for which dispersal between subpopulations is crucial. Under a metapopulation approach, a translocation may contribute to a global conservation success despite being a local establishment failure. Therefore, the ability to disperse and to select a habitat in which to settle with minimum assistance should be considered alongside survival and reproduction for the assessment of both the short- and long-term viability of translocated populations.

Box 5.3 The use of population modelling in management of post-release dispersal and habitat selection issues in translocations

Population models taking into account animal behaviour may help to forecast and to quantify risks of translocation failure due to post-release dispersal, prior to actual release. Recently, two modelling studies investigated the consequences of generic and widespread mechanisms at play during habitat selection – including conspecific attraction on post-release dispersal during the establishment phase (Mihoub *et al.*, in preparation) and the persistence phase (Mihoub *et al.*, 2009) of reintroduction. The first warns that a translocation may fail as early as the establishment phase for species using social cues. In such cases, released animals may prefer to settle in a remnant population of conspecifics rather than to settle within the envisaged release area (a so-called 'vacuum effect'; see Mihoub *et al.*, in preparation). This phenomenon has been observed and reported from past translocation monitoring (see the example of griffon vulture reintroduction in Box 5.2). The second highlights the fact that failure can occur during the persistence phase in heterogeneous environments, despite the establishment phase being achieved. Conspecific attraction behaviours may cause the population to aggregate within poor habitat (a so-called ecological trap), despite more suitable habitat being available nearby (Mihoub *et al.*, 2009).

Although these modelling analyses are mostly theoretical and use generic behavioural rules, they can allow: (i) anchoring observed behavioural patterns that lead to translocation failure in a theoretical framework and (ii) quantifying a priori probabilities of extinction

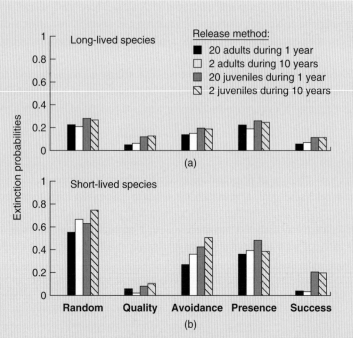

Figure 5.3   Extinction probabilities for populations of (a) long-lived and (b) short-lived reintroduced species according to four release strategies (adapted from Figure 2 in Mihoub *et al.*, 2009). Five breeding habitat selection strategies were considered. Demographic rates for long-lived species are taken from a reintroduced population of griffon vulture (Sarrazin *et al.*, 1996) whereas demographic rates for short-lived species are taken from a population of willow flycatchers (*Empidonax traillii*) (Sedgwick, 2004). Plotted extinction probabilities are generated from 1000 Monte Carlo simulations, during 10 generations, resulting in the age-structured female life cycle demographic model, under global environmental stochasticity in an extremely heterogeneous habitat. This modelling work clearly showed that, for short lived species, releasing adults is less risky than releasing juveniles, especially if individuals use the breeding performance of their conspecifics to select their habitat.

during different translocation phases (i.e. population establishment or persistence) with regard to the habitat selection behaviour. In turn, such modelling approaches may increase the efficiency and the success of translocation programmes by providing a way to help optimize management effort or structure hypotheses that can be tested with targeted management (Armstrong and Reynolds, this volume, Chapters 6 and Nichols and Armstrong, this volume, Chapter 7). For example, as seen in Figure 5.3, different release management strategies can be confronted by modelling in order to identify the age class, the frequency and the number of individuals released per year that will maximize the persistence of a reintroduced population while accounting for various patterns of habitat selection behaviour.

Acknowledgements

The authors owe special thanks to Brad Griffith for sharing his database on reintroduction outcomes. They are grateful to Pierre Migot, Pierre-Yves Quenette and Michel Catusse from the 'ONCFS, équipe Ours' for permission to use the figure on brown bear reintroduction (Figure 5.2) and Dr Linus Svensson, Oikos Editorial Board, for permission to use Figure 5.3 adapted from Mihoub et al. (2009).

References

Allen, S.H. & Sargeant, A.B. (1993) Dispersal patterns of red foxes relative to population density. *Journal of Wildlife Management*, 57, 526–533.

Angelstam, P. & Sandegren, F. (1982) A release of pen-reared capercaillie in central Sweden – survival, dispersal and choice of habitat. In *Proceedings of the Second International Symposium on Grouse at Dalhousie Castle*, Edinburgh, Scotland, March 1981, ed. T.W.I. Lovel, pp. 204–209. World Pheasant Assocation, Exning, Suffolk.

Armstrong, D.P. (1995) Effects of familiarity on the outcome of translocations, II. A test using New Zealand robins. *Biological Conservation*, 71, 281–288.

Armstrong, D.P. & Craig, J.L. (1995) Effects of familiarity on the outcome of translocations, I. A test using saddlebacks *Philesturnus carunculatus rufusater*. *Biological Conservation*, 71, 133–141.

Armstrong, D.P. & Reynolds, M.H. (2011) Modelling reintroduced populations: the state of the art and future directions. In *Reintroduction Biology: integrating science* and management, eds J.G. Ewen, D.P. Armstrong, K.A. Parker & P.J. Seddon, Chapter 6. Wiley-Blackwell, Oxford, UK.

Armstrong, D.P. & Seddon, P.J. (2008) Directions in reintroduction biology. *Trends in Ecology and Evolution*, 23, 20–25.

Banko, P.C., Burgett, J., Conry, P.J. *et al.* (2009) *Revised Recovery Plan for the 'Alalà (Corvus hawaiiensis)*. US Fish and Wildlife Service, Portland, Oregon.

Banks, P.B., Norrdahl, K. & Korpimäki, E. (2002) Mobility decisions and the predation risks of reintroduction. *Biological Conservation*, 103, 133–138.

Bergerud, A.T. & Mercer, W.E. (1989) Caribou introductions in eastern North America. *Wildlife Society Bulletin*, 17, 111–120.

Bremmer-Harrison, S., Prodohl, P.A. & Elwood, R.W. (2004) Behavioral trait assessment as a release criterion: boldness predicts early death in reintroduction programme of captive-bred swift fox (*Vulpes velox*). *Animal Conservation*, 7, 313–320.

Bright, P.W. & Morris, P.A. (1994) Animal translocation for conservation: performance of dormice in relation to release methods, origin and season. *Journal of Applied Ecology*, 31, 699–708.

Buccholz, R. (2007) Behavioural ecology: an effective and relevant conservation tool. *Trends in Ecology and Evolution*, 22, 401–407.

Calvete, C. & Estrada, R. (2004) Short term survival and dispersal of translocated European wild rabbits: improving the release protocol. *Biological Conservation*, 120, 507–516.

Campbell, L. & Croft, D.B. (2001) Comparison of hard and soft release of hand reared Eastern kangaroos. In *Veterinary Conservation Biology, Wildlife Health and Management in Australasia*, eds A. Martin & L. Vogelnest, Proceedings of the International Joint Conference, Taronga Zoo, Sydney, Australia.

Caro, T. (2007). Behavior and conservation: a bridge too far? *Trends in Ecology and Evolution*, 22, 394–400.

Castellon, T.D. & Sieving, K.E. (2006) An experimental test of matrix permeability and corridor use by an endemic understory bird. *Conservation Biology*, 20, 135–145.

Castro, I., Alley, J.C., Empson, R.A. *et al.* (1994) Translocation of the hihi or stitchbird *Notiomystis cincta* to Kapiti Island, New Zealand: transfer techniques and comparison of released strategies. In *Reintroduction biology of Australian and New Zealand fauna*, ed. M. Serena. Surrey Beatty & Sons, Chipping Norton, UK.

Clarke, R.H., Boulton, R.L. & Clarke, M.F. (2002) Translocation of the socially complex black-eared miner *Manorina melanotis*: a trial using hard and soft release techniques. *Pacific Conservation Biology*, 8, 223–234.

Clobert, J., Danchin, E., Dhondt, A.A. *et al.* (2001) *Dispersal*. Oxford University Press, Oxford, UK.

Couzin, I.D., Krause, J., Franks, N.R. *et al.* (2005) Effective leadership and decision-making in animal groups on the move. *Nature*, 434, 513–516.

Danchin, E., Giraldeau, L.-A., Valone, T.J. *et al.* (2004) Public information: from nosy neighbors to cultural evolution. *Science*, 205, 487–491.

Danchin, E., Heg, D. & Doligez, B. (2001) Public information and breeding habitat selection. In *Dispersal*, eds J. Clobert, E. Danchin, A.A. Dhondt & J.D. Nichols, pp. 243–258. Oxford University Press, Oxford, UK.

Danielson, B.J. & Gaines, M.S. (1987) The influences of conspecific and heterospecific residents on colonization. *Ecology*, 68, 1778–1784.

Day, J.I. (1985) *Javelina Research and Management in Arizona.* Arizona Game and Fish Department, Phoenix, Arizona.

Dickens, M., Delehanty, D.J., Reed, J.M. *et al.* (2009) What happens to translocated game birds that 'disappear'? *Animal Conservation*, 12, 418–425.

Diffendorfer, J.E., Gaines, M.S. & Holt, R.D. (1995) Habitat fragmentation and movements of three small mammals (*Sigmodon*, *Microtus* and *Peromyscus*). *Ecology*, 76, 827–839.

Dingemanse, N.J., Both, C., Van Noordwijk, A. *et al.* (2003) Natal dispersal and personalities in great tits (*Parus major*). *Proceedings of the Royal Society Biological Sciences Series B*, 270, 741–747.

Deredec, A. & Courchamp, F. (2007) Importance of the Allee effect for reintroductions. *Ecoscience*, 14(4), 440–451.

Dolev, A., Saltz, D., Bar-David, S. *et al.* (2002) Impact of repeated releases on space-use patterns of Persian fallow deer. *Journal of Wildlife Management*, 66, 737–746.

Doligez, B., Thomson, D.L. & van Noordwijk, A.J. (2004) Using large-scale data analysis to assess life history and behavioural traits: the case of the reintroduced white stork *Ciconia ciconia* population in the Netherlands. *Animal Biodiversity and Conservation*, 27, 387–402.

Dunham, K.M. (2000) Dispersal pattern of mountain gazelles *Gazella gazella* released in central Arabia. *Journal of Arid Environments*, 44, 247–258.

Dzialak, M.R., Lacki, M.J., Larkin, J.L. *et al.* (2005) Corridors affect dispersal initiation in reintroduced peregrine falcons. *Animal Conservation*, 8, 421–430.

Erickson, D.W. & McCullough, C.R. (1987) Fates of translocated river otters in Missouri. *Wildlife Society Bulletin*, 15, 511–517.

Ferreras, P., Delibes, M., Palomares, F. *et al.* (2004) Proximate and ultimate causes of dispersal in the Iberian lynx *Lynx pardinus*. *Behavioral Ecology*, 15, 31–40.

Finlayson, G.R. & Moseby, K.E. (2004) Managing confined populations: the influence of density on the home range and habitat use of reintroduced burrowing bettongs (*Bettongia lesueur*). *Wildlife Research*, 31, 457–463.

Fischer, J. & Lindenmayer, D.B. (2000) An assessment of the published results of animal relocations. *Biological Conservation*, 96, 1–11.

Forsman, J.T., Hjernquist, M.B., Taipale, J. *et al.* (2008) Competitor density cues for habitat quality facilitating habitat selection and investment decisions. *Behavioural Ecology*, 19, 539–545.

Frair, J.L., Merrill, E.H., Allen, J.R. *et al.* (2007) Know thy enemy: experience affects elk translocation success in risky landscapes. *Journal of Wildlife Management*, 71, 541–554.

Fritts, S.H., Mack, C.M., Smith, D. *et al.* (2001) Outcomes of hard and soft releases of reintroduced wolves in central Idaho and the Greater Yellowstone Area. In *Large Mammal Restoration*, eds D.S. Maehr, R.F. Noss & J.L. Larkin, pp. 125–147. Island Press, Washington, DC.

Fritts, S.H., Paul, W.J. & Mech, L.D. (1984) Movements of translocated wolves in Minnesota. *Journal of Wildlife Management*, 48, 709–721.

Gardner, R.H. & Gustafson, E.J. (2004) Simulating dispersal of reintroduced species within heterogeneous landscapes. *Ecological Modelling*, 171, 339–358.

Gerber, L.R., Seabloom, E.W., Burton, R.S. *et al.* (2003) Translocation of an imperilled woodrat population: integrating spatial and habitat patterns. *Animal Conservation*, 6, 1–8.

Griffith, B., Comly, L.M., Chilelli, M. *et al.* (1998) Annotated bibliography of wildlife translocations, http://users.iab.uaf.edu/~brad_g_riffith/translocations.shtml.

Griffith, B., Scott, J.M., Carpenter, J. *et al.* (1989) Translocation as a species conservation tool: status and strategy. *Science*, 245, 477–480.

Guisan, A. & Thuillier, W. (2005) Predicting species distribution: offering more than simple habitats models. *Ecology Letters*, 8, 993–1009.

Hardman, B. & Moro, D. (2006) Optimising reintroduction success by delayed dispersal: is the release protocol important for hare-wallabies? *Biological Conservation*, 128, 403–411.

Hellstedt, P. & Kallio, E.R. (2005) Survival and behaviour of captive-born weasels (*Mustela nivalis nivalis*) released in nature. *Journal of Zoology*, 266, 37–44.

Hester, J.M., Price, S.J. & Dorcas, M.E. (2008) Effects of relocation on movements and home ranges of eastern box turtles. *Journal of Wildlife Management*, 72, 772–777.

Hirzel, A.H., Posse, B., Oggier, P.-A. *et al.* (2004) Ecological requirements of reintroduced species and the implications for the release policy: the case of the bearded vulture. *Journal of Applied Ecology*, 41, 1103–1116.

Hromada, M., Antczak, M., Valone, T.J. *et al.* (2008) Settling decisions and heterospecific social information use in shrikes. *PLoS ONE*, 3, e3930.

Hunter, L. (1998) Early post-release movements and behaviour of reintroduced cheetahs and lions, and technical considerations in large carnivore restoration. In *Proceedings of a Symosium on Cheetahs as Game Ranch Animals*, pp. 72–82.

Wildlife Group of the South African Veterinary Association, Onderstepoort, South Africa.

Imms, R.A. & Hjermann, D.O. (2001) Condition-dependent dispersal. In *Dispersal*, eds J. Clobert, E. Danchin, A.A. Dhondt & J.D. Nichols, pp. 203–216. Oxford University Press, Oxford, UK.

IUCN (1998) *IUCN Guidelines for Re-introductions*. UICN/SSC Re-introduction Specialist Group, Gland, Switzerland.

Jamieson, I.G. & Wilson, G.C. (2003) Immediate and long-term effects of translocations on breeding success in takahe *Porphyrio hochstetteri*. *Bird Conservation International*, 13, 299–306.

King, S.R.B. & Gurnell, J. (2005) Habitat use and spatial dynamics of takhi introduced to Hustai National Park, Mongolia. *Biological Conservation*, 124, 277–290.

Knisley, C.B., Hill, J.M. & Scherer, A.M. (2005) Translocation of threatened tiger beetle *Cicindela dorsalis dorsalis* (Coleoptera: Cicindelidae) to Sandy Hook, New Jersey. *Annals of the Entomological Society of America*, 98, 552–557.

Kress, S.W. (1997) Using animal behaviour for conservation: case studies in seabird restoration from the Maine Coast, USA. *Journal of the Yamashina Institute for Ornithology*, 29, 1–26.

Kress, S.W. & Nettleship, D.N. (1988) Re-establishment of Atlantic puffins (*Fratercula arctica*) at a former breeding site in the Gulf of Maine. *Journal of Field Ornithology*, 59, 161–170.

Kurzejeski, E.W. & Root, G. (1988) Survival of reintroduced ruffed grouse in north Missouri. *Journal of Wildlife Management*, 52, 248–252.

Lambin, X., Aars, J. & Piertney, S.B. (2001) Dispersal, intraspecific competition, kin competition and kin facilitation: a review of the empirical evidence. In *Dispersal*, eds J. Clobert, E. Danchin, A.A. Dhondt & J.D. Nichols, pp. 110–122. Oxford University Press, Oxford, UK.

Lance, A.N. (1974) Releases of pen-reared red grouse (*Lagopus l. scoticus*) to restock breeding populations in Ireland. *International Congress of Game Biology*, 11, 225–229.

Larkin, J.L., Cox, J.J., Wichrowski, M.W. *et al.* (2004) Influences on release-site fidelity of translocated elk. *Restoration Ecology*, 12, 97–105.

Legendre, S., Clobert, J., Møller, A.P. *et al.* (1999) Demographic stochasticity and social mating system in the process of extinction of small populations: the case of passerines introduced to New Zealand. *American Naturalist*, 153, 449–463.

Le Gouar, P., Robert, A., Choisy, J.P. *et al.* (2008) Roles of survival and dispersal in reintroduction success of Griffon vulture (*Gyps fulvus*). *Ecological Applications*, 18, 859–872.

Lenain, D.M. & Warringthon, S. (2001) Is translocation an effective tool to remove predatory foxes from a desert area? *Journal of Arid Environment*, 48, 205–209.

Letty, J., Marchandeau, S., Clobert, J. *et al.* (2000) Improving translocation success: an experimental study of antistress treatment and release method for wild rabbits. *Animal Conservation*, 3, 211–219.

Letty, J., Marchandeau, S., Reitz, F. *et al.* (2002) Survival and movements of translocated wild rabbits (*Oryctolagus cuniculus*). *Game and Wildlife Science*, 19, 1–23.

Lindsey, P.A., Alexander, R.R., du Toit, J.T. *et al.* (2005) The potential contribution of ecotourism to African wild dog conservation in South Africa. *Biological Conservation*, 123, 339–348.

Linnell, J.D.C., Aanes, R. Swenson, J.E. *et al.* (1997) Translocation of carnivores as a method for managing problems animals: a review. *Biological Conservation*, 6, 1245–1257.

Lorimer, H. (2006) Herding memories of humans and animals. *Environment and Planning D-Society and Space*, 24, 497–518.

Lovegrove, T.G. (1996) Island releases of saddlebacks *Philesturnus carunculatus* in New Zealand. *Biological Conservation*, 77, 151–157.

Macdonald, D.W. & Johnson, D.D.P. (2001) Dispersal in theory and practice: consequences for conservation biology. In *Dispersal*, eds J. Clobert, E. Danchin, A.A. Dhondt & J.D. Nichols, pp. 358–372. Oxford University Press, Oxford, UK.

Mathews, F., Orros, M., McLaren, G. *et al.* (2005) Keeping fit on the ark: assessing the suitability of captive-bred animals for release. *Biological Conservation*, 121, 569–577.

Matthews, K.R. (2003) Response of mountain yellow-legged frogs, *Rana muscosa*, to short distance translocation. *Journal of Herpetology*, 37, 621–626.

Maxwell, J.M. & Jamieson, I.G. (1997) Survival and recruitment of captive-reared and wild-reared takahe in Fiordland, New Zealand. *Conservation Biology*, 11, 683–691.

McPhee, E. (2003) Generations in captivity increases behavioral variance: considerations for captive breeding and reintroduction programs. *Biological Conservation*, 115, 71–77.

Mihoub, J.B., Le Gouar, P., Robert, A. *et al.* (in preparation) Post-release dispersal from release areas following translocation attempts: social attraction and the 'vacuum effect'.

Mihoub, J.B., Le Gouar, P. & Sarrazin, F. (2009) Breeding habitat selection behaviours in heterogeneous environments: implications for modelling reintroduction. *Oikos*, 118, 663–674.

Moehrenschlager, A. & Macdonald, D.W. (2003) Movement and survival parameters of translocated and resident swift foxes *Vulpes velox*. *Animal Conservation*, 6, 199–206.

Mönkkönen, M., Härdling, R., Forsman, J.T. *et al.* (1999) Evolution of heterospecific attraction: using other species as cues in habitat selection. *Evolutionary Ecology*, 13, 91–104.

Mootnick, A., Baker, E. & Sheeran, L. (2005) Familiarity during immaturity: implications for the captive propagation of Gibbons. *International Journal of Primatology*, 26, 1417–1433.

Morgart, J.R., Smith, D.R. & Krausman, P.R. (1987) An evaluation of two methods used to reintroduce desert bighorn sheep. *Transactions of the Congress of the International Union of Game Biology*, 18, 126.

Nappee, C. (1982) Capercaillie and black grouse breeding in the Parc National des Cevennes and first release results. In *Proceedings of the Second International Symposyum on Grouse at Dalhousie Castle, Edinburgh, Scotland. March 1981*, ed. T.W.I. Lovel, pp. 218–228. World Pheasant Association, Exning, Suffolk.

Nichols, J.D. & Armstrong, D.P. (2011) Monitoring for reintroductions. In *Reintroduction Biology: integrating science and management*, eds J.G. Ewen, D.P. Armstrong, K.A. Parker & P.J. Seddon, Chapter 7. Wiley-Blackwell, Oxford, UK.

Nilsen, E.B., Milner-Gulland, E.J., Schofield, L. *et al.* (2007) Wolf reintroduction to Scotland: public attitudes and consequences for red deer management. *Proceedings of the Royal Society B*, 274, 995–1002.

Nussey, D.H., Pemberton, J., Donald, A. *et al.* (2006) Genetic consequences of human management in an introduced island population of red deer (*Cervus elaphus*). *Heredity*, 97, 56–65.

Osborne, P.E. & Seddon, P.J. (2011) Selecting suitable habitats for reintroductions: variation, change and the role of species distribution modelling. In *Reintroduction Biology: integrating science and management*, eds J.G. Ewen, D.P. Armstrong, K.A. Parker & P.J. Seddon, Chapter 3. Wiley-Blackwell, Oxford, UK.

Ostro, L.E.T., Silver, S.C., Koontz, F.W. *et al.* (2000) Habitat selection by translocated black howler monkeys in Belize. *Animal Conservation*, 3, 175–181.

Reed, J.M. & Dobson, A.P. (1993) Behavioural constraints and conservation biology: conspecific attraction and recruitment. *Trends in Ecology and Evolution*, 8, 253–256.

Reed, J.M. & Levine, S.H. (2005) A model for behavioral regulation of metapopulation dynamics. *Ecological Modelling*, 183, 411–423.

Rittenhouse, C.D., Millspaugh, J.J., Hubbard, M.W. *et al.* (2007) Movements of translocated and resident three-toed box turtles. *Journal of Herpethology*, 41, 115–121.

Robert, A. (2009) Captive breeding genetics and reintroduction success. *Biological Conservation*, 142, 2915–2922.

Saltz, D., Rowen, M. & Rubenstein, D.I. (2000) The effect of space-use patterns of reintroduced Asiatic wild ass on effective population size. *Conservation Biology*, 14, 1852–1861.

Sarrazin, F. (2007) Introductory remarks: a demographic frame for reintroduction. *Ecoscience*, 14, iii–v.

Sarrazin, F. & Barbault, R. (1996) Re-introductions: challenges and lessons for basic ecology. *Trends in Ecology and Evolution*, 11, 474–478.

Sarrazin, F., Bagnolini, C., Pinna, J.L. *et al.* (1996) Breeding biology during establishment of a reintroduced griffon vulture (*Gyps fulvus*) population. *Ibis*, 138: 315–325.

Schadt, S., Revilla, E., Wiegand, T. *et al.* (2002) Assessing the suitability of central European landscape for the reintroduction of Eurasian lynx. *Journal of Applied Ecology*, 39, 189–203.

Scherzinger, W.T. (1987) Reintroduction of the Ural owl in the Bavarian National Park, Germany. U.S. Forest Service General Technical Report RM 142, pp. 75–80.

Seddon, P.J., Armstrong, D.P. & Maloney, R.F. (2007) Developing the science of reintroduction biology. *Conservation Biology*, 21, 303–312.

Sedgwick, J.A. (2004) Site fidelity, territory fidelity, and natal philopatry in willow flycatchers (*Empidonax traillii*). *Auk*, 121, 1103–1121.

Shandruk, L.J. (1986) Elk Island National Park Trumpeter Swan Transplant Pilot Project – Final Report. In *Proceedings Papers of the Tenth Trumpeter Swan Society Conference*, ed. D. Compton, pp. 66–67. Grand Prairie, Alberta, Canada, 3–6 September 1986.

Shier, D.M. & Owings, D.H. (2006) Effects of predator training on behavior and post-release survival of captive prairie dogs (*Cynomys ludovicianus*). *Biological Conservation*, 132, 126–135.

Short, J., Bradshaw, S.D., Giles, J. *et al.* (1992) Reintroduction of macropods (Marsupialia, Macropodoidea) in Australia: a review. *Biological Conservation*, 62, 189–204.

Snyder, N.F.R. & Wallace, M.P. (1987) Reintroduction of the thick-billed parrot in Arizona. In *Proceedings of the Jean Delacour/IFCB Symposium on Breeding Birds in Captivity*, ed. A. Risser, pp. 360–384, Los Angeles, California.

Spinola, R.M., Serfass, T.L. & Brooks, R.P. (2008) Survival and post-release movements of river otters translocated to western New York. *Northeastern Naturalist*, 15, 13–24.

Stamps, J. & Swaisgood, R.R. (2007) Someplace like home: experience, habitat selection and conservation biology. *Applied Animal Behaviour Science*, 102, 392–409.

Stanley-Price, M.R. (1989) *Animal Reintroductions: the Arabian oryx in Oman*. Cambridge University Press, New York, USA.

Steffens, K.E., Seddon, P.J., Mathieu, R. *et al.* (2005) Habitat selection by South Island saddlebacks and Stewart Island robins reintroduced to Ulva Island. *New Zealand Journal of Ecology*, 29, 221–229.

Stoinski, T.S., Beck, B.B., Bloomsmith, M.A. *et al.* (2003) A behavioral comparison of captive-born, reintroduced golden lion tamarins and their wild-born offspring. *Behaviour*, 140, 137–160.

Stout, I.J., Doonan, T.J., Roberts, R.E. *et al.* (1989) Comparisons of results of three gopher tortoise relocations in central and southwest Florida. In *Proceedings of a Symposium of Gopher Tortoise Relocation, Florida Nongame Wildlife Program,* eds J.E. Diemer, D.R. Jackson, J.L. Landers, J.N. Layne & D.A. Wood, pp. 15–42. Technical Report 5.

Sutherland, W.J., Armstrong, D., Butchart, S.H.M. *et al.* (2010) Standards for documenting and monitoring bird reintroduction projects. *Conservation Letters,* 3, 229–235.

Szulkin, M. & Sheldon, B.C. (2007) Dispersal as a means of inbreeding avoidance in a wild bird population. *Proceedings of the Royal Society B,* 275, 703–711.

Teixeira, C.P., de Azevedo, C.S., Mendl, M. *et al.* (2007) Revisiting translocation and reintroduction programmes: the importance of considering stress. *Animal Behaviour,* 73, 1–13.

Toepfer, J.E. (1988) *The ecology of the greater prairie chicken as related to reintroductions.* Thesis, Montana State University, Bozeman, Montana, USA.

Trakhtenbrot, A., Nathan, R., Perry, G. *et al.* (2005) The importance of long-distance dispersal in biodiversity conservation. *Diversity and Distributions,* 11, 173–181.

Tuberville, T.D., Clark, E.E., Buhlmann, K.A. *et al.* (2005) Translocation as a conservation tool: site fidelity and movement of repatriated gopher tortoises (*Gopherus polyphemus*). *Animal Conservation,* 8, 349–358.

Tweed, E.J., Foster, J.T., Woodworth, B.L. *et al.* (2003) Survival, dispersal and home-range establishment of reintroduced captive-reared puaiohi, *Myadestes palmeri*. *Biological Conservation,* 111, 1–9.

van Heezik, Y., Maloney, R.F. & Seddon, P.J. (2009) Movements of translocated captive-bred and released critically endangered kaki (black stilts) *Himantopus novaezelandiae* and the value of long-term post-release monitoring. *Oryx,* 43, 639–647.

VanVuren, D., Kuenzi, A.J., Loredo, I. *et al.* (1997) Translocation as a nonlethal alternative for managing California ground squirrels. *Journal of Wildlife Management,* 61, 351–359.

VanZant, J.L. & Wooten, M.C. (2003) Translocation of Choctawhatchee beach mice (*Peromyscus polionotus allophrys*): hard lessons learned. *Biological Conservation,* 112, 405–413.

Wallace, M.P. (2000) Retaining behaviour in captivity for re-introduction. In *Behaviour and Conservation,* eds M. Gosling & W. Sutherland, pp. 315–329. Cambridge University Press, Cambridge, UK.

Watters, J.V. & Meehan C.L. (2007) Different strokes: can managing behavioural types increase post-release success? *Applied Animal Behaviour Science,* 102, 364–379.

Wichrowski, M.W., Maehr, D.S., Larkin, J.L. *et al.* (2005) Activity and movements of reintroduced elk in south eastern Kentucky. *Southeastern Naturalist*, 4, 365–374.

Wielebnowski, N. (1998) Contributions of behvioural studies to captive management and breeding of rare and endangered mammals. In *Behavioral Ecology and Conservation Biology*, ed. T. Caro, pp. 130–162. Oxford University Press, New York, USA.

Wolf, M.C., Griffith, B., Reed, C. *et al.* (1996) Avian and mammalian translocations: update and reanalysis of 1987 survey data. *Conservation Biology*, 10, 1142–1154.

Zwank, P.J., Geaghan, J.P. & Dewhurst, D.A. (1988) Foraging differences between native and released Mississippi sandhill cranes: implications for conservation. *Conservation Biology*, 2, 386–390.

6

Modelling Reintroduced Populations: The State of the Art and Future Directions

Doug P. Armstrong[1] and Michelle H. Reynolds[2]

[1]Massey University, Palmerston North, New Zealand
[2]US Geological Survey Pacific Island Ecosystems Research Center, Hawaii, United States of America

'... the obvious purpose for building quantitative models for reintroduced populations is to make predictions that can be used to inform management. It needs to be emphasized that the purpose is not to mimic reality, hence Box's (1976) famous quote that "all models are wrong, but some are useful". Making models useful inevitably means embracing the KISS principle to some extent, i.e. "keep it simple, stupid".'

Page 175

Introduction

Our understanding of any system constitutes a *model* of that system. In a reintroduction, the systems of interest are usually the source population, the reintroduced population, the ecosystem in the release area and any meta-populations that may include the source and/or reintroduced populations. For reintroduction to be sensible, the proponents probably need to believe that: (a) the species can persist at the release area; (b) the reintroduction will not have undue negative impacts on the ecosystem there; (c) the impacts on the source populations are outweighed by the benefits of the reintroduction; and (d) the release area is unlikely to be colonized naturally in the future. The

Reintroduction Biology: Integrating Science and Management. First Edition.
Edited by John G. Ewen, Doug P. Armstrong, Kevin A. Parker and Philip J. Seddon.
© 2012 Blackwell Publishing Ltd. Published 2012 by Blackwell Publishing Ltd.

proponents therefore need to believe they have a reasonable understanding of each of the above systems, meaning they are using models of some form.

Despite this, until recently most papers in the field of reintroduction biology make no mention of models whatsoever (Seddon *et al.*, 2007). The reason is that the term 'model' is normally used specifically for quantitative models, whereas most papers describe models verbally (e.g. 'exotic predators were thought to have caused the original extirpation, so these predators were eradicated before reintroduction'). In our experience, most decisions about reintroduction are made through round-the-table (or 'round-the-internet') discussion of intuitive verbal models. These intuitive models may reflect a good understanding of the systems involved and lead to reasonable decisions. However, development of more explicit conceptual models can substantially clarify thinking about the systems involved, and subsequent quantitative models can be used to guide a range of decisions about reintroduction programmes, as will be illustrated in this chapter.

Several other chapters in this volume also discuss quantitative models applied to reintroduction biology. In particular, Osborne & Seddon (Chapter 3) discuss application of species distribution modelling to choosing reintroduction sites, Le Gouar *et al.* (Chapter 5) discuss behavioural models for predicting post-release dispersal, Ewen *et al.* (Chapter 9) and Sainsbury *et al.* (Chapter 10) discuss risk analyses with respect to disease transmission in reintroduction and Keller *et al.* (Chapter 11), Groombridge *et al.* (Chapter 12) and Jamieson & Lacy (Chapter 13) all discuss models used to predict loss of genetic variation in reintroduced populations. The current chapter is the first in a series of three chapters focusing on the application of population modelling to reintroduction, i.e. the use of models predicting establishment, growth, persistence and/or spread of populations.

We focus on the creation of population or meta-population models using available data, and predictions from these models that can be used to guide reintroduction programmes. In this volume, Nichols & Armstrong (Chapter 7) then focus on the monitoring needed to provide crucial data and McCarthy *et al.* (Chapter 8) focus on the strategic improvement of models over time through adaptive management. The theme of the latter two chapters is structured decision making, where objectives are clearly identified and monitoring and management decisions are adaptive and optimized with respect to those objectives and their associated uncertainties (see Nichols & Armstrong, this volume, Chapter 7, for a more detailed explanation of structured decision making). The ordering of the chapters may appear to be reversed, i.e. it would be logical to first define objectives and design an adaptive management programme, then collect data and finally begin fitting models to those data. However, while population models have been used to guide many reintroduction programmes, to date the decisions made about reintroductions have been largely intuitive. The current chapter therefore focuses on the state-of-the-art for applying population modelling to reintroduction, whereas the following two chapters provide a theoretical framework for improving the logic and rigour of reintroduction programmes in the future.

The first part of the current chapter is an overview of rationale and principles underlying population models for reintroduction programmes and the steps involved in building reliable models. This is followed by a review of published literature on the use of quantitative population models in reintroduction programmes, focusing on the types of models used and their applications to management. Finally, we point to future directions that we believe will improve our capacity to guide reintroduction programmes through population modelling, providing a link to the material covered in the following two chapters. We hope that this chapter will empower people to get started in creating population models for enhancing their programmes as well as providing direction for more experienced modellers. However, while the chapter is meant to be as accessible as possible, we assume that readers are familiar with basic concepts in population ecology. For readers with little or no background in population ecology, we recommend White's (2000a) book in the chapter on population modelling, Akçakaya et al.'s (1999) introductory textbook on applied population ecology and/or Williams et al.'s (2002) more advanced textbook on analysis and management of wildlife populations.

What should models be used to predict?

Population models can potentially be used to predict future population sizes, rates of growth and spread, and probabilities of short-term *establishment* (Box 6.1) and longer-term *persistence*. The term *projections* is often used to describe predicted population trends over time, and we will adopt that usage here. The appropriate focus of predictions depends on the degree of uncertainty about habitat suitability and the amount of habitat available. For example, if a species is reintroduced to a small island following eradication of an exotic predator that clearly caused the initial extinction, and no new threats have impacted the island, the reintroduced population will be expected to grow in the short term but its long-term viability may be unknown. In such cases it may make sense to focus on persistence and to ensure that models incorporate factors relevant to long-term persistence of small populations. These include *inbreeding depression* (see this volume, Chapters 11 to 13), *Allee effects* (Box 6.2), *demographic stochasticity* (variation due to random fates of individuals), *environmental stochasticity* (variation due to environmental conditions) and catastrophes (Caughley, 1994). These issues are also critical to modelling impacts of reintroductions on source populations, as the harvesting needs to be sustainable in the context of the dynamics of those source populations. However, with many reintroduced populations it is uncertain whether the habitat can support the species and therefore uncertain whether the population will grow at all. Consequently, for at least the first few years after reintroduction, it usually makes sense to focus on predicting the growth of the reintroduced population in relation to factors likely to affect that growth. On a more immediate time frame, it may be important to predict the probability of establishment in relation to the release strategy used (including the number of individuals released) and the likely *post-release effects* (i.e. temporary effects of the translocation process on survival, reproduction or dispersal rates; see Box 6.1.)

Box 6.1 **Post-release effects and population establishment**

Translocated animals face numerous stresses, so it is not surprising that there is often elevated mortality immediately after release (Dickens *et al.*, 2010). Similar effects may occur in plant reintroductions (e.g. Morgan, 1999) although there may be a greater capacity to minimize post-planting mortality through careful management. Hamilton *et al.* (2010) defined the period of elevated mortality as the *acclimation period* and defined survival over this period as *post-release survival.*

Post-release effects are changes in survival or other vital rates attributable to translocation stress. Tavecchia *et al.* (2009) described the mortality attributable to translocation stress as the *cost of release*, which is given by

$$1 - \phi'_i/\phi_i$$

where ϕ'_i is survival over some post-release interval (which must be the same or longer than the acclimation period) and ϕ_i is the survival expected over that interval in normal circumstances. The approximate standard error based on the delta method (see text) is given by

$$SE(1 - \hat{\phi}'_i/\hat{\phi}_i) = \frac{\hat{\phi}'_i}{\hat{\phi}_i}\sqrt{\left(\frac{SE(\hat{\phi}'_i)}{\hat{\phi}'_i}\right)^2 + \left(\frac{SE(\hat{\phi}_i)}{\hat{\phi}_i}\right)^2 - \frac{2cov(\hat{\phi}'_i, \hat{\phi}_i)}{\hat{\phi}'_i\hat{\phi}_i}}$$

Tavecchia *et al.* (2009) estimated ϕ_i using data from established individuals present at the start of the same interval, i.e. individuals that had not been recently translocated. However, similar inference can be made from changes in survival probability over time. For example, if 45% of translocated individuals survived the first month ($\hat{\phi}'_i = 0.45$), but monthly survival subsequently increased to 90% and remained at that level ($\hat{\phi}_i = 0.9$), then the estimated cost of release would be 50% mortality (0.5). However, it is important to be aware of possible confounding factors such as age, season and habitat changes.

There may also be post-release effects on dispersal and/or reproduction. It is common for animals to disperse away from the release area soon after translocation (Le Gouar *et al.*, this volume, Chapter 5), so there may often be a cost of release in terms of dispersal as well as mortality. Separate effects can potentially be estimated if dispersal and mortality can be distinguished, and this will be important if estimates are to be used for predictions at other sites, especially if the sites differ greatly in connectivity. However, when making predictions for the same site where data are being collected, it may be sufficient to estimate post-release effects on *apparent survival* without distinguishing dispersal from mortality. Post-release effects on reproductive success are also expected in some circumstances (Dickens *et al.*, 2010), and have been speculated in several cases where reproduction of a reintroduced population was unexpectedly low (Nolet & Baveco, 1996; Lloyd & Powlesland, 1994; Ortiz-Catedral & Brunton, 2008). Clear effects are more difficult to infer than those on mortality and dispersal due to the longer time frames involved, but can be estimated if data are available from multiple subsequent years, especially if translocations take place in two or more years (Armstrong & Ewen, 2001a).

It is essential to consider post-release effects when modelling reintroduced populations, for two reasons. The first reason is that vital rate estimates including post-release effects will result in misleading projections, which may in turn result in poor management decisions. Projections may be misleading simply because vital rates are overly pessimistic (Armstrong & Ewen, 2001a), but also because low initial vital rates may be misinterpreted as Allee effects (see Box 6.2). The simplest approach to this problem is to ignore any data collected during the likely acclimation period when estimating vital rates. If it is unclear whether low initial vital rates are attributable to post-release effects, it is sensible to compare predictions of alternative models that do or do not incorporate post-release effects (Nolet & Baveco, 1996). However, it is important in these situations not to simply assume that vital rates will improve in the future.

The second reason it is essential to consider post-release effects is that they affect establishment probability. The *establishment phase* of a reintroduction refers to the period when the population is susceptible to threats that will disappear if the population survives this phase (Armstrong & Seddon, 2008). These threats include demographic stochasticity and Allee effects (see Box 6.2) that are associated with small initial population size, but also post-release effects. Any modelling of establishment probability as a function of the number of individuals released therefore needs to incorporate post-release effects. It is therefore useful to have estimates of post-release effects available from previous studies (e.g. Tavecchia *et al.*, 2009), including experimental studies comparing post-release survival under different release strategies (e.g. van Heezik *et al.*, 1999; Hardman & Moro, 2006; Hamilton *et al.*, 2010).

Box 6.2 **Allee effects**

Allee effects, the suppression of vital rates at low population densities, can occur through several mechanisms, including ineffective predator defence, ineffective foraging and low contact rates of potential mates or gametes (Courchamp *et al.*, 2008). Allee effects are clearly relevant to reintroduction, given that reintroduced populations are usually initially at low densities, and Allee effects can potentially result in reintroduced populations failing to establish (see Box 6.1) despite the habitat being suitable. However, Allee effects have so far received relatively little attention in the reintroduction literature. Deredec & Courchamp (2007) asserted that Allee effects are ubiquitous in small populations and urged reintroduction practitioners to give them greater consideration. They particularly emphasized the benefits of making initial population sizes as large as possible.

We agree with Deredec & Courchamp (2007) that Allee effects are probably ubiquitous in small populations and that they need to be considered in reintroduction programmes. However, Allee effects should not be assumed to always have a significant impact and it is essential to be aware of the costs involved in avoiding potential Allee effects. In particular, releasing large numbers of individuals may be costly to source populations. More importantly, it could be dangerous to uncritically attribute reintroduction failures to Allee effects, as this could lead to subsequent larger releases to sites where previous reintroduction attempts failed due to poor habitat. Armstrong & Wittmer (2011) advocated the use of structured decision-making frameworks (Nichols & Armstrong, this volume, Chapter 7) for considering Allee effects in reintroduction strategies, and illustrated how this could be done using a stochastic population model including Allee effects. As these frameworks would involve decisions about how many individuals to release, it would be sensible to consider genetic issues and Allee effects simultaneously (Jamieson & Lacy, this volume, Box 13.3).

Whether or not Allee effects should be included in models for reintroduced populations depends on the situation. Although Allee effects may be ubiquitous in small populations, it is important to distinguish trivial Allee effects from those likely to significantly reduce population growth. Following the KISS principle (see text), it is best to keep models as simple as possible, excluding details not needed to capture the population's dynamics. Figure 6.1 shows an example of a trivial Allee effect where a female Laysan teal failed to reproduce in the second breeding season after reintroduction to Midway Atoll due to isolation from males in the low-density population. This had little effect on total productivity since she was one of 18 potentially breeding females in the population that year; hence Allee effects were not considered in the model used to predict future population growth (see Box 6.4).

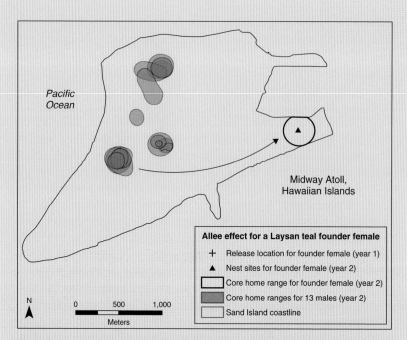

Figure 6.1   **An example of a trivial Allee effect in a reintroduction population.**
A female Laysan teal translocated to Midway Atoll (see Box 6.4) took up
residence in an otherwise uninhabited part of Sand Island (part of Midway
Atoll) in the second breeding season. She produced infertile eggs that year,
presumably due to lack of contact with males, resulting in a slight reduction
in total productivity of the 18 females present that season, and potentially a
slight reduction in the variance in genetic contributions among the founders.

In contrast to this trivial Allee effect, data from 21 translocations of
black-faced impala (*Aepyceros melampus petersi*) in Namibia strongly
suggested that small translocated populations had a lower probability of
persistence due to vulnerability to predation from cheetahs (*Acinonyx
jubatus*) (Matson *et al.*, 2004). In this situation, it would clearly be
sensible to include Allee effects in population models used to guide
subsequent management.

   Although it may be clear from the species and system whether
Allee effects are likely to be important, it may be sensible to test for Allee

effects after release using model selection procedures (see text). This can address not only whether Allee effects should be included in population models but also which vital rates they apply to and what functional forms should be used (see Boukal & Berec, 2002, for a useful review of models applicable to Allee effects). The caveat here is that it is easy to infer spurious Allee effects due to an imperfect understanding of the system since various factors are confounded with population density. For example, while Angulo *et al.* (2007) concluded that positive density dependence in apparent survival of island foxes was an Allee effect caused by eagle predation, a more comprehensive analysis by Bakker *et al.* (2009) suggested that the positive density dependence was most likely due to emigration. A further challenge of modelling Allee effects is that because they are likely to be most relevant early in reintroduction programmes, it may be necessary to fit models based on meta-analyses of data from other species and sites. In such circumstances it will be particularly important to have a clear theoretical framework in place to anticipate the circumstances where different forms of Allee effects may be expected and be wary of spurious correlations between vital rates and population density.

These different foci are closely related to Caughley's (1994) distinction between the *small population paradigm*, which means focusing on the effects of population smallness, and *declining population paradigm*, which means focusing on factors affecting rates of growth or decline. Whereas the immediate establishment of reintroduced populations and long-term persistence of island populations might largely fall into the small population paradigm, the prediction of population growth of an intermediate time frame largely falls into the declining population paradigm. However, Caughley's (1994) key point was that we need to avoid thinking solely within either of these paradigms, and we need to be aware that the effects of growth rate and smallness are interrelated. For example, although small populations are more vulnerable to demographic stochasticity, demographic stochasticity is unlikely to be important in any population that can grow rapidly at low density (Figure 6.2).

Assuming a reintroduction attempt survives the *establishment phase* (Box 6.1), we suggest that the initial focus should be estimating its population

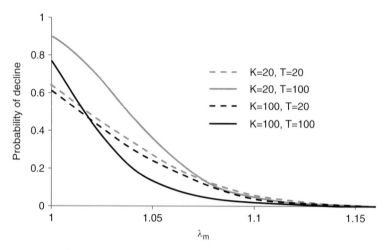

Figure 6.2 **Probability of decline due to demographic stochasticity in hypothetical female-only populations. Populations start with 10 individuals and the probability of decline from this initial number is given over 20- and 100-year time frames (T). The number of survivors and new recruits each year are initially selected from binomial distributions based on probabilities s (survival) and f (fecundity), but the populations are constrained to remain at or below K (carrying capacity). The parameter λ_m is the initial rate of increase expected and is given by $1 + s + f$. Note that the effect of demographic stochasticity is most pronounced when λ_m is close to 1 (i.e. slow growth expected) and K is low, but also depends on the time frame considered.**

growth rate and putting a confidence or credible interval around that estimate. Given that the number of individuals released is usually well below the expected carrying capacity, positive growth is essential. However, initial population trends may give a misleading indicator of future growth, not only because of post-release effects (Box 6.1) but because the initial population may not have a stable age structure or sex ratio. For example, the first year's growth of a population of mainly breeding-age females may give an unrealistically optimistic impression of the population's future. Such misleading inferences can be avoided by estimating the population's finite rate of increase (λ), which is growth rate predicted when the sex and age distribution stabilizes (Leslie, 1945; Williams *et al.*, 2002, section 8.4.2). The finite rate of increase is the proportional change in population size expected over some time interval (usually one year); hence the most essential criterion for reintroduction success is that λ > 1. However, as shown in Figure 6.2, if λ is close to 1 then trajectories of small populations will also be strongly influenced by demographic stochasticity. It is therefore essential in such circumstances to account for demographic stochasticity when making population projections (McCarthy *et al.*, 1994). Once populations are clearly growing, it may be appropriate to incorporate *density dependence* into models to project long-term population trends (e.g. Nicoll *et al.*, 2003; Armstrong *et al.*, 2005; Saether *et al.*, 2007) and then to assess potential threats to long-term persistence if the population is likely to remain small and isolated.

Models for reintroduced populations are most immediately useful if the predictions can be used to adjust release-site management such as predator control (Armstrong *et al.*, 2006) or supplementary feeding (Armstrong *et al.*, 2007), giving an ideal scenario for adaptive management (McCarthy *et al.*, this volume, Chapter 8). However, it may also be useful simply to predict population persistence under current conditions to assess whether future management is likely to be required, or develop models in order to guide future reintroductions. Regardless of how the model is applied, it ought to be possible to at least envisage how it will be used in management decisions if the purpose of the model development is thought to be conservation related.

Although it is often suggested that population models should be used to compare relative responses to different management strategies (e.g. Beissinger & Westphal, 1998; Reed *et al.*, 2002; McCarthy *et al.*, 2003), the absolute estimates may be equally important. For example, while it is useful to know that λ will be greater under strategy A than strategy B, the decisions taken may be quite different if both are > 1 (adopt the cheapest strategy if growth is the only objective) than if both are < 1 (try a new strategy or move the population to another site). Reluctance to focus on absolute estimates at least partially reflects the realization that estimated probabilities of extinction are subject to tremendous uncertainty unless those probabilities are close to 0 or 1 or the time frames considered are extremely short (Fieberg & Ellner, 2000). However, rather than ignoring absolute estimates, the better approach is to quantify uncertainty around predictions and account for this uncertainty in decision making.

Principles of model building

As noted above, the obvious purpose for building quantitative models for reintroduced populations is to make predictions that can be used to inform management. It needs to be emphasized that the purpose is not to mimic reality – hence Box's (1976) famous quote that 'all models are wrong, but some are useful'. Making models useful inevitably means embracing the KISS principle to some extent, i.e. 'keep it simple, stupid'. This point is especially important for conservation biologists, as we tend to be fascinated by diversity and complexity, so are unsatisfied using models that are gross simplifications of reality even if those models are useful for management decisions.

KISS is not only common sense but is supported by information theory (Akaike, 1973) showing that the best predictions come from models with optimal compromise between simplicity (few parameters) and fit to the data (high likelihood). Alternative models can therefore be fitted to the data, with the best models for making subsequent predictions selected based on information-theoretic criteria (Burnham & Anderson, 2002) or related Bayesian methods (Spiegelhalter *et al.*, 2002). Although there has been considerable heat generated by philosophical differences between frequentists (including the information-theoretic approach) and Bayesians, these differences are largely irrelevant to practical conservation and both approaches can be extremely useful. White (2000a), Williams *et al.* (2002), Maunder (2004) and Conroy & Carroll (2009) provide further information on principles for building effective models for conservation of populations.

Building models for reintroduced populations

The steps for constructing models for reintroduced populations are no different from any other population subject to management, and suggested steps can be found in most introductions to population modelling. Armstrong *et al.* (2002) reviewed steps taken for modelling population dynamics of reintroduced forest birds on New Zealand islands, and the steps below are based on that paper. However, Bakker *et al.* (2009) suggested a similar set of steps with specific reference to the island fox (*Urocyon littoralis*) conservation programme, which also includes reintroductions. Here we suggest 10 steps as follows:

1. Identify key questions.
2. Identify possible management actions.
3. Create conceptual models.
4. Collect relevant data under one or more management actions.

5. Select best model(s) for vital rates based on data.
6. Construct deterministic population models based on models for vital rates.
7. Incorporate uncertainty in parameter estimation and model selection.
8. Incorporate demographic stochasticity.
9. Incorporate environmental stochasticity and catastrophes.
10. Predict distributions of outcomes under possible management scenarios.

We now briefly outline each of these steps, with specific reference to reintroduction.

1. Identify key questions

The key questions surrounding a reintroduction programme should be sorted out well before the reintroduction takes place, as these questions will not only dictate the type of monitoring and management that takes place but whether the reintroduction takes place at all. The obvious question is whether the population is likely to persist under one or more possible management scenarios, but there are many other possible questions. Armstrong & Seddon (2008) proposed 10 general questions that should be considered when planning any reintroduction (Figure 6.3), and some of these are reviewed by McCarthy *et al.* (this volume, chapter 8) in the context of adaptive management.

2. Identify possible management actions

Possible management actions follow directly from the questions identified, but are also constrained by feasibility, affordability and social acceptability. For example, if the key question for a proposed reintroduction was whether the species could co-exist with an exotic predator, management actions could include predator control or no predator control, or different methods or intensities of predator control.

3. Create conceptual models

Conceptual models are explicit ideas about how the systems of interest work based on knowledge of that system as well as general theory. Conceptual models may also be called *hypotheses* (Hilborn & Mangel, 1997), a term that

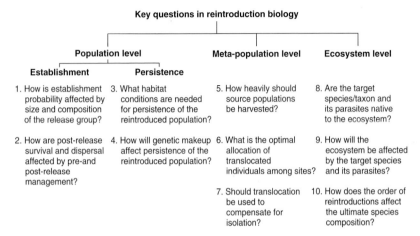

Figure 6.3 **Ten key questions for reintroduction biology, divided into questions at the population, metapopulation and ecosystem level. (Reproduced from Armstrong & Seddon, 2008).**

is used in the next two chapters. The conceptual model-building stage should ideally involve all stakeholders (Holling, 1978); hence models need to be developed in an accessible format. Conceptual models for populations will focus on the factors likely to affect a population's *vital rates*, i.e. survival, reproduction, dispersal and sex allocation (the probability of an individual being male or female). For example, the life cycle diagram in Figure 6.4 represents hypotheses about a population reintroduced to a site with an exotic predator that is controlled by ongoing management. In this scenario the stakeholders believe that the survival and reproduction rates of females are unlikely to depend on the numbers of males present, so they use a female-only model. They also believe that age effects in these rates can be captured by making a simple distinction between first-year and older females. They know that reproduction rates will depend on the density of the predator. However, they are uncertain whether adult or juvenile survival will be affected by the predator, so the diagram represents two or more alternative hypotheses.

Questions to consider at this stage include:

- Should we use a female-only model or include both sexes?
- Should we categorize individuals by age or other life stages?
- Should we include Allee effects and/or other forms of density dependence?

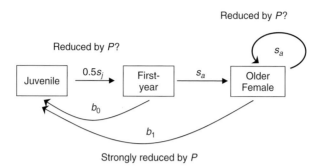

Figure 6.4   A hypothetical female-only life-cycle model proposed before reintroduction of a species affected by an exotic predator. This model assumes a pre-breeding census, meaning the numbers of first-year and older females are counted just before breeding occurs. It is known that reproduction rates (b) will depend on predator density (P), but it is unclear whether the survival rates (s) will be affected by the predator. Females are expected to breed in their first year, but first-year and older females are distinguished because they are expected to have different mean reproduction rates.

- Should we use an individual-based model?
- Should we model dispersal as well as survival and reproduction?
- Should we include metapopulation or group structure?
- Should we use a spatially explicit model?
- Should we include temporal variation (environmental stochasticity)?
- Should we include catastrophes such as cyclones or biological invasions?
- Should we include random individual variations in vital rates?
- How are vital rates likely to be affected by proposed management?
- Are vital rates likely to decline with inbreeding and/or loss of heterozygosity?

Answers to these questions will depend on the system and management problem. However, there will always be uncertainties, and these can potentially be resolved by creating alternative models and confronting these with data (step 5).

4. Collect relevant data under one or more management actions

There are all sorts of data that can potentially be collected, so monitoring should be tailored to address uncertainties (Nichols & Williams, 2006; Nichols

& Armstrong, this volume, Chapter 7). The obvious data to collect are potential indicators of habitat quality at proposed reintroduction sites before release, and then after release continue to monitor key habitat variables as well as state variables of the reintroduced population (e.g. survival, reproduction and abundance). However, data available for other sites and/or species may also be relevant, and models based on such data can be used to make prior population projections for proposed sites with data available or generate probability distributions applicable to any site (Holland *et al.*, 2009). Where it is possible to vary management actions at release sites, decisions need to be made about the timing or spatial distribution of those actions. However, prior indication of the effect of management actions can also be inferred from other available data. The following two chapters deal with these issues in depth so we will not discuss them further here.

5. Select best model(s) for vital rates based on data

Population models are usually comprised of models for the various vital rates, which are also called *demographic rates*. This does not need to be a separate step from constructing a population model (steps 6 to 9 below). In fact, the ideal approach is probably to model all data simultaneously and apply model selection procedures to the population model as a whole rather than to individual components. Barker *et al.* (2009) recently described how all data available for a reintroduced population could be modelled simultaneously using Bayesian updating in WinBUGS (Spiegelhalter *et al.*, 2003), and illustrated how this allowed correlations and uncertainties to be better accounted for than when vital rates were modelled separately. However, less experienced analysts will find it less daunting to model the vital rates first and then combine these submodels into a population model.

Confronting models with data requires converting our conceptual models into mathematical functions (Hilborn & Mangel, 1997). For example, Table 6.1 shows possible models for adult survival, juvenile survival and reproductive success based on the conceptual model shown in Figure 6.4. Such models can be fitted to the available data, allowing three things to happen. First, goodness-of-fit procedures can be used to check that the model gives a reasonable fit to the data and can therefore be used to make reliable inferences. Second, assuming the model has a reasonable fit, the estimates and associated standard errors for the parameters in the model allow predictions to be made.

Table 6.1 **A possible set of generalized linear models for vital rates of a rein-troduced population, and corresponding steps in a female-only population model incorporating demographic stochasticity. In this scenario, mean reproduction rates differ for first-year and older females, and all rates are affected by a management treatment. Alternative models could differ in the parameters included and in the forms of the equations. The log and logit link functions shown here are standard for reproduction and survival respectively as they constrain the vital rates to sensible values (> 0 for reproduction, 0–1 for survival).**

Vital rate	Model for vital rate*	Step in stochastic population model[†]
Mean juveniles produced per female	$\ln(b_A) = \alpha_b + \beta_{bA} * A + \beta_{bM} * M$	$J_{A,t} \sim \text{poisson}\,(F_{A,t} * b_A)$
Pr. juvenile survives to breeding	$\text{logit}(sj) = \alpha_{sj} + \beta_{sjM} * M$	$N_{0,t+1} \sim \text{binomial}\,(J_{0,t} + J_{1,t}, sj)$
Pr. a recruit is female	$\theta = 0.5$	$F_{0,t+1} \sim \text{binomial}\,(N_{0,t+1}, \theta)$
Pr. adult female survives 1 year	$\text{logit}(sa) = \alpha_{sa} + \beta_{saM} * M$	$F_{1,t+1} \sim \text{binomial}\,(F_{0,t} + F_{1,t}, sa)$

*α_b, reproduction parameter for first-year females ($A = 0$) with no management ($M = 0$); β_{bA}, increase in reproduction for older females ($A = 1$); β_{bM}, effect of management on reproduction; α_{sj}, juvenile survival parameter with no management; β_{sjM}, effect of management on juvenile survival; α_{sa}, adult female survival parameter with no management; β_{saM}, effect of management on adult female survival.

[†]$J_{A,t}$, number of juveniles produced by first-year females ($A = 0$) or older females ($A = 1$) in breeding season t; $N_{0,t+1}$, number of first-year individuals next breeding season; $F_{A,t+1}$, number of first-year ($A = 0$) or older ($A = 1$) females next breeding season.

Fitting the models in Table 6.1 to data would give estimates for the seven parameters shown (α_b, β_{bA}, β_{bM}, α_{sj}, β_{sjM}, α_{sa} and β_{saM}), allowing survival and reproduction rates to be estimated with and without management. Third, fitting enables comparison of alternative models, which might have parameters added or removed (e.g. to change the number of age categories) or could have a different functional form.

Models can be compared using Akaike's Information Criterion (AIC) or similar methods that have been derived to maximize predictive value (Akaike,

1973). Spiegelhalter *et al.* (2002) developed a Bayesian analogue called the Deviance Information Criterion (DIC), which allows prior information and random effects to be included in model comparisons. Where the best model is ambiguous, model averaging can be used to obtain parameter estimates that take the relative support for alternative models into account. Burnham & Anderson (2002) provide a comprehensive treatment of model selection and inference, and Johnson & Ormland (2004) give a brief introduction to these concepts.

Fitting models to data usually means finding maximum likelihood estimates for the parameters, i.e. finding the values that maximise the probability of getting the observed data. A variety of statistical software packages are available for doing this, with R (R Development Core Team, 2007) and WinBUGS providing particularly powerful and flexible platforms for model fitting. However, the standard package for analysing survival data is Programme MARK (White & Burnham, 1999), which is freely available and supported by an exceptionally clear online book and other useful information (see http://www.cnr.colostate.edu/~gwhite/mark and http://www.phidot.org). MARK can be used to fit models to many types of mark-recapture data, but the standards for modelling survival are the *live encounters* and *known fate* procedures, both of which require data to be collected in discrete encounter occasions (Nichols & Armstrong, this volume, Chapter 7). It is also possible to use MARK to estimate recruitment (number of new individuals added per individual alive at the last time step). Data on numbers of offspring per female or pair can be modelled using standard statistical packages, but it is best to be able to specify a distribution that is realistic for reproduction data rather than a normal distribution (the Poisson is the standard distribution for modelling reproduction because it constrains the values to be whole non-negative numbers; see appendix E of Williams *et al.*, 2002). It is also useful to be able to fit mixed models incorporating random variation among individual females and/or years. Bayesian inference is particularly amenable to modelling such random effects (King *et al.*, 2010); hence accessible Bayesian software such as WinBUGS is a powerful tool for improving our modelling of reintroduced populations.

The models shown in Table 6.1 include possible effects of age and management, and these are obvious factors to consider when modelling survival and reproduction of any population, as are differences between sexes. However, post-release effects (Box 6.1) also always need to be accounted for when making predictions for reintroduced populations, and here we are assuming that the data used to fit the models in Table 6.1 were not confounded by post-release effects. Depending on the time frame, monitoring frequency and circumstances of the reintroduction, a range of other factors may need to considered, including seasonal variation, random individual variation, inbreeding depression and density dependence including Allee effects (Box 6.2).

6. Construct deterministic population models based on models for vital rates

A *deterministic* population model has no uncertainty included, and therefore makes fixed predictions based on current best estimates of population parameters. When combining vital rates into a population model, it makes sense to start with a deterministic model and then gradually build up complexity by incorporating uncertainties (Ferriere *et al.*, 1996). This stepwise approach is not strictly necessary and may be impossible if using a population modelling package (Box 6.3). However, gradually building complexity can make the process less daunting, make it easier to spot mistakes and helps identify sources of uncertainty in predictions. We illustrate this approach using the hypothetical reintroduction scenario described above (Figure 6.4 and Table 6.1), while Box 6.4 summarizes a real modelling exercise for the reintroduced population of Laysan teal (*Anas laysanensis*) on Midway Atoll.

Box 6.3 Population viability analysis packages

When creating a simulation model for a reintroduced population, the first question may be whether to create purpose-specific code or to use 'canned' software. Several specialized population viability analysis (PVA) packages have been developed, and all of these are potentially applicable to reintroduced populations. VORTEX (Lacy *et al.*, 2005) is the most commonly used PVA package for modelling reintroduced populations, but other packages used include RAMAS Age (Ferson & Akçakaya, 1991), RAMAS GIS (Akçakaya & Root, 2002a), RAMAS Metapop (Akçakaya & Root, 2002b), RISKMAN (Taylor *et al.*, 2006), ULM (Legendre & Clobert, 1995) and ALEX (Possingham & Davies, 1995). Brook *et al.* (1997, 1999) and Keedwell (2004) have made useful comparisons of PVA software, but it is sensible to check the latest online information when comparing packages because they are constantly evolving. For example, while early versions of VORTEX imposed a fixed model structure and had no facility for assessing effects of uncertainty in parameter estimates, the current version allows flexible functions for all rate parameters and has an automated sensitivity testing module.

There are pros and cons to using canned software. The obvious benefits are that it may allow modelling that is beyond the technical capabilities of the users and otherwise may save time by avoiding the need to reinvent the wheel for each modelling exercise. Packages should also be relatively error-free, given that they are designed by experts and rigorously checked. On the other hand, the process of model building clarifies thinking about the problem and is often at least as useful as the predictions made. More importantly, packages can easily be misused if the users do not fully understand the modelling taking place. Conroy & Carroll (2009) provide a detailed discussion of the dangers of using PVA software, and it is notable that Williams *et al.* (2002) make no mention of PVA software in their comprehensive book on analysis and management of wildlife populations.

The key issue with PVA packages is to ensure that the horse is pulling the cart. That is, the models used for reintroduced populations should be developed using the principles and procedures outlined in this chapter, and not on particular features of a software package. Letting software dictate the modelling may lead to essential aspects of the model being omitted or, perhaps more frequently, unnecessary complexity. On the other hand, PVA packages can be extremely useful if they provide a suitable platform for running models appropriate for the situation.

Box 6.4 **Population prediction for Laysan teal reintroduced to Midway Atoll**

In 2004 wild Laysan teal (Figure 6.5) were translocated to Midway Atoll, which is within the species' presumed prehistoric range in the Hawaiian Islands. To assess early population establishment or failure, all founders and their first-generation offspring were monitored using

Figure 6.5   A reintroduced Laysan teal and her first generation brood on Midway Atoll (USGS Photo: J. Breeden).

radio telemetry for two years after the first release to estimate survival and reproduction rates (Table 6.2).

Reynolds *et al.* 2008 used a deterministic matrix model to project population growth from 2007 to 2010 based on the data from the

Table 6.2   **Estimates and standard errors for parameters measured from monitoring data for Laysan teal on Midway Atoll from 2004 to 2006. Post-release effects (Box 6.1) on reproduction were likely during the first breeding season, so reproductive rates are based on the second breeding season only.**

Parameter	Definition	Estimate	SE
b_0	Mean numbers of fledglings per first-year female	0.60	0.30
b_1	Mean numbers of fledglings per older female	3.20	0.41
s_p	Probability post-fledgling survives the next year	0.82	0.03
s_a	Probability adult survives the next year	0.92	0.03

first two years. The combination of radio telemetry monitoring of all individuals and the use of the unbiased Kaplan–Meier estimates of survival parameters greatly reduced the uncertainty associated with parameter estimates. The model assumed a post-breeding census, recognized two age classes (post-fledglings and adults) and applied only to the female portion of the population. A post-breeding census means that the annual time steps in the model run from the end of the breeding season, as is sensible for waterfowl because this is when they are easiest to survey. This conceptual model (Figure 6.6) was converted to a quantitative matrix model using the estimated survival and reproduction rates.

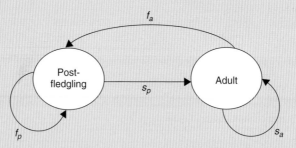

Figure 6.6 A female-only life-cycle model for the reintroduced population of Laysan teal on Midway Atoll. This model assumes a post-breeding census in contrast to the example in Figure 6.4, which assumes a pre-breeding census.

Converting to a matrix model means calculating the fecundities (f) or net reproduction rates for the two age classes (Figure 6.6). Assuming annual mortality takes place before breeding, the fecundity rates are obtained by multiplying the reproduction and survival rates and dividing by two (since only half the fledglings are female). This gives the projection matrix

$$\begin{bmatrix} 0.25 & 1.47 \\ 0.82 & 0.92 \end{bmatrix}$$

which gives a finite rate of increase (λ) of 1.73 and predicts that the population will increase from 56 females in 2006 to about 500 by 2010 (Figure 6.7).

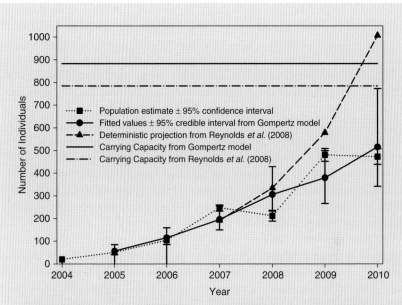

Figure 6.7 **Predicted growth of the reintroduced Laysan teal on Midway Atoll from 2006 to 2010 based on a deterministic model, assuming the population is double the number of females. This is compared to abundance estimates based on numbers of marked and unmarked ducks recorded in subsequent surveys (Reynolds *et al.*, 2011) and a Gompertz function fitted to those estimates.**

The observed growth was lower than that predicted by the matrix model from 2008 onwards. This may have been due to one or more unusual mortality events (181 carcasses were recovered during a botulism outbreak in late 2008) and/or density dependence, which would be expected if the maximum density were similar to the source population on Laysan Island. However, the deterministic matrix projections did not include uncertainty in parameter estimates, nor did they allow for stochasticity. Incorporating uncertainty in parameter estimates gives a 95% prediction interval of 329–826 females in 2010, and adding demographic stochasticity changes this interval to 305–830. Doubling these shows that the 2010 population estimate is not much below the range of projections that would have been made if estimation uncertainty and demographic stochasticity had been included.

Even if a modelling package is going to be used, it is sensible to start with a deterministic model created in a simple format such as a Microsoft Excel® spreadsheet. Spreadsheets are a natural format for population modelling, and for simple scenarios have advantages over other approaches; i.e. they are transparent, easy to code and accessible to most people (White, 2000a). Although Excel® has well-known numerical deficiencies compared to more specialized software (McCullough & Wilson, 2005), the level of accuracy concerned is too fine to affect predictions for reintroduced populations. The main limitations of spreadsheet models are that they are slow to run and unsuitable for complex scenarios such as spatially explicit dynamics. However, spreadsheet modelling can be surprisingly versatile, especially with the additional functions in the PopTools (Hood, 2010) and @RISK (Palisade, 2000) Add-Ins. Here we show how a spreadsheet model can be developed for the scenario shown in Figure 6.4 and Table 6.1.

In a deterministic model, estimates for vital rates are treated as fixed values and there is no uncertainty in population projections. The vital rates are often derived parameters based on two or more other parameters. For example, the estimated mean reproduction rates of first-year females (b_0) and older females (b_1) in Table 6.1 are calculated from the parameters α_b, β_{bA} and β_{bM}. The function shown would allow these rates to be calculated for any level of management intensity (M). For example, when $M = 1$ the estimated mean number of juveniles per first-year female is given by

$$\hat{b}_0 = \exp(\hat{\alpha}_b + \hat{\beta}_{bM})$$

For illustration we will assume that these calculations gave estimates of 1 juvenile per first-year female, 2 juveniles per older female, a 0.4 probability of a juvenile surviving to the next breeding season and a 0.7 probability of an adult surviving one year (Table 6.3). Assuming an initial population of 10 first-year females and 10 older females, a deterministic model (Table 6.3, rows 9 to 13) gives projections of 20.0, 20.8, 21.5, 22.3 and 23.1 females over the next five years. Examining these numbers shows that the model population's rate of increase quickly converges to 1.035.

The estimates could also be expressed as a matrix model, but this requires parameters to be measured over the same time frame. This means combining reproduction and juvenile survival rates into *fecundity* rates, which are the net reproduction rates over a year. With a pre-breeding census (Figure 6.4), these are obtained by multiplying the number of juveniles per female (1 or 2 depending on age class) by the juvenile survival rate (0.4) and dividing by two

Table 6.3 A spreadsheet model in Microsoft Excel, with code for both deterministic and stochastic projections. This quantitative population model is based on the life cycle shown in Figure 6.4, with hypothetical estimates and standard errors given for the four vital rates. The deterministic model (rows 8 to 13) makes fixed projections based on the estimates alone. Uncertainty in estimates is incorporated (rows 15 to 20) by sampling the four rates from distributions each time the model is run, with mean reproduction rates taken to be lognormally distributed (cells D3:G4) and survival probabilities taken to be logit-normally distributed (cells D5:G6). Demographic stochasticity in survival, reproduction and sex allocation is incorporated (rows 22 to 27) by sampling from binomial distributions (cells C25:C26) and gamma approximations of the Poisson distributions (cells C23:C24; the PopTools function dpoissondev can also be used). Models can be run multiple times by pressing the F9 key on the keyboard and distributions of outcomes can be obtained by writing a macro that activates the Calculate function multiple times and collates the results. Only the first year is shown for simplicity, but cells C8 to C27 can be dragged to the right to get projections for as many years as needed.

	A	B	C	D	E	F	G
				REAL		TRANSFORMED	
1	PARAMETER						
2		Est.	SE	Random	Est.	SE	Random
3	Mean juvs per 1st-yr fem.	1.00	0.10	=EXP(G3)	=LN(B3)	=C3/B3	=NORMINV(RAND(),E3,F3)
4	Mean juvs per older fem.	2.00	0.20	=EXP(G4)	=LN(B4)	=C4/B4	=NORMINV(RAND(),E4,F4)
5	Pr. juv. surv. to breeding	0.40	0.08	=EXP(G5)/(1+EXP(G5))	=LN(B5/(1-B5))	=C5/(B5*(1-B5))	=NORMINV(RAND(),E5,F5)
6	Pr. ad. fem. surv. 1 year	0.70	0.10	=EXP(G6)/(1+EXP(G6))	=LN(B6/(1-B6))	=C6/(B6*(1-B6))	=NORMINV(RAND(),E6,F6)
7							

(Continued)

Table 6.3 *(Cont'd)*

	A	B	C	D	E	F	G
8	**DETERMINISTIC**	*Year 0*	*Year 1*				
9	No. fledged by 1st-yr. fem.		=B11*B3				
10	No. fledged by older fem.		=B12*B4				
11	No. fem. recruits	10	=SUM(C9:C10)*B5*0.5				
12	No. ad. fem. survivors	10	=B13*B6				
13	Total females	=SUM(B11:B12)	=SUM(C11:C12)				
14							
15	**UNCERTAINTY**	*Year 0*	*Year 1*				
16	No. fledged by 1st-yr. fem.		=B18*D3				
17	No. fledged by older fem.		=B19*D4				
18	No. fem. recruits	10	=SUM(C16:C17)*D5*0.5				
19	No. ad. fem. survivors	10	=B20*D6				
20	Total females	=SUM(B18:B19)	=SUM(C18:C19)				
21							
22	**DEM. STOCH.**	*Year 0*	*Year 1*				
23	No. fledged by 1st-yr. fem.		=IF(B25*D3>0,ROUND(GAMMAINV(RAND(),B25*D3,1),0),0)				
24	No. fledged by older fem.		=IF(B26*D4>0,ROUND(GAMMAINV(RAND(),B26*D4,1),0),0)				
25	No. female recruits	10	=CRITBINOM(SUM(C23:C24),D5*0.5,RAND())				
26	No. ad. fem. survivors	10	=CRITBINOM(B27,D6,RAND())				
27	Total females	=SUM(B25:B26)	=SUM(C25:C26)				

(assuming half of the surviving juveniles are female). This yields the projection matrix

$$\begin{bmatrix} 0.2 & 0.4 \\ 0.7 & 0.7 \end{bmatrix}$$

where the fecundity rates are shown on the top (first-year females on the left and older ones on the right) and the adult survival probabilities are shown on the bottom. The finite rate of increase (λ) is given by the dominant eigenvalue of the matrix, which can be confirmed to be 1.035 using the EigVal function in PopTools.

7. Incorporate uncertainty in parameter estimation and model selection

Population projections are full of uncertainties, so while we recommend starting the model-building process with a deterministic model, we caution that it is dangerous to use deterministic models for management decisions. Uncertainty can be divided into two forms: *stochasticity*, which is the inherent variability of the system, and *incertitude*, which is uncertainty attributable to our imperfect understanding of that system. Incertitude can in turn be divided into uncertainty in parameter estimates and uncertainty in model selection.

A simple way to incorporate uncertainty in parameter estimates is to sample parameter values randomly from distributions defined by their means and standard errors (White, 2000b). The second spreadsheet model in Table 6.3 (rows 16 to 20) illustrates how this can work, as the vital rates are sampled from distributions in each run of the model. The standard errors for these distributions (Table 6.3, cells C3:C6) were assigned arbitrarily, but represent fairly precise estimation.

Standard errors for derived parameters are easy to obtain if using Bayesian inference to analyse the data (McCarthy, 2007; King *et al.*, 2010). Otherwise, the delta method (Williams *et al.*, 2002, appendix F) can be used to calculate approximate standard errors for derived parameters based on the estimates, standard errors and covariances of the original parameters. Under the delta method, the approximate standard error for a function of two parameters x and y is given by

$$SE(g[x,y]) = \sqrt{\left(SE(x)\frac{\partial g[x,y]}{\partial x}\right)^2 + \left(SE(y)\frac{\partial g[x,y]}{\partial y}\right)^2 + 2cov(x,y)\frac{\partial g[x,y]}{\partial x}\frac{\partial g[x,y]}{\partial y}}$$

As shown above, when the management intensity (M) is set to 1 in our hypothetical scenario, the reproduction rate of first-year females (b_0) is derived from parameters α_b and β_{bM}. Applying the delta method gives

$$SE(\hat{b}_0) = \exp(\hat{\alpha}_b + \hat{\beta}_{bM})\sqrt{(SE(\hat{\alpha}_b))^2 + (SE(\hat{\beta}_{bM}))^2 + 2cov(\hat{\alpha}_b, \hat{\beta}_{bM})}$$

It is actually better to do the delta approximation before backtransforming the vital rates (e.g. calculate the standard error for $\ln(\hat{b}_0)$ rather than $\hat{b}_0$ in this example). However, showing real values for vital rates on the spreadsheet makes it easier to explain the model to other people.

Uncertainty in model selection can be incorporated using some form of model averaging (Burnham & Anderson, 2002). This can be done in two different ways although the results should be similar. The first approach is to incorporate model averaging into the estimates and standard errors for the vital rates. For example, the estimates and standard errors in Table 6.3 could already have been obtained using model averaging rather than being based on a single model. The second approach is to average the projections from more than one population model. For example, if population model A had twice the weight of model B based on model selection procedures, then projections could be obtained where there was a two-third chance of model A being used in each run and a one-third chance of model B being used.

8. Incorporate demographic stochasticity

Demographic stochasticity refers to the inherent uncertainty in births, deaths and sex allocation. For example, while exactly 14 out of 20 adults survive the first year of the deterministic version of the spreadsheet model in Table 6.3, adding demographic stochasticity would mean that each individual has a 70% chance of surviving the year. Demographic stochasticity in survival is modelled by sampling the numbers of adult and juvenile survivors from binomial distributions each year, and the number of females among the surviving juveniles is sampled in the same way (Tables 6.1 and 6.3). Modelling demographic stochasticity in reproduction is less straightforward, but the standard method is to sample the number of offspring from a Poisson distribution (Tables 6.1 and 6.3).

Demographic stochasticity is the easiest form of uncertainty to incorporate into population models, mainly because no data are required to model it.

However, we recommend focusing on uncertainty in parameter estimation first because that step is essential for obtaining realistic distributions of predictions. The effects of demographic stochasticity on population projections will be relatively minor in comparison to parameter estimation error in some circumstances. For example, demographic stochasticity has a relatively minor effect on projections for the reintroduced Laysan teal on Midway Atoll due to the rapid growth of that population (Box 6.4). However, it is critical to include demographic stochasticity when modelling small populations where λ is close to 1 (Figures 6.2 and 6.8).

9. Incorporate environmental stochasticity

Environmental stochasticity refers to uncertainty caused by variation in mean vital rates due to environmental conditions, and can include both spatial and temporal variation. Environmental stochasticity can be added to population models by sampling vital rates from distributions each year rather than considering rates to be fixed over time (White, 2000a). This is similar to the way that parameter uncertainty is incorporated (Table 6.3) except that

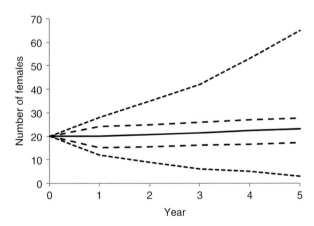

Figure 6.8   Uncertainty in population projections from the spreadsheet models in Table 6.3. The solid line shows the projection from the deterministic model, the long-dash lines shows the 95% prediction intervals when uncertainty in parameter estimates is added and the short-dash lines show the 95% predictions intervals when demographic stochasticity is also added (prediction intervals are based on 10 000 runs of the model).

the random sampling occurs each year rather than at the start of each run. In some systems it is important to model *catastrophes*, which are infrequent but extreme forms of stochasticity that dramatically impact populations (Caughley, 1994). For example, populations restricted to low-lying coastal habitats may be vulnerable to tsunamis, small islands with busy airports or harbours are vulnerable to exotic invasions and isolated populations may be particularly vulnerable to new pathogens. Historical disaster occurrences and subsequent extinction rates or other known population impacts can be applied.

Although adding environmental stochasticity is technically straightforward, the data requirements for estimating random temporal or spatial variations are onerous. It is essential to distinguish *process variation* caused by environmental stochasticity from *sampling variation* (White, 2000a, 2000b), and while the methods for doing so have advanced greatly in recent years (King *et al.*, 2010), the data requirements have not. As it is unlikely that random temporal or spatial variation can be estimated in most reintroduced populations, it will usually be necessary to base variance estimates on meta-analysis of long-term datasets.

10. Predict distributions of outcomes under possible management scenarios

The types of predictions made from models will depend on the situation, as discussed above. However, the essential feature of predictions used for management decisions is that they incorporate uncertainty. For example, the projection from a deterministic model (Table 6.3, rows 9 to 13) that a reintroduced population would increase from 20 to 23 females over five years does not capture the necessary information. Adding parameter estimation uncertainty (Table 6.3, rows 16 to 20) gives a 95% prediction interval of 17–28 females after 5 years, and adding demographic stochasticity (Table 6.3, rows 23 to 27) widens this interval to 3–66 females (Figure 6.8). Management decisions would therefore need to be based on predictions incorporating both these forms of uncertainty. This hypothetical scenario focused on management of an exotic predator likely to threaten the reintroduced population (Figure 6.4), so predictions under different levels of management intensity (M) would be made by deriving estimates of vital rates (and associated standard errors) using the models shown in Table 6.1.

The modelling scenario we have chosen here is deliberately simple, with just enough detail to make it realistic. However, similar spreadsheet models have been used to predict viability of reintroduced populations as a function of predator control (Armstrong *et al.*, 2006) and supplementary feeding (Armstrong *et al.*, 2007), and the spreadsheets were only slightly more complicated than those shown here. In many cases, creating the population model (steps 6 to 10) is a minor part of the modelling process, and is much less complex and time consuming than collecting the data (step 4) and modelling the vital rates (step 5). However, in other cases the population models will need to be much more complex, especially if individual-based and/or spatially explicit modelling is needed (see step 3), and these are the situations in which specialized software is most likely to be useful (Box 6.3).

Modellers may also prefer to make predictions using a purely analytical approach (e.g. Holland *et al.*, 2009) rather than the simulation approach employed here or in PVA packages (Box 6.3). Analytical models are more elegant and more transparent, at least to people with sufficient mathematical knowledge to understand them. However, simulations can model processes that are mathematically intractable, e.g. where dynamics may depend on spatially explicit individual dispersal behaviour, and are generally more accessible.

Published literature on modelling reintroduced populations

Although the majority of models involving reintroductions have probably not been published, we expect the published literature to capture significant new developments and therefore to represent the state-of-the-art up to about two years ago. We searched web-of-science up to the end of 2010 for the terms. . .

reintroduc* OR re-introduc* OR translocat*
AND
population model* OR population viability*

and then repeated this search with the term 'metapopulation' substituted for 'population'. We then examined all of the papers and retained those where: (1) models were fitted to data; (2) these models were used to predict establishment, growth, persistence or spread of either existing or prospective populations; and (3) at least one of the populations was reintroduced or otherwise subjected to translocation to or from the population. We did not include papers that modelled vital rates without combining them into a population model or that modelled captive populations only. We also excluded papers using hypothetical scenarios to demonstrate theory relevant to modelling reintroduced populations (Robert et al., 2004; Tenhumberg et al., 2004; Crone et al., 2007; Rout et al., 2007, 2009; Mihoub et al., 2009), although the distinction is not straightforward since all of these authors used real data in their examples. These papers should clearly be taken into account in future modelling.

Of the 89 papers found (Table 6.4), 83% were published in the last decade (2001–2010), indicating a new and developing literature. Most papers focused on translocation programmes in Europe (32), North America (23) or New Zealand (14), with others focused on programmes in Australia (6), South Africa (6), the Middle East (4), China (1), Hawaii (1), Madagascar (1) and Mongolia (1). The majority involved mammals (46) or birds (29), but other papers involved lizards (3), tuatara (1), tortoises (2), frogs (2), fish (2), butterflies (1) or plants (3).

The selection of papers nicely illustrates the ways in which quantitative population models can be used to guide reintroduction programmes. The majority of papers focused on predicting growth, persistence and/or spread of existing reintroduced populations, whereas a smaller number focused on prospective reintroduced populations or impacts to source populations. Many of the papers modelled the effects of key habitat factors, such as food availability, predation risk and human persecution, and/or modelled the effects of potential management actions, such as food supplementation, predator control, ongoing translocation to compensate for habitat isolation or alternative release strategies. Surprisingly only one paper (Armstrong et al., 2007) used an active adaptive management approach whereby management was deliberately manipulated over time to gain information to improve predictions, and no papers reported use of passive adaptive management (McCarthy et al., this volume, Chapter 8). Rout et al. (2009) also demonstrated how formal active adaptive management could be applied in a hypothetical scenario.

The majority of papers (46) used non-spatial single-population stochastic models, but there were also 28 papers using models that were spatially explicit and/or had meta-population or group structure. A smaller number (5) estimated population growth rates (r or λ) without explicitly simulating dynamics, but several of the papers using stochastic models also derived point estimates or distributions for growth rates. There were also simpler single-population deterministic models, one diffusion model and one model predicting occupancy of quadrats. The population models included a mix of age- or stage-based models and individual-based models. Many models

Table 6.4 Summary of published papers up to 2010 using population models to make predictions about reintroduced populations or other populations subject to translocation. Papers are listed in alphabetical order, with * indicating papers that incorporated uncertainty in parameter estimation into the model predictions.

Paper	Species/Region	Model	Application
Akçakaya et al., 1995	Helmeted honeyeater (Australia)	Spatially explicit meta-population model using RAMAS GIS	Evaluate effectiveness of translocation as management strategy
Armstrong & Ewen, 2001a	North Island robin (New Zealand)	Single-population model using VORTEX	Assess impact of poison drop (for rat eradication) on reintroduced population
Armstrong & Ewen, 2001b*	North Island robin (New Zealand)	Single-population model using VORTEX	Post hoc evaluation of usefulness of follow-up translocation
Armstrong & Ewen, 2002*	North Island robin (New Zealand)	Single-population model using VORTEX	Predict viability of small island population
Armstrong et al., 2002*	Forest birds (New Zealand)	Single-population stochastic models	Summary of use of population models for management of reintroduced pops
Armstrong et al., 2005	North Island saddleback (New Zealand)	Single-population stochastic spreadsheet model	Understand population regulation
Armstrong & Davidson, 2006	North Island saddleback (New Zealand)	Single-population stochastic spreadsheet model	Project growth of new population as function of predator control based on data from other populations
Armstrong et al., 2006*	North Island robin (New Zealand)	Single-population stochastic spreadsheet model	Population projections as function of predator control

(*Continued*)

Table 6.4 (*Cont'd*)

Paper	Species/Region	Model	Application
Armstrong et al., 2007 *	Hihi (New Zealand)	Single-population stochastic spreadsheet model	Population projections under alternative management strategies involving mite control and supplementary feeding
Bach et al., 2010	African wild dog (South Africa)	Single-population model using VORTEX	Predict persistence at proposed reintroduction site based on other data, considering inbreeding depression
Bakker et al., 2009*	Island fox (USA)	Single-population stochastic model	Projections for both remnant and reintroduced populations as function of eagle predation
Bar-David et al., 2008	Persian fallow deer (Israel)	Spatially explicit individual-based model	Predict effect of planned development on expansion of population
Bretagnolle & Inchausti, 2005	Little bustard (France)	Spatially explicit metapopulation model using RAMAS Metapop	Predict effect of restocking
Brook et al., 2002	Carpentarian rock-rat (Australia)	Single-population stochastic spreadsheet model	Predict viability in relation to fire management
Bustmante, 1996	Bearded vulture (Spain)	Single-population model using VORTEX	Predict risk to captive source population, considering inbreeding depression
Cade, 2007	California condor (USA)	Single-population deterministic model	Predict impact of lead shot
Carroll et al., 2003	Grey wolf (USA)	Spatially explicit metapopulation model	Develop regional reintroduction strategy

Chapron et al., 2009	Pyrenean brown bear (France)	Single-population stochastic model using ULM	Predict effect of adjusting sex ratio through translocation
Clark & Eastridge, 2006	Black bear (USA)	Single-population model using RISKMAN	Impact of translocation and hunting on source population
Cramer & Portier, 2001	Florida panther (USA)	Spatially explicit individual-based model	Assess potential reintroduction sites
Curtis & Naujokaitis-Lewis, 2008*	Sand lizard (Sweden)	Meta-population model using RAMAS Metapop	Predict extinction probability as function of spatial parameters and vital rates
Dimond & Armstrong, 2007	North Island robin (New Zealand)	Single-population stochastic spreadsheet model	Project size of source population as function of harvesting
Eastridge & Clark, 2001	Black bear (USA)	Single-population deterministic model using RISKMAN	Project effect of follow-up translocation
Evans et al., 2009	White-tailed eagle (Scotland)	Single-population model using VORTEX	Predict growth as function of number wild and released birds in population
Fernandez et al., 2006	Wild boar (Denmark)	Spatially explicit individual-based model	Assess reintroduction sites with respect to disease risk from piggeries
Gilad et al., 2008	Arabian oryx (Israel, Oman)	Metapopulation model using STELLA	Validate model through retrospective predictions and project growth for prospective reintroductions
Green et al., 1996	White-tailed eagle (Scotland)	Single-population stochastic model	Project growth

(Continued)

Table 6.4 (*Cont'd*)

Paper	Species/Region	Model	Application
Gusset *et al.*, 2009	African wild dog (South Africa)	Individual-based pack-structured model	Predict persistence as function of translocation to add packs and harvest for reintroduction elsewhere
Haig *et al.*, 1993	Red-cockaded woodpecker (USA)	Single-population individual-based model using VORTEX	Predict effect of supplementation on viability under hypothetical inbreeding depression
Haines *et al.*, 2006	Ocelot (USA)	Spatially explicit model in RAMAS GIS	Predict impact of supplementation on viability of existing meta-population
Haydon *et al.*, 2008	Manitoban elk (Ontario)	Spatially explicit individual-based stochastic model	Predict growth and spread as function of movement behaviour
Holland *et al.*, 2009*	European wild boar (Europe)	Estimate growth rate from meta-analysis of vital rates	Distribution of population trajectories for a random reintroduction site
Howells & Edwards-Jones, 1997	Wild boar (Scotland)	Single-population model using VORTEX	Assess feasibility of reintroduction to Scotland based on data from Poland
Kauffman *et al.*, 2003*	Peregrine falcon (USA)	Estimate population growth rate	Estimate growth in urban versus rural habitats
Kauffman *et al.*, 2004	Peregrine falcon (USA)	Matrix models connected by dispersal	Predict effect of source-sink dynamics on the meta-population

Reference	Species	Model	Aim
Kinley & Newhouse, 2008	American badger (Canada)	Population growth rate estimated from vital rates	Translocated animals to assess whether a local decline due to temporary variation in habitat quality
Kohlmann et al., 2005	Island fox (USA)	Single-population model using VORTEX	Predict impact of translocation on source population
Kramer-Schadt et al., 2005	European lynx (Germany)	Spatially explicit individual-based model	Predict viability of potential reintroduction
Leaper et al., 1999	Wild boar (Scotland)	Single-population model using RAMAS Age	Predict probability of persistence at prospective release sites as function of number released
Leech et al., 2007	Rifleman (New Zealand)	Single-population deterministic model	Estimate growth rate
Lennartsson, 2002	Gentianella campestris (Sweden)	Single-population stochastic models	Viability of six reintroduced populations as function of inbreeding depression
León-Cortés et al., 2003	Hamearis lucina (Wales)	Meta-population model	Predict cause of original extinction as function of meta-population dynamics and hence potential for reintroduction
Liu et al., 2004	Wild rice (China)	Quadrat spread model	Predict growth of reintroduced population as function of habitat characteristics
Lloyd et al., 2009*	Brown-headed nuthatch, eastern bluebird (USA)	Population growth rate estimated using Pradel model in MARK	Estimate population growth

(Continued)

Table 6.4 (*Cont'd*)

Paper	Species/Region	Model	Application
Macdonald et al., 2000	European beaver (Scotland)	Spatially explicit individual-based stochastic model (same as used by South et al., 2000)	Predict spread of reintroduced populations based on data from literature
Mackey et al., 2009	African elephant (South Africa)	Population growth rate estimated from vital rates	Predict reduction in growth rate through immunocontraception
Marshall & Edwards-Jones, 1998	Capercaillie (Scotland)	Single-population model using VORTEX	Estimate area needed to support a viable population if species reintroduced
McCallum et al., 1995	Bridled nailtail wallaby (Australia)	Single-population stochastic model	Predict effect of release number on persistence under a random quota predation
McCleery et al., 2005	Key Largo woodrat (USA)	Single-population stochastic model using RAMAS Metapop	Predict persistence as function of supplementation
Meriggi et al., 2007	Grey and red-legged partridges (Italy)	Single-population models using VORTEX	Predict persistence as function of release strategy, immigration and population parameters
Mitchell et al., 2010	Tuatara (New Zealand)	Single-population stochastic model using VORTEX	Probability of extinction due to skewed sex ratio expected with climate change
Münzbergová et al., 2005	*Succisa pratensis* (Sweden)	Spatially explicit meta-population model	Assess optimal reintroduction strategy for maintaining meta-population

Muths & Dreitz, 2008	Wyoming toad (USA)	Single-population deterministic model	Estimate number released per year to achieve target population
Nolet & Baveco, 1996	European beaver (Netherlands)	Single-population stochastic model	Predict persistence with and without a post-release effect on reproduction
Osbourne, 2005	Great bustard (England)	Single-population deterministic model	Project growth as function of number of eggs sourced to reintroduce population
Pedrono et al., 2004	Ploughshare tortoise (Madagascar)	Meta-population model using ULM	Predict persistence of wild-captive meta-population as function of reintroductions from captive stock
Pergams et al., 2000	Anacapa deer mouse (USA)	Meta-population model using VORTEX	Develop reintroduction strategy in case population eradicated, considering inbreeding depression
Perkins et al., 2008*	Florida grasshopper sparrow (USA)	Spatially explicit meta-population model using RAMAS GIS	Estimate probability of extinction under different management options
Reed et al., 2009	Desert tortoise (USA)	Single-population deterministic model	Project growth under alternative management actions including restocking
Reynolds et al., 2008	Laysan teal (Hawaii)	Single-population deterministic model	Project growth of reintroduced population
Roth et al., 2008	Red wolf (USA)	Single-population stochastic model	Predict effect of coyote control on prospective red wolf reintroductions
Saether et al., 2002*	Ibex (Switzerland)	Single-population stochastic models fitted to population estimates	Project dynamics of a reintroduced population

(Continued)

Table 6.4 (*Cont'd*)

Paper	Species/Region	Model	Application
Saether *et al.*, 2007	Ibex (Switzerland)	Single-population stochastic population models fitted to population estimates	Project dynamics of 23 reintroduced populations
Saltz *et al.*, 2006	Asiatic wild ass (Israel)	Single-population stochastic model	Predict effect of increased environmental stochasticity expected under global climate change
Schadt *et al.*, 2002	European lynx (Europe)	Spatially explicit individual-based model	Predict viability of potential reintroduction sites
Schaub *et al.*, 2004	White stork (Switzerland)	Single-population stochastic model	Estimating λ
Schaub *et al.*, 2009	Bearded vulture (Europe)	Single-population stochastic population	Population projections with and without ongoing follow-up translocations
Slotta-Bachmayr *et al.*, 2004	Przewalski's horse (Mongolia)	Single-population model using VORTEX	Predict persistence and optimize release strategies
Smart *et al.*, 2010	Red kite (Scotland)	Single-population model using VORTEX	Project population growth with and without illegal killing
Somers, 1997	Warthog (South Africa)	Single-population model using VORTEX	Predict impact on source population from translocation
South *et al.*, 2000	European beaver (Scotland)	Spatially explicit individual-based stochastic model in C; meta-population model using VORTEX	Predict persistence as function of release strategy based on data from literature

Reference	Species (location)	Model type	Purpose
South et al., 2001	European beaver (England)	Spatially explicit individual-based stochastic model (same as used by South et al., 2000)	As for above but for new site in England
Southgate & Possingham, 1995	Bilby (Australia)	Meta-population model using ALEX	Predict persistence as function of spatial configuration of reintroduction sites
Steury & Murray, 2004	Canada lynx (USA)	Single-population deterministic and stochastic models	Predict persistence at potential reintroduction sites as function of hare density release strategy
Strauss, 2002	Arabian oryx (Saudi Arabia)	Single-population and meta-population models using VORTEX	Estimate probability of persistence
Swart & Lawes, 1996	Samango monkey (South Africa)	Spatially explicit stochastic meta-population model	Predict effect of translocation or corridors to mitigate fragmentation
Tocher et al., 2006	Hamilton's frog (New Zealand)	Single-population deterministic model	Balance risk to source population and potential reintroduced population based on numbers and age distribution
Todd et al., 2002*	Eastern barred bandicoot (Australia)	Single-population stochastic model	Estimate of removals on a source population that went extinct
Todd et al., 2004	Trout cod (Australia)	Single-population stochastic model	Population projection including effects of potential supplementation strategies
Towns & Ferreira, 2001	Four skink species (New Zealand)	Single-population deterministic model	Project growth as function of number released

(Continued)

Table 6.4 (*Cont'd*)

Paper	Species/Region	Model	Application
Towns et al., 2003	Suter's skink (New Zealand)	Single-population models using VORTEX	Predict persistence as function of rat predation
Vandel et al., 2006	European lynx (France)	Spatially explicit stochastic population model	Retrospective predictions of model to observed spread
Vincenzi et al., 2008*	Marble trout (Slovenia)	Single-population stochastic model	Project persistence of translocated populations
Wakamiya & Roy, 2009	Peregrine falcon (USA)	Meta-population model using RAMAS GIS	Compare population predictions and costs under several proposed reintroduction scenarios
Watson & Chadwick, 2007	Cape mountain zebra (South Africa)	Diffusion model fitted to population estimates	Effect of translocation among three isolated populations
Wear et al., 2005	Black bear (USA)	Single-population stochastic model using RISKMAN	Estimate λ and effect of supplementation and poaching
Wiegand et al., 2004a*	Brown bear (Austria/Slovenia)	Spatially explicit individual-based model	Project growth and spread
Wiegand, et al., 2004b*	Brown bear (Austria/Slovenia)	Spatially explicit individual-based model	Calibrate predictions with population-level data to reduce uncertainty
Wood et al., 2007	Tree squirrels (general)	Single-population model using VORTEX	Predict effect of release on establishment probability of hypothetical populations

were individual-based by default simply because they used the VORTEX package, which runs individual-based simulations to facilitate modelling of inbreeding depression (Lacy *et al.*, 2005). However, at least 14 papers made use of individual-based frameworks to model inbreeding or spatial and/or social dynamics, and these are labelled 'individual-based' in Table 6.4. In general, these differences in model type reflected sensible choices with respect to the problem, i.e. the questions being asked, characteristics of the species and release area, and limitations of the data.

Close to half (34) of the papers used software specifically designed for population or meta-population viability analysis (VORTEX, RAMAS, RISKMAN, ULM, ALEX; see Box 6.3), with VORTEX used most commonly. Other software used to model population dynamics included Microsoft Excel®, STELLA® (High Performance Systems, 2001), SMALLTALK (Baveco & Smeulders, 1994), SAS® (SAS Institute, 2005), MATLAB (MathWorks, 1992), Mathematica (Wolfram Research, 1996), R and C++, although software was often not mentioned. Program MARK (White & Burnham, 1999) was frequently used for modelling survival, whereas reproduction was usually modelled with standard statistical software such as SAS or R. Choice of software obviously reflected differences among authors in modelling experience and familiarity with particular packages, but in most cases the software used allowed construction of models that appeared sensible for the problem.

The papers varied greatly in the apparent reliability of methods used for model selection and parameter estimation, and in the degree to which they accounted for uncertainty. In many of the earlier studies, a particular model structure was assumed a priori without clear justification and/or parameter values were plugged in based on the literature or educated guesses. Such models can play a useful role, but should be regarded as semi-quantitative exploratory exercises rather than predictive quantitative models. In contrast, most recent studies have had a clear theoretical framework, and used model selection and parameter estimation procedures along the lines recommended in this chapter.

Uncertainty in parameter estimation was explicitly incorporated into model predictions in 18 of the 89 papers, and these are indicated in Table 6.4. A few papers used a semi-quantitative approach where they compared predictions under parameter values deemed to be low, medium or high with respect to the plausible ranges (e.g. South *et al.*, 2000; Brook *et al.*, 2002; Tocher *et al.*, 2006). In addition, many papers assessed sensitivity of predictions to perturbation of vital rates (e.g. by changing the estimated values by 10%) but without reference to the standard errors of the estimates (e.g. Gilad *et al.*, 2008; Reynolds *et al.*, 2008; Gusset *et al.*, 2009; Bach *et al.*, 2010). Among the papers that explicitly incorporated estimation uncertainty into their predictions, Wiegand *et al.* (2004b), Vincenzi *et al.* (2008), Curtis & Naujokaitis-Lewis (2008), Bakker *et al.* (2009) and Holland *et al.* (2009) provide particularly sophisticated methodologies. Curtis & Naujokaitis-Lewis' (2008) paper is especially notable, as they developed software for generated random input parameters (GRIP) in conjunction with the PVA package RAMAS Metapop, providing an accessible methodology for incorporating uncertainty in spatial parameters as well as vital rates.

The literature summarized in Table 6.4 provides a wealth of information on methods for modelling reintroduced populations and using these models to address important questions. However, there is a striking lack of cross-referencing among these papers, and some authors even made statements to the effect that reintroduced populations were rarely monitored and rarely modelled, and failed to cite any of the earlier papers. We hope that this summary will encourage researchers to build on this existing body of literature when constructing future models for reintroduction.

Future directions

Future directions in the modelling of reintroduced populations will partially reflect advances in population modelling in general, as well as shifts in focus of species recovery programmes. We particularly expect to see greater emphasis on integrated Bayesian models, where all data are modelled simultaneously (e.g. Barker *et al.*, 2009), and increased used of Bayesian approaches to population modelling in general (King *et al.*, 2010). We also expect a widening taxonomic and life history breadth in population modelling. For example, Menges (2008) has advocated a much greater role of population modelling in plant reintroduction programmes.

We also see future modelling directions specific for improving reintroduction programmes and will conclude this chapter with five suggested directions in which to focus further research.

1. Long-term genetic effects on reintroduced populations

As illustrated in Chapters 11 to 13 of this volume, reintroductions that appear successful in the short term may not be viable in the long term due to inbreeding and loss of genetic diversity. Some of the papers reviewed in this chapter have modelled the consequences of inbreeding depression, but in most cases the effects of inbreeding were set to default values (e.g. Haig *et al.*, 1993; Bustmante, 1996; Bach *et al.*, 2010) or based on related species (Pergrams *et al.*, 2000). Lennartsson's (2002) research on reintroductions of the herb *Gentianella campestris* in Sweden appears to be the only case where inbreeding depression was empirically estimated (using pollinaton experiments) and this estimate was then used to model the population consequences. Jamieson & Lacy (this volume, Chapter 13) also cite an unpublished example involving North Island robins (*Petroica longipes*) in New Zealand. However, there is clearly more scope for long-term modelling of reintroduced populations using empirically reliable estimates of inbreeding depression. The difficulties associated with long-term population predictions (Beissinger & Westphal, 1998) mean that particular care will need to be taken in modelling uncertainties in these scenarios.

2. Integrated modelling for whole reintroduction programmes

The papers reviewed here tend to focus on predicting impacts on source populations, predicting outcomes of reintroductions at the planning stage or predicting future outcomes after data have been collected from the reintroduced population. There appears to be a lack of modelling that integrates these elements, although the piecemeal nature of scientific publication may exaggerate this impression. There is obvious scope to produce models at the planning stage that are later updated through Bayesian inference when post-release data become available, and this approach should facilitate greater use of adaptive management (McCarthy *et al.*, this volume, Chapter 8).

3. Integrated modelling using data from multiple programmes

Reintroduction biology has largely consisted of isolated case studies until recently (Seddon *et al.*, 2007), and this may be reflected in the lack of cross-referencing in the modelling papers reviewed here (Table 6.4). Integrating information among projects not only allows understanding to increase over time but also allows predictions to be made when data for the current reintroduction are sparse or absent. Papers modelling reintroductions at the planning stage have compensated for the lack of data by using data from the literature and other sources, but generally the approaches have been somewhat ad hoc and the validity of the predictions impossible to assess. In contrast, Holland *et al.* (2009) showed how meta-analysis of data from wild boars (*Sus scrofa*) could be used to generate reliable probability distributions applicable to any new reintroduction. McCarthy *et al.* (this volume, Chapter 8) show how data from multiple species can be integrated in a meta-analysis with species and case study included as random effects.

There is also potential to integrate different data types. For example, while most of the modelling studies reviewed here used data on vital rates, in a few cases models were fitted to abundance estimates only (Saether *et al.*, 2002, 2007; Watson & Chadwick, 2007). It is possible to model these two types of data simultaneously (Brooks *et al.*, 2004), meaning that meta-analyses could include different types of monitoring data. There may also be potential to integrate vital rate data with species distribution modelling based on occupancy data (Osborne & Seddon, this volume, Chapter 3) or studies modelling reintroduction success as a binary variable (e.g. Griffith *et al.*, 1989).

4. Modelling conservation introductions

Most of the papers reviewed in this chapter have involved reintroductions, i.e. attempts to re-establish species within their historic ranges (Seddon *et al.*, this volume, Chapter 1). An exception is Vincenzi *et al.*'s (2008) study, which involved modelling marble trout *Salmo marmoratus* introductions to previously fishless streams. Although motivated by conservation of a threatened species, we would expect such introductions to be controversial given the possible ecosystem impacts. Species may also be introduced completely outside their historic ranges for conservation purposes. Although this is only justifiable in particular circumstances, these circumstances may be becoming more common. In particular, proposals to introduce species outside their historic ranges in response to climate change have recently become topical and controversial (Dawson *et al.*, 2011). Sensible decisions about such controversial actions require reliable model predictions, but making predictions about populations introduced into changing climates and ecosystems outside their historic ranges is going to be extremely challenging.

5. Modelling within structured decision frameworks

The last future direction we want to emphasize is that models increasingly need to be done within the context of structured decision frameworks that explicitly define objectives, describe uncertainty, and weigh costs and benefits of possible outcomes. This is particularly important for controversial decisions such as proposed conservation introductions in response to climate change.

However, decision frameworks also make it possible explicitly to integrate the range of factors that need to be considered when making decisions concerning all translocations. This includes cost effectiveness, which Wakamiya & Roy (2009) considered when modelling reintroduction strategies for peregrine falcons (*Falco peregrinus*). However, other issues include ethical costs, risks to other species, populations and ecosystems, and allocation of resources between management and monitoring. The next two chapters provide a framework for addressing such issues.

Acknowledgements

The authors thank Hannah Kokko, Ali Chauvenet, Mick McCarthy, Gillian Dennis and Brady Mattsson for comments on the manuscript, Andrew McClung for assisting with draft revisions and Paul Berkowitz for producing Figure 6.1. The initial draft was prepared as part of the 2008 Graeme Caughley Travelling Fellowship to DPA from the Australian Academy of Science. Use of product names is for descriptive purposes and does not imply endorsement by the USGS.

References

Akaike, H. (1973) Information theory as an extension of the maximum likelihood principle. In *Second International Symposium on Information Theory*, eds B.N. Petrov & F. Csaki, pp. 267–281. Akademiai Kiado, Budapest, Hungary.

Akçakaya, H.R. & Root, W. (2002a) *RAMAS GIS: linking spatial data with population viability analysis (version 4.0)*. Applied Biomathematics, Setauket, New York, USA.

Akçakaya, H.R. & Root, W. (2002b) *RAMAS Metapop: viability analysis for stage-structured metapopulations (version 4.0)*. Applied Biomathematics, Setauket, New York, USA.

Akçakaya, H.R., Burgman, M.A. & Ginzburg, L.V. (1999) *Applied population Ecology*, 2nd edition. Sinauer, Sunderland, Massachusetts, USA.

Akçakaya, H.R., McCarthy, M.A. & Pearce, J.L. (1995) Linking landscape data with population viability analysis – management options for the helmeted honeyeater *Lichenostomus melanops cassidix*. *Biological Conservation*, 73, 169–176.

Angulo, E, Roemer, G.W., Berec, L. *et al.* (2007) Double Allee effects and extinction in the island fox. *Conservation Biology*, 21, 567–577.

Armstrong, D.P. & Davidson, R.S. (2006) Developing population models for guiding reintroductions of extirpated bird species back to the New Zealand mainland. *New Zealand Journal of Ecology*, 30, 73–85.

Armstrong, D.P. & Ewen, J.G. (2001a) Assessing the value of follow-up translocations: a case study using New Zealand robins. *Biological Conservation*, 101, 239–247.

Armstrong, D.P. & Ewen, J.G. (2001b) Estimating impacts of poison operations using mark-recapture analysis and population viability analysis: an example with New Zealand robins (*Petroica australis*). *New Zealand Journal of Ecology*, 25, 29–38.

Armstrong, D.P. & Ewen, J.G. (2002) Dynamics and viability of a New Zealand robin population reintroduced to regenerating fragmented habitat. *Conservation Biology*, 16, 1074–1085.

Armstrong, D.P. & Seddon, P.J. (2008) Directions in reintroduction biology. *Trends in Ecology and Evolution*, 23, 20–25.

Armstrong, D.P. & Wittmer, H.U. (2011) Incorporating Allee effects into reintroduction strategies. *Ecological Research*, 26, 687–695.

Armstrong, D.P., Castro, I. & Griffiths, R. (2007) Using adaptive management to determine requirements of re-introduced populations: the case of the New Zealand hihi. *Journal of Applied Ecology*, 44, 953–962.

Armstrong, D.P., Davidson, R.S., Dimond, W.J. *et al.* (2002) Population dynamics of reintroduced forest birds on New Zealand islands. *Journal of Biogeography*, 29, 609–621.

Armstrong, D.P., Davidson, R.S., Perrott, J.K. *et al.* (2005) Density-dependent population growth in a reintroduced population of North Island saddlebacks. *Journal of Animal Ecology*, 74, 160–170.

Armstrong, D.P., Raeburn, E.H., Lewis, R.M. *et al.* (2006) Estimating the viability of a reintroduced New Zealand robin population as a function of predator control. *Journal of Wildlife Management*, 70, 1020–1027.

Bach, L.A., Pedersen, R.B.F., Hayward, M.W. *et al.* (2010) Assessing re-introductions of the African wild dog (*Lycaon pictus*) in the Limpopo Valley Conservancy, South Africa, using the stochastic simulation program VORTEX. *Journal for Nature Conservation*, 18, 237–246.

Bakker, V.J., Doak, D.F., Roemer, G.W. *et al.* (2009) Incorporating ecological drivers and uncertainty into a demographic population viability analysis for the island fox. *Ecological Monographs*, 79, 77–108.

Bar-David, S., Saltz, D., Dayan, T. *et al.* (2008) Using spatially expanding populations as a tool for evaluating landscape planning: the reintroduced Persian fallow deer as a case study. *Journal for Nature Conservation*, 16, 164–174.

Barker, R.J., Schofield, M.R., Armstrong, D.P. *et al.* (2009) Bayesian hierarchical models for inference about population growth. In *Modelling Demographic Processes in Marked Populations*, eds D.L. Thomson, E.G. Cooch & M.J. Conroy, pp. 3–17. Springer, New York, USA.

Baveco, J.M. & Smeulders, A.M.W. (1994) Objects for simulation: smalltalk and ecology. *Simulation*, 62, 42–57.

Beissinger, S.R. & Westphal, M.I. (1998) On the use of demographic models of population viability in endangered species management. *Journal of Wildlife Management*, 62, 821–841.

Boukal, D.S. & Berec, L. (2002) Single-species models of the Allee effect: extinction boundaries, sex ratios and mate encounters. *Journal of Theoretical Biology*, 218, 375–394.

Box, G.E.P. (1976) Science and statistics. *Journal of the American Statistical Association*, 71, 791–799.

Bretagnolle, V. & Inchausti, P. (2005) Modelling population reinforcement at a large spatial scale as a conservation strategy for the declining little bustard (*Tetrax tetrax*) in agricultural habitats. *Animal Conservation*, 8, 59–68.

Brook, B.W., Cannon, J.R., Lacy, R.C. *et al.* (1999) Comparison of the population viability packages GAPPS, INMAT, RAMAS and VORTEX for the whooping crane (*Grus americana*). *Animal Conservation*, 2, 23–32.

Brook, B.W., Griffiths, A.D. & Puckey, H.L. (2002) Modelling strategies for the management of the critically endangered Carpentarian rock-rat (*Zyzomys palatalis*) of northern Australia. *Journal of Environmental Management*, 65, 355–368.

Brook, B.W., Lim, L., Harden, R. *et al.* (1997) Does population viability analysis software predict the behaviour of real populations? A retrospective study on the Lord Howe Island woodhen. *Biological Conservation*, 82, 119–128.

Brooks, S. P., King, R. & Morgan, B.J.T. (2004) A Bayesian approach to combining animal abundance and demographic data. *Animal Biodiversity and Conservation*, 27, 515–529.

Burnham, K.P. & Anderson, D.R. (2002) *Model Selection and Multimodal Inference: a practical information-theoretic approach*, 2nd edition. Springer-Verlag, New York, USA.

Bustmante, J. (1996) Population viability analysis of captive and released bearded vulture populations. *Conservation Biology*, 10, 822–831.

Cade, T.J. (2007) Exposure of California condors to lead from spent ammunition. *Journal of Wildlife Management*, 71, 2125–2133.

Carroll, C., Phillips, M.K., Schumaker, N.H. *et al.* (2003) Impacts of landscape change on wolf restoration success: planning a reintroduction program based on static and dynamic spatial models. *Conservation Biology*, 17, 536–548.

Caughley, G. (1994) Directions in conservation biology. *Journal of Animal Ecology*, 63, 215–244.

Chapron, G., Wielgus, R., Quenette, P.Y. *et al.* (2009) Diagnosing mechanisms of decline and planning for recovery of an endangered brown bear (*Ursus arctos*) population. *PLOS ONE*, 4 (10), article e7568.

Clark, J.D. & Eastridge, R. (2006) Growth and sustainability of black bears at White River National Wildlife Refuge, Arkansas. *Journal of Wildlife Management*, 70, 1094–1101.

Conroy, M.J. & Carroll, J.P. (2009) *Quantitative Conservation of Vertebrates*. Wiley-Blackwell, Chichester, UK.

Courchamp. F., Berec, L. & Gascoigne, J. (2008) *Allee Effects in Ecology and Conservation*. Oxford University Press, New York, USA.

Cramer, P.C. & Portier, K.M. (2001) Modeling Florida panther movements in response to human attributes of the landscape and ecological settings. *Ecological Modelling*, 140, 51–80.

Crone, E.E., Pickering, D. & Schultz, C.B. (2007) Can captive rearing promote recovery of endangered butterflies? An assessment in the face of uncertainty. *Biological Conservation*, 139, 103–112.

Curtis, J.M.R. & Naujokaitis-Lewis, I. (2008) Sensitivity of population viability to spatial and nonspatial parameters using grip. *Ecological Applications*, 18, 1002–1013.

Dawson, T.P., Jackson, S.T., House, J.I. *et al.* (2011) Beyond predictions: biodiversity conservation in a changing climate. *Science*, 332, 53–58.

Deredec, A. & Courchamp, F. (2007) Importance of the Allee effect for reintroductions. *Ecoscience*, 4, 440–451.

Dickens, M.J., Delehanty, D.J. & Romero, L.M. (2010) Stress: an inevitable component of animal translocation. *Biological Conservation*, 143, 1329–1341.

Dimond, W.J. & Armstrong, D.P. (2007) Adaptive harvesting of source populations for translocation: a case study with New Zealand robins. *Conservation Biology*, 21, 114–124.

Eastridge, R. & Clark, J.D. (2001) Evaluation of two soft-release techniques to reintroduce black bears. *Wildlife Society Bulletin*, 29, 1163–1174.

Evans, R.J., Wilson, J.D., Amar, A. *et al.* (2009) Growth and demography of a re-introduced population of white-tailed Eagles *Haliaeetus albicilla*. *Ibis*, 151, 244–254.

Ewen, J.G., Acevedo-Whitehouse, K., Alley, M. *et al.* (2011) Empirical consideration of parasites and health in reintroduction. In *Reintroduction Biology: integrating science and management*, eds J.G. Ewen, D.P. Armstrong, K.A. Parker & P.J. Seddon, Chapter 9. Wiley-Blackwell, Oxford, UK.

Fernandez, N., Kramer-Schadt, S. & Thulke, H.H. (2006) Viability and risk assessment in species restoration: planning reintroductions for the wild boar, a potential disease reservoir. *Ecology and Society*, 11, 6.

Ferriere, R., Sarrazin, F., Legendre, S. *et al.* (1996) Matrix population models applied to viability analysis and conservation: theory and practice using the ULM software. *Acta Oecologica – International Journal of Ecology*, 17, 629 656.

Ferson, S. & Akçakaya, H.R. (1991) *RAMAS/age User Manual: modelling fluctuations in age-structured populations*. Exeter Software/Applied Biomathematics, New York, USA.

Fieberg, J. & Ellner, S.P. (2000) When is it meaningful to estimate an extinction probability? *Ecology*, 81, 2040–2047.

Gilad, O., Grant, W.E. & Saltz, D. (2008) Simulated dynamics of Arabian oryx (*Oryx leucoryx*) in the Israeli Negev: effects of migration corridors and post-reintroduction changes in natality on population viability. *Ecological Modelling*, 210, 169–178.

Green, R.E., Pienkowski, M.W. & Love, J.A. (1996) Long-term viability of the reintroduced population of the white-tailed eagle *Haliaeetus albicilla* in Scotland. *Journal of Applied Ecology*, 33, 357–368.

Griffith, B., Scott, J.M., Carpenter, J.W. *et al.* (1989) Translocation as a species conservation tool – status and strategy. *Science*, 245, 477–480.

Groombridge, J.J., Raisin, C., Bristol, R. *et al.* (2011) Genetic consequences of reintroductions and insights from population history. In *Reintroduction Biology: integrating science and management*, eds J.G. Ewen, D.P. Armstrong, K.A. Parker & P.J. Seddon, Chapter 12. Wiley-Blackwell, Oxford, UK.

Gusset, M., Jakoby, O., Muller, M.S. *et al.* (2009) Dogs on the catwalk: modelling re-introduction and translocation of endangered wild dogs in South Africa. *Biological Conservation*, 142, 2774–2781.

Haig, S.M., Belthoff, J.R. & Allen, D.H. (1993) Population viability analysis for a small population of red-cockaded woodpeckers and an evaluation of enhancement strategies. *Conservation Biology*, 7, 289–301.

Haines, A.M., Tewes, M.E., Laack, L.L. *et al.* (2006) A habitat-based population viability analysis for ocelots (*Leopardus pardalis*) in the United States. *Biological Conservation*, 132, 424–436.

Hamilton, L.P., Kelly, P.A., Williams, D.F. *et al.* (2010) Factors associated with survival of reintroduced riparian brush rabbits in California. *Biological Conservation*, 143, 999–1007.

Hardman, B. & Moro, D. (2006) Optimising reintroduction success by delayed dispersal: is the release protocol important for hare-wallabies? *Biological Conservation*, 128, 403–411.

Haydon, D.T., Morales, J.M., Yott, A. *et al.* (2008) Socially informed random walks: incorporating group dynamics into models of population spread and growth. *Proceedings of the Royal Society of London B*, 275, 1101–1109.

High Performance Systems (2001) *An introduction to systems thinking*; http://www.hps-inc.com.

Hilborn, R. & Mangel, M. (1997) *The Ecological Detective: confronting models with data*. Princeton University Press, Princeton, New Jersey, USA.

Holland, E.P., Burrow, J.F., Dythama, C. *et al.* (2009) Modelling with uncertainty: introducing a probabilistic framework to predict animal population dynamics. *Ecological Modelling*, 220, 1203–1217.

Holling, C.S. (1978) *Adaptive Environmental Assessment and Management*. John Wiley & Sons, Ltd, Chichester, UK.

Hood, G. M. (2010) *PopTools version 3.2.3*; http://www.poptools.org.

Howells, O. & Edwards-Jones, G. (1997) A feasibility study of reintroducing wild boar *Sus scrofa* to Scotland: are existing woodlands large enough to support minimum viable populations? *Biological Conservation*, 81, 77–89.

Jamieson, I.G. & Lacy, R.C. (2011) Managing genetic issues in reintroduction biology. In *Reintroduction biology: integrating science and management*, eds J.G. Ewen, D.P. Armstrong, K.A. Parker & P.J. Seddon, Chapter 13. Wiley-Blackwell, Oxford, UK.

Johnson, J.B. & Ormland, K.S. (2004) Model selection in ecology and evolution. *Trends in Ecology and Evolution*, 19, 101–108.

Kauffman, M.J., Frick, W.F. & Linthicum, J. (2003) Estimation of habitat-specific demography and population growth for peregrine falcons in California. *Ecological Applications*, 13, 1802–1816.

Kauffman, M.J., Pollock, J.F. & Walton, B. (2004) Spatial structure, dispersal, and management of a recovering raptor population. *American Naturalist*, 164, 582–597.

Keedwell, R.J. (2004) Use of population viability analysis in conservation management in New Zealand. In *Science for Conservation 243*. New Zealand Department of Conservation, Wellington, New Zealand.

Keller, L.K., Biebach, I., Ewing, S.R. *et al.* (2011) The genetics of reintroductions: inbreeding and genetic drift. In *Reintroduction Biology: integrating science and management*, eds J.G. Ewen, D.P. Armstrong, K.A. Parker & P.J. Seddon, Chapter 11. Wiley-Blackwell, Oxford, UK.

King, R., Morgan, B.J., Gimenez, O. *et al.* (2010) *Bayesian Analysis for Population Ecology*. CRC Press, Boca Raton, Florida, USA.

Kinley, T.A. & Newhouse, N.J. (2008) Ecology and translocation-aided recovery of an endangered badger population. *Journal of Wildife Management*, 72, 113–122.

Kohlmann, S.G., Schmidt, G.A. & Garcelon, D.K. (2005) A population viability analysis for the Island Fox on Santa Catalina Island, California. *Ecological Modelling*, 183, 77–94.

Kramer-Schadt, S., Revilla, E. & Wiegand, T. (2005) Lynx reintroductions in fragmented landscapes of Germany: projects with a future or misunderstood wildlife conservation. *Biological Conservation*, 125, 169–182.

Lacy, R.C., Borbat, M. & Pollak, J.P. (2005) *VORTEX: a stochastic simulation of the extinction process*. Chicago Zoological Society, Brookfield, Illinois, USA.

Leaper, R., Massei, G., Gorman, M.L. *et al.* (1999) The feasibility of reintroducing wild boar (*Sus scrofa*) to Scotland. *Mammal Review*, 29, 239–259.

Leech, T.J., Craig, E., Beaven, B. *et al.* (2007) Reintroduction of rifleman *Acanthisitta chloris* to Ulva Island, New Zealand: evaluation of techniques and population persistence. *Oryx*, 41, 369–375.

Legendre, S. & Clobert, J. (1995) ULM, a software for conservation and evolutionary biologists. *Journal of Applied Statistics*, 22, 817–834.

Le Gouar, P., Mihoub, J.-B. & Sarrazin, F. (2011) Dispersal and habitat selection: behavioural and spatial constraints for animal translocations. In *Reintroduction*

Biology: integrating science and management, eds J.G. Ewen, D.P. Armstrong, K.A. Parker & P.J. Seddon, Chapter 5. Wiley-Blackwell, Oxford, UK.

Lennartsson, T. (2002) Extinction thresholds and disrupted plant-pollinator interactions in fragmented plant populations. *Ecology*, 83, 3060–3072.

León-Cortés, J.L., Lennon, J.J. & Thomas, C.D. (2003) Ecological dynamics of extinct species in empty habitat networks. 1. The role of habitat pattern and quantity, stochasticity and dispersal. *Oikos*, 102, 449–464.

Leslie, P.H. (1945) On the use of matrices in certain population mathematics. *Biometrika*, 33, 183–212.

Liu, G.H., Zhou, J., Huang, D.S. *et al.* (2004) Spatial and temporal dynamics of a restored population of *Oryza rufipogon* in Huli Marsh, South China. *Restoration Ecology*, 12, 456–463.

Lloyd, B.D. & Powlesland, R.G. (1994) The decline of kakapo *Strigops habroptilus* and attempts at conservation by translocation. *Biological Conservation*, 69, 75–85.

Lloyd, J.D., Slater, G.L. & Snow, S. (2009) Demography of reintroduced eastern bluebirds and brown-headed nuthatches. *Journal of Wildife Management*, 73, 955–964.

Macdonald, D.W., Tattersall, F.H., Rushton, S. (2000) Reintroducing the beaver (*Castor fiber*) to Scotland: a protocol for identifying and assessing suitable release sites. *Animal Conservation*, 3, 125–133.

Mackey, R.L., Page, B.R., Grobler, D. *et al.* (2009) Modelling the effectiveness of contraception for controlling introduced populations of elephant in South Africa. *African Journal of Ecology*, 47, 747–755.

Marshall, K. & Edwards-Jones, G. (1998) Reintroducing capercaillie (*Tetrao urogallus*) into southern Scotland: identification of minimum viable populations at potential release sites. *Biodiversity and Conservation*, 7, 275–296.

MathWorks (1992) *MATLAB Reference Guide*. MathWorks, Natick, Massachussets, USA.

Matson T., Goldizen, A.W. & Jarman, P.J. (2004) Factors affecting the success of translocations of the black-faced impala in Namibia. *Biological Conservation*, 116, 359–365.

Maunder, M.N. (2004) Population viability analysis based on combining Bayesian, integrated, and hierarchical analyses. *Acta Oecologica*, 26, 85–94.

McCallum, H., Timmers, P. & Hoyle, S. (1995) Modelling the impact of predation on reintroductions of bridled nailtail wallabies. *Wildlife Research*, 22, 163–171.

McCarthy, M.A. (2007) *Bayesian Methods for Ecology*. Cambridge University Press, Cambridge, UK.

McCarthy, M.A., Andelman, S.A. & Possingham, H.P. (2003) Reliability of relative predictions in population viability analysis. *Conservation Biology*, 17, 982–989.

McCarthy, M.A., Armstrong, D.P. & Runge, M.C. (2011) Adaptive management of reintroduction. In *Reintroduction Biology: integrating science and management*, eds

J.G. Ewen, D.P. Armstrong, K.A. Parker & P.J. Seddon, Chapter 8. Wiley-Blackwell, Oxford, UK.

McCarthy, M.A., Franklin, D.C. & Burgman, M.A. (1994) The importance of demographic uncertainty: an example from the helmeted honeyeater *Lichenostomus melanops cassidix*. *Conservation Biology*, 67, 135–142.

McCleery, R.A., Lopez, R.R., Silvy, N.J. *et al.* (2005) Effectiveness of supplemental stockings for the endangered Key Largo woodrat. *Biological Conservation*, 124, 27–33.

McCullough, B.D. & Wilson, B. (2005) On the accuracy of statistical procedures in Microsoft Excel 2003. *Computational Statistics and Data Analysis*, 49, 1244–1252.

Menges, E.S. (2008) Restoration demography and genetics of plants: when is a translocation successful? *Australian Journal of Botany*, 56, 187–196.

Meriggi, A., Della Stella, R.M., Brangi, A. *et al.* (2007) The reintroduction of grey and red-legged partridges (*Perdix perdix* and *Alectoris rufa*) in central Italy: a metapopulation approach. *Italian Journal of Zoology*, 74, 215–237.

Mihoub, J.B., Le Gouar, P. & Sarrazin, F. (2009) Breeding habitat selection behaviors in heterogeneous environments: implications for modeling reintroduction. *Oikos*, 118, 663–674.

Mitchell, N.J., Allendorf, F.W., Keall, S.N. *et al.* (2010) Demographic effects of temperature-dependent sex determination: will tuatara survive global warming? *Global Change Biology*, 16, 60–72.

Morgan, J.W. (1999) Have tubestock plantings successfully established populations of rare grassland species into reintroduction sites in western Victoria? *Biological Conservation*, 89, 235–243.

Münzbergová, Z., Milden, M., Ehrlen, J. *et al.* (2005) Population viability and reintroduction strategies: a spatially explicit landscape-level approach. *Ecological Applications*, 15, 1377–1386.

Muths, E. & Dreitz, V. (2008) Monitoring programs to assess reintroduction efforts: a critical component in recovery. *Animal Biodiversity and Conservation*, 31, 47–56.

Nichols, J.D. & Williams, B.K. (2006) Monitoring for conservation. *Trends in Ecology and Evolution*, 21, 668–673.

Nichols, J.D. & Armstrong, D.P. (2011) Monitoring for reintroductions. In *Reintroduction Biology: integrating science and management*, eds J.G. Ewen, D.P. Armstrong, K.A. Parker & P.J. Seddon, Chapter 7. Wiley-Blackwell, Oxford, UK.

Nicoll, M.A.C., Jones, C.G. & Norris, K. (2003) Declining survival rates in a reintroduced population of the Mauritius kestrel: evidence for non-linear density dependence and environmental stochasticity. *Journal of Animal Ecology*, 72, 917–926.

Nolet, B.A. & Baveco, J.M. (1996) Development and viability of a translocated beaver *Castor fiber* population in the Netherlands. *Biological Conservation*, 75, 125–137.

Ortiz-Catedral, L. & Brunton, D.H. (2008) Clutch parameters and reproductive success of a translocated population of red-crowned parakeet (*Cyanoramphus novaezelandiae*). *Australian Journal of Zoology*, 56, 389–393.

Osborne, P.E. (2005) Key issues in assessing the feasibility of reintroducing the great bustard *Otis tarda* to Britain. *Oryx*, 39, 22–29.

Osborne, P.E. & Seddon, P.J. (2011) Selecting suitable habitats for reintroductions: variation, change and the role of species distribution modelling. In *Reintroduction Biology: integrating science and management*, eds J.G. Ewen, D.P. Armstrong, K.A. Parker & P.J. Seddon, Chapter 3. Wiley-Blackwell, Oxford, UK.

Palisade (2000) *@RISK: advanced risk analysis for spreadsheets*. Palisade Corporation, Newfield, New York, USA.

Pedrono, M., Smith, L.L., Clobert, J. *et al.* (2004) Wild-captive metapopulation viability analysis. *Biological Conservation*, 119, 463–473.

Pergams, O.R.W., Lacy, R.C. & Ashley, M.V. (2000) Conservation and management of Anacapa Island deer mice. *Conservation Biology*, 14, 819–832.

Perkins, D.W., Vickery, P.D. & Shriver, W.G. (2008) Population viability analysis of the Florida grasshopper sparrow (*Ammodramus savannarum floridanus*): testing recovery goals and management options. *Auk*, 125, 167–177.

Possingham, H. P. & Davies, I. (1995) ALEX: a model for the viability analysis of spatially structured populations. *Biological Conservation*, 73, 143–150.

R Development Core Team (2007) *R: a language and environment for statistical computing*. R Foundation for Statistical Computing, Vienna, Austria; http://www.R-project.org.

Reed, J.M., Fefferman, N. & Averill-Murray, R.C. (2009) Vital rate sensitivity analysis as a tool for assessing management actions for the desert tortoise. *Biological Conservation*, 142, 2710–2717.

Reed, J.M., Mills, L.S., Dunning, J.B. *et al.* (2002) Emerging issues in population viability analysis. *Conservation Biology*, 16, 7–19.

Reynolds, M.H., Brinck, K.W. & Laniawe, L. (2011) *Population estimates and monitoring guidelines for endangered Laysan teal, Anas Laysanensis, at Midway Atoll: pilot study results 2008–2010*, Hawaii Cooperative Studies Unit Technical Report HCSU-021. University of Hawaii, Hilo, Hawaii, USA.

Reynolds, M.H., Seavy, N.E., Vekasy, M.S. *et al.* (2008) Translocation and early post-release demography of endangered Laysan teal. *Animal Conservation*, 11, 160–168.

Robert, A., Sarrazin, F., Couvet, D. *et al.* (2004) Releasing adults versus young in reintroductions: interactions between demography and genetics. *Conservation Biology*, 18, 1078–1087.

Roth, J.D., Murray, D.L. & Steury, T.D. (2008) Spatial dynamics of sympatric canids: modeling the impact of coyotes on red wolf recovery. *Ecological Modelling*, 214, 391–403.

Rout, T.M., Hauser, C.E. & Possingham, H.P. (2007) Minimise long-term loss or maximise short-term gain? Optimal translocation strategies for threatened species. *Ecological Modelling*, 201, 67–74.

Rout, T.M., Hauser, C.E. & Possingham, H.P. (2009) Optimal adaptive management for the translocation of a threatened species. *Ecological Applications*, 19, 515–526.

Saether, B.E., Engen, S., Filli, F. *et al.* (2002) Stochastic population dynamics of an introduced Swiss population of the ibex. *Ecology*, 83, 3457–3465.

Saether, B.E., Lillegard, M., Grotan, V. *et al.* (2007) Predicting fluctuations of reintroduced ibex populations: the importance of density dependence, environmental stochasticity and uncertain population estimates. *Journal of Animal Ecology*, 76, 326–336.

Sainsbury, A.W., Armstrong, D.P. & Ewen, J.G. (2011) Methods of disease risk analysis for reintroduction programmes. In *Reintroduction Biology: integrating science and management*, eds J.G. Ewen, D.P. Armstrong, K.A. Parker & P.J. Seddon, Chapter 10. Wiley-Blackwell, Oxford, UK.

Saltz, D., Rubenstein, D.I. & White, G.C. (2006) The impact of increased environmental stochasticity due to climate change on the dynamics of asiatic wild ass. *Conservation Biology*, 20, 1402–1409.

SAS Institute (2005) *SAS/STAT User's Guide. Release 8.02.* SAS Institute, Cary, North Carolina, USA.

Schadt, S., Revilla, E., Wiegand, T. *et al.* (2002) Assessing the suitability of central European landscapes for the reintroduction of Eurasian lynx. *Journal of Applied Ecology*, 39, 189–203.

Schaub, M., Pradel, R. & Lebreton, J.D. (2004) Is the reintroduced white stork (*Ciconia ciconia*) population in Switzerland self-sustainable? *Biological Conservation*, 119, 105–114.

Schaub, M., Zink, R., Beissmann, H. *et al.* (2009) When to end releases in reintroduction programmes: demographic rates and population viability analysis of bearded vultures in the Alps. *Journal of Applied Ecology*, 46 (1), 92–100.

Seddon, P.J., Armstrong, D.P. & Maloney, R.F. (2007) Developing the science of reintroduction biology. *Conservation Biology*, 21, 303–312.

Slotta-Bachmayr, L., Boegel, R., Kaczensky, P. *et al.* (2004) Use of population viability analysis to identify management priorities and success in reintroducing Przewalski's horses to southwestern Mongolia. *Journal of Wildlife Management*, 68, 790–798.

Smart, J., Amar, A., Sim, I.M.W. *et al.* (2010) Illegal killing slows population recovery of a re-introduced raptor of high conservation concern – the red kite *Milvus milvus*. *Biological Conservation*, 143, 1278–1286.

Somers, M.J. (1997) The sustainability of harvesting a warthog population: assessment of management options using simulation modelling. *South African Journal of Wildlife Research*, 27, 37–43.

South, A., Rushton, S. & Macdonald, D. (2000) Simulating the proposed reintroduction of the European beaver (*Castor fiber*) to Scotland. *Biological Conservation*, 93, 103–116.

South, A.B., Rushton, S.P., Macdonald, D.W. *et al.* (2001) Reintroduction of the European beaver (*Castor fiber*) to Norfolk, UK: a preliminary modelling analysis. *Journal of Zoology*, 254, 473–479.

Southgate, R. & Possingham, H. (1995) Modelling the reintroduction of the greater bilgy *Macrotis lagotis* using the metapopulation model Analysis of the Likelihood of EXtinction (ALEX). *Biological Conservation*, 73, 151–160.

Spiegelhalter, D.J., Best, N.G., Carlin, B.P. *et al.* (2002) Bayesian measures of model complexity and fit (with discussion). *Journal of the Royal Statistics Society B*, 64, 583–640.

Spiegelhalter, D., Thomas, A., Best, N. *et al.* (2003) *WinBUGS User Manual, Version 1.4.3*; http://www.mrc-bsu.cam.ac.uk/bugs.

Steury, T.D. & Murray, D.L. (2004) Modeling the reintroduction of lynx to the southern portion of its range. *Biological Conservation*, 117, 127–141.

Strauss, W.M. (2002) Towards the effective management of the Arabian oryx *Oryx leucoryx* in the Kingdom of Saudi Arabia. *Zeitschrift für Jagdwissenschaft*, 48, 7–16.

Swart, J. & Lawes, M.J. (1996) The effect of habitat patch connectivity on samango monkey (*Cercopithecus mitis*) metapopulation persistence. *Ecological Modelling*, 93, 57–74.

Tavecchia, G., Viedma, C., Martinez-Abrain, A. *et al.* (2009) Maximizing re-introduction success: assessing the immediate cost of release in a threatened waterfowl. *Biological Conservation*, 142, 3005–3012.

Taylor, M.K., Obbard, M., Pond, B. *et al.* (2006) *A Guide to Using RISKMAN: stochastic and deterministic population modeling RISK MANagement decision tool for harvested and unharvested populations, Version 1.9.003*. The Queen's Printer for Ontario, Ontario, Canada.

Tenhumberg, B., Tyre, A.J., Shea, K. *et al.* (2004) Linking wild and captive populations to maximize species persistence: optimal translocation strategies. *Conservation Biology*, 18, 1304–1314.

Tocher, M.D., Fletcher, D. & Bishop, P.J. (2006) A modelling approach to determine a translocation scenario for the endangered New Zealand frog *Leiopelma hamiltoni*. *Herpetological Journal*, 16, 97–106.

Todd, C.R., Jenkins, S. & Bearlin, A.R. (2002) Lessons about extinction and translocation: models for eastern barred bandicoots (*Perameles gunnii*) at Woodlands Historic Park, Victoria, Australia. *Biological Conservation*, 106, 211–223.

Todd, C.R., Nicol, S.J. & Koehn, J.D. (2004) Density-dependence uncertainty in population models for the conservation management of trout cod, *Maccullochella macquariensis*. *Ecological Modelling*, 171, 359–380.

Towns, D.R. & Ferreira, S.M. (2001) Conservation of New Zealand lizards (Lacertilia: Scincidae) by translocation of small populations. *Biological Conservation*, 98, 211–222.

Towns, D.R., Parrish, G.R. & Westbrooke, I. (2003) Inferring vulnerability to introduced predators without experimental demonstration: case study of Suter's skink in New Zealand. *Conservation Biology*, 17, 1361–1371.

Vandel, J.M., Stahl, P., Herrenschmidt, V. *et al.* (2006) Reintroduction of the lynx into the Vosges mountain massif: from animal survival and movements to population development. *Biological Conservation*, 131, 370–385.

van Heezik, Y., Seddon, P.J. & Maloney, R.F. (1999) Helping reintroduced houbara bustards avoid predation: effective anti-predator training and the predictive value of pre-release behaviour. *Animal Conservation*, 2, 155–163.

Vincenzi, S., Crivelli, A.J., Jesensek, D. *et al.* (2008) Potential factors controlling the population viability of newly introduced endangered marble trout populations. *Biological Conservation*, 141, 198–210.

Wakamiya, S.M. & Roy, C.L. (2009) Use of monitoring data and population viability analysis to inform reintroduction decisions: peregrine falcons in the Midwestern United States. *Biological Conservation*, 142, 1767–1776.

Watson, L.H. & Chadwick, P. (2007) Management of Cape Mountain zebra in the Kammanassie Nature Reserve, South Africa. *South African Journal of Wildlife Research*, 37, 31–39.

Wear, B.J., Eastridge, R. & Clark, J.D. (2005) Factors affecting settling, survival, and viability of black bears reintroduced to Felsenthal National Wildlife Refuge, Arkansas. *Wildlife Society Bulletin*, 33, 1363–1374.

White, G.C. (2000a) Modelling population dynamics. In *Ecology and Management of Large Mammals in North America*, eds S. Demarais & P.R. Krausman, pp. 85–107. Prentice-Hall, Englewood Cliffs, New Jersey, USA.

White, G.C. (2000b) Population viability analysis: data requirements and essential analyses. In *Research Techniques in Animal Ecology: controversies and consequences*, eds L. Boitani & T.K. Fuller, pp. 287–331. Columbia University Press, New York, USA.

White, G.C. & Burnham, K.P. (1999) Program MARK: survival estimation from populations of marked animals. *Bird Study*, 46 (Suppl.), 120–138.

Wiegand, T., Knauer, F., Kaczensky, P. *et al.* (2004a) Expansion of brown bears (*Ursus arctos*) into the eastern Alps: a spatially explicit population model. *Biodiversity and Conservation*, 13, 79–114.

Wiegand, T., Revilla, E. & Knauer, F. (2004b) Dealing with uncertainty in spatially explicit population models. *Biodiversity and Conservation*, 13, 53–78.

Williams, B.K., Nichols, J.R. & Conroy, M.J. (2002) *Analysis and Management of Animal Populations*. Academic Press, New York, USA.

Wolfram Research (1996) *Mathematica, Version 3.0*. Wolfram Research, Champaign, Illinois, USA.

Wood, D.J.A., Koprowski, J.L. & Lurz, P.W.W. (2007) Tree squirrel introduction: a theoretical approach with population viability analysis. *Journal of Mammalogy*, 88, 1271–1279.

7

Monitoring for Reintroductions

James D. Nichols[1] and Doug P. Armstrong[2]

[1]Patuxent Wildlife Research Center, U.S. Geological Survey,
United States of America
[2]Massey University, Palmerston North, New Zealand

'*Our rationale for this initial focus on the entire reintroduction programme as a decision process, rather than on monitoring per se, is that monitoring is only useful because of its contribution to the larger decision process of which it is a part.*'

Page 224

Introduction

Monitoring is viewed by many as an important component of reintroduction programmes. Nonetheless, many reintroduction programmes have been criticized for a failure either to conduct adequate monitoring or to report the results of such monitoring (IUCN, 1987, 1998; Lyles & May, 1987; Griffith *et al.*, 1989). We hold the view that monitoring of reintroduced populations is important for the success of both the focal reintroduction program and other future programmes. In addition, pre-release monitoring of predators, competitors and various other aspects of habitat (for cover, potential food resources, etc.) can be likewise important to determine the appropriateness of an area for reintroduction (Osborne & Seddon, this volume, Chapter 3). However, monitoring must be done strategically to ensure we get a good return for the monitoring effort invested (Ewen & Armstrong, 2007).

Reintroduction Biology: Integrating Science and Management. First Edition.
Edited by John G. Ewen, Doug P. Armstrong, Kevin A. Parker and Philip J. Seddon.
© 2012 Blackwell Publishing Ltd. Published 2012 by Blackwell Publishing Ltd.

Structured decision making (SDM) provides a framework for making such strategic decisions about monitoring and other aspects of reintroduction programmes. We provide a brief outline of the components of SDM processes, with special reference to reintroduction programmes. Adaptive management (AM) can be viewed as a subset of SDM developed for recurrent decisions for which uncertainty exists. Many reintroduction programmes require recurrent decisions (potential releases at multiple points in time, at multiple locations, and of similar species at multiple times and locations), providing an opportunity to use adaptive management (AM) as a means of learning and reducing uncertainty over time. The scope for AM to improve reintroduction programmes is covered in depth in the following chapter (McCarthy *et al.*, this volume, Chapter 8), but here we briefly review the AM process in order to show the broader context within which monitoring takes place. Our rationale for this initial focus on the entire reintroduction programme as a decision process, rather than on monitoring per se, is that monitoring is only useful because of its contribution to the larger decision process of which it is a part (Yoccoz *et al.*, 2001; Nichols & Williams, 2006). Under this view, monitoring programmes inherit their designs from the larger processes to which they contribute, in this case reintroduction programmes.

We consider three central questions important to the design of monitoring programmes: why, what and how? The 'why' question concerns the role of monitoring in a SDM program directed at reintroduction. The 'what' question addresses the selection of state variables and vital rates to monitor. 'State variables' (e.g. population size, proportion of habitat patches occupied by a species, species richness, etc.) refer to quantities that provide information about the status or health of a system. 'Vital rates' refer to the rate parameters that bring about changes in state variables (e.g. vital rates for population size are rates of survival, reproductive recruitment, emigration and immigration). The 'how' question focuses on issues of sampling and inference. Finally, we discuss special challenges for monitoring posed by reintroduction efforts and emphasize the importance of integrating monitoring into overall programmes designed to make reintroduction decisions.

Structured decision making (SDM)

Structured decision making is a process that entails breaking a decision into specific components, focusing on one component at a time, and then reassembling the components to develop a solution (e.g. Clemen & Reilly,

2001; Martin *et al.*, 2009; McCarthy *et al.*, this volume, Chapter 8). The idea of SDM is to provide a transparent and logical approach to identify decisions that are optimal with respect to specified objectives, available actions and knowledge of system behaviour. In conservation, components of decision-making problems often include: objectives, potential management actions, model(s) of system behaviour (in particular models that predict the system response to the different management options), a monitoring program to track the system state and finally a method to identify the solution (Williams *et al.*, 2002; Dorazio & Johnson, 2003; McCarthy & Possingham, 2007). These components are typically characterized by substantial uncertainties that should be accommodated in the optimization process when possible.

The specification of objectives is a critical component of any decision-making process. Objectives are developed by relevant stakeholders and may represent compromises among competing views. Objectives associated with reintroduction programmes might be expressed in terms of a desired population size, a minimal probability of local extinction, non-negative population growth rate and restoration of an ecosystem function. Competing objectives can be accommodated in any of several ways, including common currency and use of constraints. In reintroduction programmes, for example, population objectives may be constrained by monetary costs, impact on source populations and risk of spreading unwanted pathogens to the release area. Formal techniques to select appropriate objectives may be especially useful when dealing with diverse stakeholder groups (Clemen & Reilly, 2001; Burgman, 2005). Once objectives and constraints have been selected, they can be formalized mathematically into an objective function. The objective function quantifies the benefit (or return) obtained by implementing specific decisions at each time step, accumulated over the time horizon of the decision problem (Lubow, 1995; Williams *et al.*, 2002; Fonnesbeck, 2005). In general, objectives should drive any management programme and provide the yardstick for judging management success or failure.

Decisions in SDM usually involve selecting the appropriate management action from a set of such potential actions. Alternative actions in reintroduction programmes may include releases of different sizes, compositions (age, sex, size) or origins (wild caught, captive reared) of release groups, different release methods (e.g. immediate versus held at release site), parasite and pathogen management (screening/quarantine and treatment), and different types of release area management (e.g. predator control, other types of habitat management, no management). Objectives and potential actions should reflect the value judgements of stakeholders (Nichols & Williams, 2006). For example, predator control may be an acceptable management action to some stakeholder groups but not others and will generally be more controversial if the predators are native species.

All informed decisions require predictions of system response to each of the potential management actions, and these predictions are based on models (Armstrong & Reynolds, this volume, Chapter 6). Models are therefore essential in SDM, as there is no basis for selecting one management action or another without predictions. In conservation, there is often substantial uncertainty about system responses to management actions. In some cases this uncertainty can be expressed as uncertainty about a key parameter (e.g. a single parameter defining the relationship between annual survival probability and density) in a model of system behaviour. In other cases, such uncertainty is expressed as multiple discrete models representing different hypotheses about how the system works (e.g. one model expressing relatively strong density dependence in survival and another depicting relatively weak density dependence). Optimal decisions may therefore require weighting, with model weights reflecting our relative faith in the predictive abilities of the different models. In contrast to management objectives, which are driven by human values, models of system behaviour and measures of confidence in the models (model 'weights') should preferably be based on a scientific approach (Nichols, 2001; Nichols & Williams, 2006).

Monitoring provides estimates of key system state variables needed for making state-dependent decisions and assessing the degree to which objectives have been met. For example, reintroduction is only considered if the habitat at the release area is believed suitable for the species (Osborne & Seddon, this volume, Chapter 3), and monitoring is used to obtain this information about habitat status. For recurrent decision problems (i.e. those for which decisions are made periodically), monitoring facilitates learning about system behaviour in responses to management. Specifically, model-based predictions are compared with estimates from monitoring as a means of modifying model weights. In the case of reintroduction programmes, learning might focus on the efficacy of different release methods, for example. Such learning can be based not only on estimates of state variables but also on estimates of associated vital rates, and these are a product of monitoring programmes as well.

The final component of SDM is some method for identifying the solution and selecting the management action that best meets management objectives. This final component is frequently an optimization algorithm appropriate to the specific decision problem. In the case of single isolated decisions (e.g. do we attempt to reintroduce animals to this area or not?), optimization may involve simple approaches to maximization or minimization of a function for a single time step (Williams *et al.*, 2002). In the case of recurrent decision problems, decisions made at one time step drive the system to a new state that can influence and constrain the next decision, making optimization a more difficult problem. The optimization must now consider the entire sequence of decisions over the time horizon of the management programme. Stochastic dynamic programming is an algorithm that provides optimal decisions for Markovian systems (Bellman, 1957; Lubow, 1995; Williams *et al.*, 2002). Markovian systems have the property that the probability distribution of the system state at $t + 1$ depends on the system state at t and the decision at t, and the dynamics of most reintroduced populations can be characterized in this way.

Adaptive management (AM)

Adaptive management is a subset of SDM designed for recurrent decisions with uncertainty (Walters, 1986; Williams *et al.*, 2007; McCarthy *et al.*, this volume, Chapter 8). Williams (1997) classified uncertainty in natural resource management into four categories. *Environmental variation* refers to the various aspects of the environment that influence the managed resource and that exhibit variation (weather, habitat, predator populations). *Partial observability* refers to the inability directly to observe system state variables and vital rates. Such attributes must be estimated, and uncertainty is associated with such inference. *Partial controllability* refers to our usual inability to implement management actions directly and precisely. For example, we may conduct predator control prior to reintroduction, but may not know what fraction of predators has been removed or the abundance of predators following control. *Structural uncertainty* refers to uncertainty about how the managed system responds to management actions. We may represent this uncertainty using multiple discrete models, as described above, or as uncertainty in a key parameter of a single, general model. AM was developed to identify good decisions in the face of these sources of uncertainty and is especially focused on the reduction of structural uncertainty or, stated differently, on learning about system response to management actions.

The process of AM requires all of the components of SDM. AM begins with a deliberative or 'set-up phase', during which stakeholders are engaged and come to agreement on both objectives and a set of potential management actions (Williams *et al.*, 2007). The set-up phase also includes the initial development of models and a monitoring programme, both the primary responsibilities of scientists and managers. Following this developmental stage, the iterative phase of AM is initiated. At each decision point of the iterative phase, some process (e.g. an optimization algorithm) is used to select a potential action that is good (or even optimal) with respect to objectives, current system state and current state of knowledge about the system response to actions as encoded in the model set and respective model weights. Conditional on the selected action, each model predicts values of the key system state variables at the next time step. Data from the monitoring programme provides information about the subsequent system state. The comparison of the estimated system state against the predictions of each model provides the basis for updating the model weights (using Bayes' theorem, Williams *et al.*, 2002), increasing the weights of models that predict well and decreasing weights for models that predict poorly. At the next decision point in the iterative phase, the objectives, potential actions (note that, in some situations, actions may include monitoring options) and models remain the same, but the monitoring programme has provided a new estimate of system state and the measures of model credibility (model weights) have been revised.

In the iterative phase described above, learning is achieved by focusing on a comparison of model-based predictions of state variables with estimates of these variables obtained via monitoring. However, competing models also make predictions about vital rates expected to occur under different actions, and monitoring programmes can be used to estimate these as well. Comparisons of predicted and estimated vital rates provide another source of information for learning, and are sometimes more useful for discriminating among competing models than comparisons involving state variables (see below).

The iterative phase proceeds in this fashion, but can be interrupted either at regular time intervals or when new information develops that one or more components of the process should be revisited. Such an interruption is referred to as double-loop learning (Williams *et al.*, 2007) and involves a re-examination of one or more of the process components developed during the set-up phase of AM. For example, changing attitudes and values may motivate a re-examination of objectives or experience with the system may lead to new ideas for either models or management actions. Following appropriate changes to process components, the AM process returns to the iterative phase.

Typically, the iterative phase should operate on a shorter time scale than double-loop learning. The above material is intended to provide a general sketch of AM and the reader is referred to Walters (1986), Williams *et al.* (2002, 2007) and McCarthy *et al.* (this volume, Chapter 8) for more detailed accounts.

Reintroduction programmes as decision problems

Reintroductions involve multiple decisions. The main objective is usually the re-establishment of a population that will persist into the foreseeable future. Decisions therefore need to be made about whether (and how) to proceed with a reintroduction attempt, but also what, if any, ongoing management is needed to increase the probability of persistence. It is therefore useful to distinguish between decisions that need to be made before or after release.

Before release

Before a release takes place, the main objective is to produce a habitat capable of supporting the focal species, as reintroduction efforts make little sense unless selected areas are suitable for the species. For example, habitat state variables can include vegetation characteristics, food resources, and predator and competitor populations (IUCN, 1998). Potential management actions before release will focus on the factors believed to be critical to the success of subsequent reintroductions, such as actions designed to increase cover or food resources, or perhaps efforts to reduce predator or competitor populations. The more scientific components of SDM, models and monitoring, will be tailored to the specified objectives as well as to the available management actions. Models used in this phase of a reintroduction programme focus on predicting: (1) responses of habitat state variables to management actions and (2) suitability for reintroduced animals as a function of habitat state variables (Osborne & Seddon, this volume, Chapter 3). Monitoring will focus on the relevant habitat state variables.

Pre-release monitoring may be restricted to a single assessment of release site suitability. If this assessment indicates that there is a high probability that the habitat can indeed support a population of the focal species, then managers may choose to proceed with a release. However, when such an initial assessment suggests that the area is not currently suitable, AM can be used to guide further management and monitoring. For example, if the key habitat state variable is thought to be predator density, then AM could involve manipulating the intensity of predator control. There will sometimes be little opportunity for pre-release AM at the proposed release area, for example if the location just needs to regenerate after deforestation. However, AM can clearly be applied to management decisions after regeneration has occurred. One of the major criticisms of conservation biology, in general, and reintroduction programmes, in particular, is that such efforts have not paid adequate attention to learning, and that current and future conservation and reintroduction efforts have much to gain from what has been learned in previous efforts.

After release

In some situations, habitat management may be continued after release. Such management may specifically aim to reduce the high 'post-release mortality' (Armstrong & Reynolds, this volume, Chapter 6) and 'post-release dispersal' (Le Gouar *et al.*, this volume, Chapter 5) that often occur immediately after release. This management is discontinued once the population is established. Alternatively, management may need to be continued indefinitely. In either case, the objectives, actions, models and monitoring of the pre-release phase become part of the larger programme after release. In addition, monitoring may reflect decisions about the release strategies used.

Objectives after release will be based on metrics such as desired population size, a maximum acceptable probability of extinction over a specified time frame or a minimum acceptable probability of species occupancy in a suitable habitat. The vital rates of the reintroduced populations, and parameters that relate them to management actions, will be key components of population models that project changes in abundance or occupancy over time (Armstrong & Reynolds, this volume, Chapter 6). This modelling involves assessing hypotheses about factors affecting vital rates, and some of these may concern the effects of different release strategies on vital rates of newly released animals. In one of the initial actual applications of AM to a translocation problem, Rout *et al.* (2009) considered releases of animals at either of two possible sites, one of which was characterized by uncertainty in the post-release survival probability.

Monitoring of the focal species usually involves estimating abundance to inform the entire decision process (e.g. Sutherland *et al.*, 2010). Because of the importance of vital rates to projections of population dynamics and responses to management, monitoring will often be directed at vital rates as well. In some cases, habitat management such as predator control may continue into this phase of management, in which case state variables such as predator densities are indexed or estimated via the monitoring programme. The immediate focus of managers may be on integrated rate parameters, such as the rate of population increase, λ (integrates survival, reproductive recruitment and movement), or rate of change in occupancy (integrates rates of local extinction and colonization), and on their relationships with management actions.

Monitoring for decision making

As noted above, monitoring programmes should inherit their designs from the larger processes to which they contribute. Yoccoz *et al.* (2001) argued that the design of any monitoring programme depends on the answers to three questions – why, what and how? We seek these answers by focusing on the rein-troduction programme that the monitoring programme is designed to inform.

Monitoring – why?

Monitoring data are used for four basic purposes in AM: making state-dependent decisions, assessing progress towards objectives, learning, and developing and refining models. Most informed decisions depend on the system state. If the system state is where we would like it to be, we might take one action, whereas if the system state is far from meeting our objectives, then we might take a very different action. Before release, the system state will include the habitat variables that are thought to be relevant to success of the reintroduction. The same state variables may be monitored after release, but it will then be useful as well to monitor attributes of the reintroduced population. Depending on the state of the population, it may be decided to increase numbers of releases, to reduce releases or to stop releases altogether.

Monitoring to assess progress towards objectives may also differ for different phases of reintroduction programmes. Pre-release monitoring will focus on habitat variables and their distance from values believed to be needed for establishment of the population. For example, estimates or indices (e.g. based on footprint tracking tunnels) of predator density may be used to decide whether their densities are low enough for the reintroduction to have a high probability of success. After release, monitoring will be directed at the abundance of the focal species, with hope for rapid population growth and attainment of objective population sizes. In some cases, this monitoring may also include occupancy, with an emphasis on the spread of the species and colonization of areas surrounding the release site(s).

Monitoring for learning, that is for discriminating among different models describing the effects of various management actions on the system state, will typically include all of the state variables monitored for making state-dependent decisions and for assessing management progress. Monitoring focused on learning may also include inference about the vital rates that are directly influenced by management actions. Models used before release will predict effects of management actions on habitat state variables, so estimates of these state variables will be useful in discriminating among the predictions of different models of effects of management actions. Changes in both habitat and focal population state often involve multiple vital rates. For example, changes in abundance are an outcome of rates of survival, reproductive recruitment and movement. However, estimates of specific vital rates may be more useful in discriminating among competing models of management effects than estimates of state variables that integrate effects of multiple vital rates. In all phases of reintroduction programmes, the selection of specific state variables and vital rates to monitor for the purpose of learning will be dictated by the specifics of the models of system response.

The final listed use of monitoring involves the process of model development and revision. Although initial models may be developed in a complete absence of data, estimates of vital rates and of parameters describing relationships between management actions and vital rates are very useful. In the double-loop phase of AM, for example, reconsiderations of model structure and parameterization can greatly benefit from estimates obtained during monitoring for the iterative phase of AM. Before release, estimates of habitat state transition probabilities and rates of mortality, reproduction and local extinction of predator/competitor populations will be useful for model revisions, as will estimates of parameters describing functional relationships between these rate parameters and management actions. Models used after release similarly require estimates of rates of mortality, reproductive recruitment and movement in and out of the release area for the focal population. As noted, estimates of vital rates (e.g. survival) for the immediate post-release period, and then for subsequent periods, are often relevant to reintroduction models. Parameters describing relationships between focal species vital rates and release strategies are also useful.

In summary, the answer to the question 'why monitor?' is not difficult to address for a formal programme of AM. The answer is inherited directly from the entire AM process and includes the general categories of making state-dependent decisions (e.g. decisions about release depend on population status at that time), assessing progress towards management objectives (do the reintroductions appear to be successful), learning (determining which models do the best job of predicting effects of management actions, as this determination permits better decisions about what action(s) to take) and developing and refining models (better estimates for revisions of models predicting management effects). For each of these general uses of monitoring data, specific data needs are dictated by other components of the decision process, such as objectives, actions and system models.

Monitoring – what?

As was the case for the previous question, the answer to the question 'what should be monitored?' is easy to answer within an AM framework. Once again the answer is dictated by the management context provided by the larger reintroduction programme. As noted in the previous section on uses of monitoring data, estimates of state variables, vital rates and parameters describing relationships between vital rates and management actions are all useful in AM (Table 7.1). Because of the close relationship between uses of monitoring and quantities to monitor, this discussion is closely related to that presented above.

Monitoring before release will focus on state variables reflecting habitat conditions. However, habitat is an elusive concept and can be characterized by an infinite number of variables. The key to characterizing habitat of release areas is to focus on those features that are hypothesized to affect the vital rates of the species to be introduced. Note that this hypothesis-driven approach differs substantially from the omnibus approach of measuring a very large number of habitat variables regardless of any hypothesized relationship to the prospects of focal species success. For many management purposes, it is possible to classify habitat into discrete states or habitat types. This approach has the advantage of being amenable to dynamic multistate modelling (e.g. Breinenger et al., 2009, 2010), an approach with which many animal ecologists are already familiar. Such models permit inference about, and projection of, responses of habitat to natural phenomena (e.g. succession) and management actions. Natural state variables for predator and/or competitor populations are abundance

Table 7.1 **Common management decisions and corresponding monitoring emphases before and after reintroduction. Numerical superscripts indicate the general role of monitoring with respect to the management decision ([1] state-dependent decision; [2] progress assessment; [3] learning). All monitoring contributes to parameter estimation for model modification.**

	Management decision	Monitoring emphasis
Before release	Proceed with release or not[1]	Relevant habitat variables at release area as predictors of suitability
	Habitat management action(s)[1,2,3]	Relevant habitat variables, their changes over time and their associated vital rates (e.g. state transition probabilities) as functions of management actions
After release	Number and composition of release group[1,2,3]	Abundance, survival, dispersal and subsequent reproduction of released animals as function of demographic state (age, sex, breeding condition, etc.)
	Release method[1,2,3]	Abundance, survival, dispersal and subsequent reproduction of released animals as function of release method
	Habitat management action(s)[1,2,3]	Relevant habitat variables, their changes over time and their associated vital rates as functions of management actions
		Abundance, survival and subsequent reproduction of released animals as functions of habitat variables

and habitat occupancy. Occupancy may be an especially useful state variable for programmes aimed at reduction of predator or competitor populations. The reason for this is that as populations are reduced to low numbers and animals become more secretive (e.g. in response to removal efforts), abundance becomes more difficult to estimate. In addition, removal programmes sometimes attempt to eliminate a species from an area. In such cases, sign-based occupancy surveys frequently provide a useful way to estimate the probability that the population is still extant in a removal area.

Monitoring of vital rates that govern habitat state change is useful both for learning and for development and revision of models. In the case of discrete habitat states, the state of the entire system will be characterized by the proportion of land area (or number of land units or habitat patches) in each of the possible habitat states. Changes in the system state are modelled using state transition probabilities, representing the probabilities that a patch in one habitat state at time t will be in any of the possible states at time $t + 1$ (e.g. see Waggoner & Stephens, 1970; Usher, 1979; Callaway & Davis, 1993; Breinenger *et al.*, 2010). For the state variable of population size (e.g. of predator or competitor populations), the relevant vital rates are survival probabilities, rates of reproductive recruitment and rates of movement in and out of the focal area. For the state variable occupancy, the vital rates are local probabilities of extinction and colonization for habitat patches or units of land. In general, monitoring that is focused on vital rates will also permit inference about the effects of management actions on them. These latter relationships are key components of models that project consequences of management actions.

Monitoring after release may focus on some of the variables and parameters monitored before release, as well as on the focal species. In addition to focal species abundance (or density), occupancy may be a relevant state variable. However, for both of these state variables, monitoring may often be directed not at the entire population but at demographic (e.g. adults) or state-based (e.g. breeders) strata within the population.

With respect to abundance, the number of animals in each age–sex class or size–sex class may be especially relevant to post-release population dynamics. The initial age and sex distributions of reintroduced populations may be quite different from the stable or equilibrium distribution, and this is always the case with plant reintroductions. Transient dynamics (patterns of population change for populations not in an equilibrium structure) of such reintroduced populations during their establishment phase are highly dependent on their structure (e.g. Caswell, 2001, 2007; Keyfitz & Caswell, 2005; Koons *et al.*, 2005, 2006, 2007). For example, releases of pre-reproductive animals will not be expected to yield high rates of population growth until the animals mature. On the other hand, release groups comprised of newly mature animals might be expected to yield higher growth rates than groups at the equilibrium structure that also contain pre-reproductive and senescent animals. Transient dynamics are not only influenced by release group structure but are also driven by stochastic and other (e.g. Allee effects) processes commonly affecting populations of low size that can last multiple generations. One metric that predicts the short-term growth potential of any establishing population is total reproductive value (e.g. Keyfitz & Caswell, 2005), and this metric is computed based on the number of animals in each age or stage class at a given point in time.

Occupancy can be criticized as an inadequate state variable for reintroduction programmes, because a key aspect of such programmes is reproduction of released animals. Survival and persistence of released animals is not enough to produce population growth and eventual reproduction is needed for a sustainable population. Estimates of the number of land units or habitat patches occupied by the species (the usual state variable of occupancy surveys) may not be adequate indicators of population health or potential for growth. Multistate occupancy surveys further categorize occupied units, for example, according to whether or not reproduction is occurring in the unit (Royle, 2004b; Royle & Link, 2005; Nichols *et al.*, 2007; MacKenzie *et al.*, 2006, 2009). The state variable in this case is the number of units in each of three states: unoccupied by the species, occupied with no reproduction and occupied with reproduction. Such stratification of the occupancy state variable is likely to be especially useful for monitoring after release.

As was true for pre-release monitoring, monitoring of vital rates after release will be useful both for learning and for development and revision of models. The same vital rates relevant for predator and competitor species are relevant for the focal species as well. However, the importance for post-release monitoring of stratification of population size and occupancy state variables applies to their associated vital rates as well. With respect to the state variable population size, rates of survival, reproductive recruitment and movement can be estimated for the relevant ages or stages used for stratification. Similarly for the state variable of multistate occupancy, the vital rates are the state-specific transition probabilities specifying the probabilities of a unit moving from one state at time t to any other state at time $t + 1$. Such stratification of vital rates is needed for models that project transient dynamics.

For programmes experimenting with different release strategies, monitoring should focus closely on subsequent survival, movement and reproductive rates of animals released according to each strategy. Such information will be most useful for reducing the uncertainty associated with predicted efficacy of different strategies and should facilitate discrimination among competing models of post-release performance. Similar reasoning leads to a focus on survival and reproductive rates of animals of wild versus captive origin, in cases where animals from both sources are available. Another necessary focus of vital rate monitoring in the release and post-release phase is immediate post-release versus later survival. In cases where this has been investigated, there is frequently evidence of low survival probability for some short period immediately following release, with increased survival for animals that survive this initial period and that are born into the population (e.g. Sarrazin & Legendre, 2000; Tavecchia *et al.*, 2009; Hamilton *et al.*, 2010). These examples again emphasize that monitoring should be tailored to the needs of the reintroduction programme. In cases where uncertainty exists about an optimal release strategy, estimation of vital rates under the different approaches will permit discrimination among competing hypotheses and learning, in this case about which approaches are likely to yield the best results. In other cases, uncertainty may still exist about the extent to which survival is depressed post-release and the length of this 'post-release' period. Resolution of this uncertainty is also important for optimal release strategies.

Monitoring – how?

Conditional on clear ideas about what should be estimated and how estimates will be used in subsequent decision making, the final question is how to monitor. Answers to the previous two monitoring questions (why and what) were dictated almost completely by the context of the AM programme for reintroduction. The answer to the 'how' question depends to a degree on this decision process context, but also on general principles of statistical inference. We will first discuss these principles in a general way and then relate them to approaches that should be especially useful for reintroduction programmes.

Inference from monitoring data: general principles

Monitoring programmes for animal and plant populations are typically based on counts. Such counts may be of animals seen from an airplane, birds seen or heard during point counts, animals captured in traps, number of surveyed patches at which an animal sign is detected, etc. Efforts to draw inferences about state variables and vital rates from such count data must account for two important sources of variation in the counts: geographic variation and detectability (probability of detection). Geographic variation refers to the variation of animal numbers and associated counts across space. If study areas are sufficiently small, as may be the case for reintroductions, then monitoring programmes may cover the entire area of interest and geographic variation is not an issue requiring much special treatment or consideration.

However, if the area of interest is too large to survey entirely, then land units to be surveyed must be selected in a manner that permits inference about the units not surveyed, and thus inference about the entire area of interest; i.e. based on counts obtained in surveyed land units, we would like to say something about the numbers of animals in units not surveyed. The key to such inference is a sampling design that permits computation of the probability of a sample unit being selected to receive survey efforts. For example, simple random sampling results in equal probability of being sampled for each unit, with that probability given by the ratio of number of units selected (sample size) to total units in the area for which inference is desired. Other probabilistic spatial sampling designs include stratified random sampling, systematic sampling, cluster sampling, adaptive sampling, double sampling and dual-frame sampling (e.g. Thompson, 2002; Pollock et al., 2002; Williams et al., 2002).

Detectability refers to the probability that an entity of interest is detected by a survey method. In surveys directed at abundance and vital rates, detectability refers to the probability that a survey method will detect an individual animal that is present. In the case of occupancy surveys, detectability is the probability of detecting the presence of the focal species in the sample unit, given that the species actually occurs in the unit. Inference about detectability is key to the use of counts to draw inference about the state variable(s) or vital rate(s) of interest. An estimate of focal species abundance in a surveyed area is obtained by dividing the count, or number of individuals detected in the area, by the estimated detection probability (Lancia et al., 1994; Borchers et al., 2002; Pollock et al., 2002; Williams et al., 2002). Similarly, an estimate of the number of sample units at which a species is present is obtained by dividing the number of units at which the species was detected, by the estimated probability of detecting the species in a unit (MacKenzie et al., 2002).

Some investigators use raw (not adjusted for detection probability) counts (of animals seen, caught, heard or of animal signs (tracks, pellets)) as 'indices' to abundance. The rationale is that, although the counts are related to true abundance in some unknown manner, comparison of counts over time, space or species can still be used for inference about *variation* in abundance. For example, if we count 10 birds in the breeding season of year 1 and 20 in the breeding season of year 2, we may not know how many birds are present at either time, but we might conclude that population size approximately doubled between the two years. If detection probability is the same for both years, then this conclusion would be justified. The difficulty with this inference is that the count represents the product of the quantity of interest (population size) and a confounding parameter (detection probability). In the example above, another plausible conclusion would be that the number of birds present in years 1 and 2 was the same, but that detection probability simply doubled between the two years. For inferences about occupancy, the number of locations at which animals (or an animal sign) is detected represents the raw count or index. Replicate samples permit inference about the number of locations that are occupied, including those at which occupancy is not detected. Whenever possible, our preference is to collect the information needed to draw separate inferences about the quantities of biological interest (e.g. abundance, occupied sites) and detection probability. Williams *et al.* (2002) provide a more detailed discussion of count statistics, estimates and indices.

Detectability is also relevant to inference about vital rates (Williams *et al.*, 2002). For example, one approach to survival estimation is to mark and release a set of animals at time t and then to return to the location at time $t + 1$ to recapture, re-sight or otherwise detect the marked animals from this original sample. Dividing the number of animals caught at $t + 1$ by the number of animals marked and released at t will usually not provide a good estimate of survival between t and $t + 1$, because not all marked animals that are still alive at the location at $t + 1$ will be captured. However, if we divide the number of marked animals caught at $t + 1$ by the estimated detection probability for $t + 1$, then we can estimate the number of marked animals that survived and did not emigrate. The ratio of this estimate to the number released at t then estimates the survival probability from t to $t + 1$. Thus, monitoring programmes that are used to estimate state variables and vital rates require assumptions about, or preferably estimates of, detection probability.

Statisticians and animal population ecologists have developed many different approaches to drawing inferences about detection probabilities for individual animals. Some approaches are based on direct observations of animals that are not individually identified, whereas others are based on identification of marked animals at multiple sampling occasions (Seber, 1982; Lancia *et al.*, 1994, 2005; Borchers *et al.*, 2002; Williams *et al.*, 2002). Observation-based approaches to inference include distance sampling, in which detection probability can be estimated because of an assumed functional relationship between detection probability of an animal and its distance from the observer (Burnham *et al.*, 1980; Buckland *et al.*, 2001, 2004). Multiple observer methods are based on surveys of a single location made by multiple observers. With two observers, numbers of individuals detected only by observer A, only by observer B and by both observers provide the information needed to estimate the detection probability for each observer and for both observers combined (Cook & Jacobson, 1979; Nichols *et al.*, 2000b). Temporal removal methods, sometimes referred to as time of detection methods, involve visual or aural surveys of a sample unit by a single observer, classifying individual animals by the time intervals during which they were first detected. The number of animals first detected in period 1, period 2, etc., provide the data required to estimate detection probability, and hence abundance (Farnsworth *et al.*, 2002). Sighting probability models (e.g. Caughley *et al.*, 1976; Samuel *et al.*, 1987) can be developed using detections and non-detections of radio-marked animals using standard survey methods. Radio signals are used to identify animals that are missed during the standard survey efforts. Covariates (e.g. habitat of observation, group size of observed animals) are obtained for animals that were and were not detected using the conventional survey methods, and these data are used to build models relating the covariates to the probability of detecting an animal. These models are then applied to standard surveys of unmarked animals in order to estimate abundance. Multiple counts (e.g. on successive days) at the same sample unit can be used to estimate abundance with the help of assumptions about the form of the spatial distribution of animals across sample units (Royle & Nichols, 2003; Royle, 2004a). Finally, we note that investigators are beginning to combine two or more of these observation-based approaches for inference about abundance (see the review of Nichols *et al.*, 2008).

The alternative to observation-based approaches using animals that are not identified individually to estimate detection probability, and hence abundance, is to use approaches that are based on individual identifications. Capture–recapture approaches are usually based on capturing animals and marking them with tags permitting individual identification at later times. However, animals of some species (e.g. tigers, jaguars, many marine mammals, some salamanders, frogs and lizards) can be identified individually by markings that can be observed by investigators. Individual animals of some species can now be identified by DNA obtained from sources such as faeces or hair samples (Woods *et al.*, 1999; Mowat & Strobeck, 2000; Gardner *et al.*, 2009). Regardless of the origin of marks used for individual identification, surveys are conducted on discrete sampling occasions. Detection histories (strings of '1's' (indicating detection) and '0's' (indicating non-detection)) of individuals across these sampling occasions permit inference about detection probabilities and abundance. Closed population capture–recapture models are based on sampling occasions that are separated by short time intervals, such that population size is not expected to change. Detection events in detection histories are modelled with a detection probability parameter and non-detection events are modelled with the complement of detection probability. Closed population models permit estimation of occasion-specific detection probability and abundance (Otis *et al.*, 1978; White *et al.*, 1982; Chao & Huggins, 2005). Open population capture–recapture models permit gains and losses of individuals to occur between sampling occasions. However, animals are known to be alive during each sampling occasion between the first detection and the final detection. The detection history data during these intervals within which animals are known to be alive are used to draw inferences about detection probabilities, and thus abundance, recruitment and survival (Jolly, 1965; Seber, 1965; Pollock *et al.*, 1990; Lebreton *et al.*, 1992; Nichols, 2005).

Some capture-based methods permit inference about detection probability without requiring individual identifications. Removal models (Otis *et al.*, 1978, White *et al.*, 1982) are based on data from multiple sampling occasions, with the relevant data being the number of animals removed (or first detected) at sampling occasions 1, 2, etc. Catch-effort models are based on similar data, with the addition of occasion-specific covariate information on sampling effort (e.g., Gould & Pollock, 1997a, 1997b). Both removal methods and catch-effort models permit inference about detection probabilities and abundance. Other approaches that use capture and removal data to draw inferences about detection and abundance are change-in-ratio approaches (Udevitz & Pollock, 1991, 1992) and trapping webs (Anderson *et al.*, 1983; Buckland *et al.*, 2001).

For the state variable of occupancy (proportion of habitat units or patches occupied) and its associated vital rates (probabilities of local extinction and colonization), monitoring programmes require repeat visits to patches within a season, a relatively short period over which occupancy does not change. At each visit to a patch, the species is detected or not. Data are again summarized as detection histories for all of the monitored patches. Information about detection probability comes primarily from patches at which the species is detected at least once during a season. At such patches, non-detection at an occasion is known to be a failure to detect, given presence, rather than an absence. Vital rates (e.g. probabilities of local extinction and colonization) are estimated based on models describing the patterns of change in occupancy for patches in the various occupancy states across seasons. Thus, inference about vital rates is based on the same approach as described for inference about occupancy itself.

In summary, there is a vast range of potential monitoring methods, and the answer to the question about 'how to monitor' depends on the entire management programme in which monitoring is embedded. However, there are also general principles that apply regardless of the management programme context. In cases where the area of interest is too large to be surveyed entirely, some sort of probabilistic sampling scheme should be used to select sample units (see Thompson, 2002; Williams *et al.*, 2002). Such selection permits valid inference about the entire area of interest based on counts at the sample units that are actually surveyed. Counts of both animals and plants almost always fail to include all individuals present in an area because of detection failures. Valid approaches to inference about population size, occupancy and their associated vital rates depend on dealing adequately with detection probability. Approaches for inference about detection probability are varied and reflect the innovation of animal ecologists and the statisticians who work with them (Seber, 1982; Lancia *et al.*, 1994, 2005; Borchers *et al.*, 2002; Pollock *et al.*, 2002; Williams *et al.*, 2002). This variety of available approaches is needed, as approaches that are suitable for one set of biological and logistical constraints may be useless in another situation.

Inference from monitoring data: reintroduction programmes

The design and conduct of monitoring for reintroduction programmes should follow the principles outlined above. Frequently, the release area may be so small that the entire area can be surveyed. If this is not true, then stratified random sampling, with strata defined by proximity of sample units to the release site(s), will often represent a reasonable approach to the sampling of space. Dealing with the issue of detectability will require any of a number of possible approaches depending on the focus of the particular monitoring survey and on associated biological and logistical constraints.

Monitoring *before release* will usually focus on habitat quality (e.g. including predator abundance, food resources, nesting sites, cover, etc.). If an initial assessment deems a habitat to be suitable, based on hypotheses about focal species needs, then pre-release monitoring may not be needed. However, if a habitat is deemed unsuitable, based on an initial assessment, then pre-release habitat management can employ an AM process that will include habitat monitoring. Detectability may or may not be an important issue in the assessment of habitat. In the case where a habitat is to be described by a vector of vegetation variables, for example, direct measurement of such variables at sample plots may be adequate. In cases for which the habitat can be classified into discrete states, and when these habitat states are relatively few in number, it may be possible for trained observers accurately to classify the habitat into states, based on direct observations. In other cases, for example when the habitat is classified via remotely sensed data, misclassification may be possible. In such cases, it is often possible to identify a subset of sites at which the truth is known (e.g. based on a more intensive ground survey effort) and to estimate misclassification probabilities directly as the fraction of the subset of sites at which classification was not correct (see Breininger *et al.*, 2010; McClintock *et al.*, 2010). Just as with detection probability, if misclassification probability can be estimated, then this source of uncertainty can be incorporated into inferences and into the larger AM decision process. In cases where the habitat is classified into discrete states, dynamic modelling requires estimates of state transition probabilities of moving from one state to another. These transition probabilities are the vital rates for dynamic habitat modelling, and they are estimated using a time series of habitat data for all sampled sites. Data for each site will be the habitat history, consisting of the site's habitat state at each time period in the history. State transition probabilities can be estimated directly from these history data, for example using multistate models (Breininger *et al.*, 2009, 2010).

Pre-release monitoring of predators and competitors will require designs that are tailored to the particular species. Initial estimates of abundance might be based on observation-based or capture-based methods for inference about detection probability, depending on the species and situation. If management includes active efforts to reduce population sizes, then monitoring may shift to an occupancy approach as the species becomes more rare. Although such occupancy modelling does not provide an estimate of abundance in most cases, it will estimate occurrence and provide an estimate of the probability that an undetected species is truly absent. Approaches to inference about vital rates will differ, depending on whether the state variable is abundance or occupancy. Inferences about survival probabilities of individual animals usually require samples of animals that can be individually identified. Open population capture–recapture models, or their analogs for use with radio-telemetry data, can then be used to estimate survival probabilities. Overall reproductive recruitment can be estimated using open-population capture–recapture models as well (e.g. Nichols & Pollock, 1990; Pradel, 1996; Schwarz & Arnason, 1996; Nichols *et al.*, 2000a). Components of reproduction can be estimated using a variety of approaches, depending on the component of interest. Nest success modelling (e.g. Mayfield, 1961; Dinsmore *et al.*, 2002; Rotella *et al.*, 2004; Shaffer, 2004; Jones & Geupel, 2007) permits inference about this parameter from nest visit data. Open population capture–recapture models and data permit inference about the probability that an individual breeds in a given year (e.g. Nichols *et al.*, 1994; Kendall *et al.*, 1997; Cam *et al.*, 1998). Rates of movement can also be estimated using open population capture–recapture data collected at multiple sites (e.g. Hestbeck *et al.*, 1991; Brownie *et al.*, 1993; Spendelow *et al.*, 1995; Lebreton *et al.*, 2009) or by following radio-tracked individuals (e.g. Richard & Armstrong, 2010). So there are many available approaches for monitoring that can be useful for assessing predator and/or competitor populations and dynamics prior to release.

It is worth noting that these inferences about demographic vital rates described above will frequently require marked individuals (or nests). Thus, even if the state variable of abundance is best estimated using an observation-based method, capture-based approaches will be needed for inference about demographic vital rates and their components. In contrast to this situation, the vital rates for occupancy dynamics (in the simplest models, rates of local extinction and colonization) are estimated using time series of the same data as used to estimate single-season occupancy (MacKenzie *et al.*, 2003, 2006, 2009). Once again, we emphasize that although the vital rates themselves will be of interest, of even more interest will be the functional relationships between these vital rates and management (e.g. control) efforts. These functional relationships can be directly estimated using the modelling approaches identified above.

Monitoring *after release* may include vegetation and predator/competitor species, as before release, but will also include the focal species being released. Abundance can be estimated using observation-based or capture-based approaches, and associated vital rates will usually require marked individuals. Capture-based approaches will require marking each released animal and then devising an approach (based on captures or direct observations) to observe these animals in future sampling occasions. Time-specific estimates of focal species abundance and rate of change in abundance should be closely related to objectives of most reintroduction programmes. Vital rates of focal species will also be important, and inference about these will usually be based on open population capture–recapture approaches. Inference models for resulting detection history data will then focus on the questions most useful to the reintroduction programme. For example, models incorporating different hypotheses about survival of newly released versus other animals and animals released using different methods will be especially useful. Stage- and age-specific variations in vital rates will be important as well because of their relevance to transient dynamics expected in the periods following initial release. If there is substantive variation in habitat quality, then modelling the functional relationship between habitat metrics and focal species vital rates will be important (Armstrong *et al.*, 2006). A problem for monitoring abundance and vital rates in reintroduction programmes is the small numbers of marked animals likely to be available for study. Even if every released animal carries an individual mark, the small numbers involved in many release efforts and the need to stratify by age and stage will make precise inference very difficult. If information obtained elsewhere is relevant and available, then formal incorporation of such prior information using Bayesian approaches may be useful (e.g. McCarthy, 2007; Link & Barker, 2010). The natural response to the problems of small sample size and imprecision by those carrying out the monitoring is to attempt to keep detection probabilities as high as possible through use of intensive sampling. Ultimately, it will be useful to incorporate costs of management actions and monitoring activities into the programme objective function, permitting optimization that accounts for the possibility of low information content of expensive monitoring, perhaps leading to decisions to reduce monitoring efforts based on cost–benefit ratios (see below).

Occupancy monitoring may be useful for inference about the geographic spread of the focal species. Even immediately following release, when the primary emphasis is on demographic monitoring for focal species abundance and vital rates, it may be useful to set up an occupancy monitoring design for neighbouring areas. Such monitoring would permit inference about the initial spread of the released population, focusing on the colonization of nearby land units. Reasonable inference models would likely model colonization in subsequent time periods as a function of both time since initial release and distance from the release site (or nearest of multiple release sites). So-called autologistic models in which colonization and extinction are modelled as a function of occupancy status of neighbouring sites may prove useful as well (Royle & Dorazio, 2008).

In summary, there is no simple recipe for specifying exactly how monitoring should be conducted in order best to inform a reintroduction programme. Instead, the monitoring programme should be tailored to the needs and logistical constraints of the specific reintroduction programme. We have suggested that monitoring should frequently focus on more than one state variable and set of vital rates, usually leading to multiple survey methods. For example, detectability may not be an important issue for monitoring some habitat state variables, but will probably be important for monitoring predator and competitor species as well as the reintroduced species. Different monitoring methods may be selected for each species, based on biology, logistics and information needs. There may also be reason to select multiple state variables, vital rates and monitoring approaches for multiple information needs with respect to a single species. For example, open population capture–recapture approaches permit inference about demographic vital rates and initial population growth of focal species. However, space–time changes in the distribution of the species as it spreads from the release area are best monitored using occupancy approaches, with special attention to rates of colonization into previously unoccupied locations.

This absence of a simple recipe for all reintroduction programme monitoring should not be viewed negatively, but instead as providing investigators with a great deal of flexibility and opportunity for success. The variety of approaches for estimating abundance and demographic vital rates (e.g. Seber, 1982; Lebreton *et al.*, 1992, 2009; Lancia *et al.*, 1994, 2005; Borchers *et al.*, 2002; Williams *et al.*, 2002) represents an impressive toolbox of methods. Although development has been much more recent, a variety of methods now exists for inference about occupancy and associated vital rates (MacKenzie *et al.*, 2006; Royle & Dorazio, 2008). Thus, the investigator has the ability to tailor monitoring approaches to specific variables, species and logistical considerations.

Discussion

Monitoring plans developed to inform reintroduction programmes face a number of challenges that are particular to such programmes. The problem of small sample sizes when monitoring the focal species was discussed briefly. Many animal monitoring programmes are purposely established at locations believed to contain large numbers of focal animals, for the sole reason that such populations are most likely to yield large samples, precise estimates and strong inferences. Reintroduction programmes represent the opposite extreme, as small numbers are frequently guaranteed, at least for the years following initial reintroductions. Intensive sampling designed to yield high detection probabilities represents a reasonable response to small numbers, but it is still likely that imprecise estimates will be characteristic of many monitoring programmes for reintroduction efforts. The small sample problem is exacerbated by the interest in transient dynamics, and thus in demographic strata (ages and stages), and the resultant need for inference models that carry different vital rate parameters for the different strata. This major source of uncertainty (partial observability) places a premium on approaches such as AM that attempt to deal with uncertainty within the overall management programme.

Another challenge for many reintroduction programmes is a lack of experience with management methods. For new reintroduction programmes, there is naturally little experience with population dynamics following releases and little or no experience with different release strategies. This lack of experience amplifies the importance of learning during the implementation of reintroduction programmes. However, the small sample sizes that characterize such programmes lead to imprecision, which serves to slow the rate of learning. The slow learning that is likely to characterize individual reintroduction programmes should call for increased efforts to exploit other reintroduction programmes, by borrowing information about related systems, species (e.g. similar to the model species concept of Jones & Merton, this volume, Chapter 2) and management programmes (e.g. MacKenzie et al., 2005; McCarthy, 2007). Therefore, if the same approaches to introduction are used in multiple areas for multiple species, then learning (updating of model weights) might use information from these multiple management efforts.

Monitoring after release will often focus on re-encountering animals released with marks (such as colour bands). Models constructed to analyse such data must account for potential post-release differences in vital rates (Armstrong & Reynolds, this volume, Chapter 6). Furthermore, vital rates of released animals may be permanently different from the vital rates of descendents, particularly if captive stock is used. Such differences may require separate inference about vital rates of wild-born animals, meaning that at least a sample of wild-born animals would need to be marked for possible re-encounter.

Despite these challenges, we believe that monitoring programmes are essential to informed management, in general, and to reintroduction programmes, in particular. Management programmes provide the context for monitoring and lead to a sharp focus on exactly the data and inferences most useful for management success. The recurrent nature of many reintroduction programmes and the existence of several important sources of uncertainty in these programmes lead to the expectation that AM will be a useful approach to such programmes. Monitoring is used for four primary purposes in AM:

making state-dependent decisions, assessing programme success, learning, and developing and refining models. This identification of the use of monitoring data facilitates the design of monitoring programmes, helping to insure that resulting data are adequate to meet programme needs. In addition to those aspects of monitoring design that are inherited directly from the larger programme, general principles of inference for ecological count data should be acknowledged as well. Specifically, monitoring methods should attempt to deal with the two key sources of variation in ecological count data, geographic variation and detectability. As outlined, there are many potentially useful approaches to deal with each of these sources of variation, with selection of an approach depending on the biological and logistical details of the system to be monitored. In summary, designs of monitoring for reintroduction programmes should be based on the information needs of the larger programmes and on basic principles of statistical inference.

In keeping with the overall theme of monitoring needs and designs being inherited from the larger management programme, we note that, in some situations, part of the decision entails funding allocation to management actions versus monitoring (e.g. Gerber *et al.*, 2005; Hauser *et al.*, 2006; Chades *et al.*, 2008). In such instances, the costs of monitoring and other management actions will be a component of the objective function, such that optimal management decisions will now include recommended monitoring effort. Two roles of monitoring, state-dependent decisions and learning, are very important determinants of management returns and success, and will figure prominently in assessing the relative value of monitoring versus other management actions. Concepts such as the expected value of perfect and partial information (e.g. see Raiffa & Schlaifer, 1961; Walters, 1986; Williams *et al.*, 2002) are relevant to this valuation of monitoring and its contributions to overall management objectives. Once again, it is knowledge of the explicit roles of monitoring in the larger management process that permits a formal treatment of this question of allocation of effort to monitoring. We believe that such formal treatment has the potential to be very useful in the design of future reintroduction programmes.

References

Anderson, D.R., Burnham, K.P., White, G.C. *et al.* (1983) Density estimation of small-mammal populations using a trapping web and distance sampling methods. *Ecology*, 64, 674–680.

Armstrong, D.P. & Reynolds, M.H. (2011) Modelling reintroduced populations: the state of the art and future directions. In *Reintroduction Biology: integrating science and management*, eds J.G. Ewen, D.P. Armstrong, K.A. Parker & P.J. Seddon, Chapter 6. Wiley-Blackwell.

Armstrong, D.P., Raeburn, E.H., Lewis, R.M. *et al.* (2006) Modelling vital rates of a reintroduced New Zealand robin population as a function of predator control. *Journal of Wildlife Management*, 70, 1028–1036.

Bellman, R. (1957) *Dynamic Programming*. Princeton University Press, Princeton, New Jersey, USA.

Borchers, D.L., Buckland, S.T. & Zucchini, W. (2002) *Estimating Animal Abundance: closed populations*. Springer, New York, USA.

Breinenger, D.R., Nichols, J.D., Carter, G.M. *et al.* (2009) Habitat-specific breeder survival of Florida scrub-jays: inferences from multistate models. *Ecology*, 90, 3180–3189.

Breinenger, D.R., Nichols, J.D. (2010) Multistate modelling of habitat dynamics: factors affecting Florida scrub transition probabilities. *Ecology*, 91, 3354–3364.

Brownie, C., Hines, J.E., Nichols, J.D. *et al.* (1993) Capture–recapture studies for multiple strata including non-Markovian transition probabilities. *Biometrics*, 49, 1173–1187.

Buckland, S.T., Anderson, D.R., Burnham, K.P. *et al.* (2001) *Introduction to Distance Sampling*. Oxford University Press, Oxford, UK.

Buckland, S.T., Anderson, D.R., Burnham, K.P. *et al.* (2004) *Advanced Distance Sampling. Estimating abundance of biological populations*. Oxford University Press, Oxford, UK.

Burgman, M. (2005) *Risks and Decisions for Conservation and Environmental Management*. Cambridge University Press, Cambridge, UK.

Burnham, K.P., Anderson, D.R. & Laake, J.L. (1980) Estimation of density from line transect sampling of biological populations. *Wildlife Monographs*, 72, 1–202.

Callaway, R.M. & Davis, F.W. (1993) Vegetation dynamics, fire, and physical environment in coastal central California. *Ecology*, 74, 1567–1578.

Cam, E., Hines, J.E., Monnat, J.-Y. *et al.* (1998) Are adult nonbreeders prudent parents? The kittiwake model. *Ecology*, 79, 2917–2930.

Caswell, H. (2001) *Matrix Population Models: construction, analysis, and interpretation*, 2nd edition. Sinauer Associates, Sunderland, Massachusetts, USA.

Caswell, H. (2007) Sensitivity analysis of transient population dynamics. *Ecology Letters*, 10, 1–15.

Caughley, G., Sinclair, R. & Scott-Kemmis, D. (1976) Experiments in aerial survey. *Journal of Wildlife Management*, 40, 290–300.

Chades, I., McDonald-Madden, E., McCarthy, M.A. *et al.* (2008) When to stop managing or surveying cryptic threatened species. *Proceedings of the National Academy of Sciences, USA*, 105, 13936–13940.

Chao, A. & Huggins, R.M. (2005) Modern closed population models. In *The Handbook of Capture–Recapture Analysis*, eds S. Amstrup, T. McDonald & B. Manly, pp. 58–86. Princeton University Press, Princeton, New Jersey, USA.

Clemen, R.T. & Reilly, T. (2001) *Making Hard Decisions with Decision Tools*. Duxbury Press, Pacific Grove, California.

Cook, R.D. & Jacobsen, J.O. (1979) A design for estimating visibility bias in aerial surveys. *Biometrics*, 35, 735–742.

Dinsmore, S.J., White, G.C. & Knopf, F.L. (2002) Advanced techniques for modelling avian nest survival. *Ecology*, 83, 3476–3488.

Dorazio, R.M. & Johnson, F.A. (2003) Bayesian inference and decision theory – a framework for decision making in natural resource management. *Ecological Applications*, 13, 556–563.

Ewen, J.G. & Armstrong, D.P. (2007) Strategic monitoring of re-introductions in ecological restoration programmes. Ecoscience, 14, 401–409

Farnsworth, G.L., Pollock, K.H., Nichols, J.D. *et al.* (2002) A removal model for estimating detection probabilities from point count surveys. *Auk*, 119, 414–425.

Fonnesbeck, C.J. (2005) Solving dynamic wildlife resource optimization problems using reinforcement learning. *Natural Resource Modelling*, 18, 1–39.

Gardner, B., Royle, J.A. & Wegan, M.T. (2009) Hierarchical models for estimating density from DNA mark-recapture studies. *Ecology*, 90, 1106–1115.

Gerber, L.R., Beger, M., McCarthy, M.A. *et al.* (2005) A theory of optimal monitoring of marine reserves. *Ecology Letters*, 8, 829–837.

Gould, W.R. & Pollock, K.H. (1997a) Catch-effort estimation of population parameters under the robust design. *Biometrics*, 53, 207–216.

Gould, W.R. & Pollock, K.H. (1997b) Catch-effort maximum likelihood estimation of population parameters. *Canadian Journal of Fisheries and Aquatic Science*, 54, 890–897.

Griffith, B., Scott, J.M., Carpenter, J.W. *et al.* (1989) Translocation as a species conservation tool: status and strategy. *Science*, 245, 477–480.

Hamilton, L.P., Kelly, P.A., Williams, D.F. *et al.* (2010). Factors associated with survival of reintroduced riparian brush rabbits in California. *Biological Conservation*, 143, 999–1007.

Hauser, C.E., Pople, A.R. & Possingham, H.P. (2006) Should managed populations be monitored every year? *Ecological Applications*, 16, 807–819.

Hestbeck, J.B., Nichols, J.D. & Malecki, R.A. (1991) Estimates of movement and site fidelity using mark-resight data of wintering Canada geese. *Ecology*, 72, 523–533.

IUCN (1987) *IUCN Position Statement on the Translocation of Living Organisms: introductions, re-introductions, and re-stocking*. IUCN, Gland, Switzerland.

IUCN (1998) *Guidelines for Reintroductions*. IUCN/SSC Re-introduction Specialist Group, IUCN, Gland, Switzerland and Cambridge, United Kingdom.

Jolly, G.M. (1965) Explicit estimates from capture–recapture data with both death and immigration-stochastic model. *Biometrika*, 52, 225–247.

Jones, C.G. & Merton, D.V. (2011) A tale of two islands: the rescue and recovery of endemic birds in New Zealand and Mauritius. In *Reintroduction Biology: integrating science and management*, eds J.G. Ewen, D.P. Armstrong, K.A. Parker & P.J. Seddon, Chapter 2. Wiley-Blackwell, Oxford, UK.

Jones, S.L. & Geupel, G.R. (eds) (2007) Beyond Mayfield: measurements of nest-survival data. *Studies in Avian Biology*, 34.

Kendall, W.L., Nichols J.D. & Hines, J.E. (1997) Estimating temporary emigration and breeding proportions using capture–recapture data with Pollock's robust design. *Ecology*, 78, 563–578.

Keyfitz, N. & Caswell, H. (2005) *Applied Mathematical Demography*, 3rd edition. Springer-Verlag, New York, USA.

Koons, D.N., Grand, J.B., Zinner, B. *et al.* (2005) Transient population dynamics: relations to life history and initial population state. *Ecological Modelling*, 185, 283–297.

Koons, D.N., Holmes, R.R. & Grand, J.B. (2007) Population interia and its sensitivity to changes in vital rates and population structure. *Ecology*, 88, 2857–2867.

Koons, D.N., Rockwell, R.F. & Grand, J.B. (2006) Population momentum: implications for wildlife management. *Journal of Wildlife Management*, 70, 19–26.

Lancia, R.A., Kendall, W.L., Pollock K.H. *et al.* (2005) Estimating the number of animals in wildlife populations. In *Research and Management Techniques for Wildlife and Habitats*, ed. C.E. Braun, pp. 106–153. The Wildlife Society, Bethesda, Maryland, USA.

Lancia, R.A., Nichols, J.D. & Pollock, K.H. (1994) Estimating the number of animals in wildlife populations. In *Research and Management Techniques for Wildlife and Habitats*, ed. T. Bookhout, pp. 215–253. The Wildlife Society, Bethesda, Maryland, USA.

Lebreton, J.-D., Burnham, K.P., Clobert, J. *et al.* (1992) Modelling survival and testing biological hypotheses using marked animals: a unified approach with case studies. *Ecological Monographs*, 62, 67–118.

Lebreton, J.-D., Nichols, J.D., Barker, R. *et al.* (2009) Modelling individual animal histories with multistate capture–recapture models. *Advances in Ecological Research*, 41, 87–173.

Le Gouar, P., Mihoub, J.-B. & Sarrazin, F. (2011) Dispersal and habitat selection: behavioural and spatial constraints for animal translocations. In *Reintroduction Biology: integrating science and management*, eds J.G. Ewen, D.P. Armstrong, K.A. Parker & P.J. Seddon, Chapter 5. Wiley-Blackwell, Oxford, UK.

Link, W.A. & Barker, R.J. (2010) *Bayesian Inference with Ecological Applications*. Academic Press, New York, USA.

Lubow, B.C. (1995) SDP: generalized software for solving stochastic dynamic optimization problems. *Wildlife Society Bulletin*, 23, 738–742.

Lyles, A.M. & May, R.M. (1987) Problems in leaving the ark. *Nature*, 326, 245–246.

MacKenzie, D.I., Nichols, J.D., Hines, J.E. *et al.* (2003) Estimating site occupancy, colonization and local extinction probabilities when a species is not detected with certainty. *Ecology*, 84, 2200–2207.

MacKenzie, D.I., Nichols, J.D., Lachman, G.B. *et al.* (2002) Estimating site occupancy when detection probabilities are less than one. *Ecology*, 83, 2248–2255.

MacKenzie, D.I., Nichols, J.D., Royle, J.A. *et al.* (2006) *Occupancy Modelling and Estimation*. Academic Press, San Diego, California, USA.

MacKenzie, D.I., Nichols, J.D., Seamans, M.E. *et al.* (2009) Dynamic models for problems of species occurrence with multiple states. *Ecology*, 90, 823–835.

MacKenzie, D.I., Nichols, J.D., Sutton, N. *et al.* (2005) Suggestions for dealing with detection probability in population studies of rare species. *Ecology*, 86, 1101–1113.

Martin, J., Runge, M.C., Nichols, J.D. *et al.* (2009) Structured decision making as a conceptual framework to identify thresholds for conservation and management. *Ecological Applications*, 19, 1079–1090.

Mayfield, H.F. (1961) Nesting success calculated from exposure. *Wilson Bulletin*, 73, 255–261.

McCarthy, M.A. (2007) *Bayesian Methods for Ecology*. Cambridge University Press, Cambridge, UK.

McCarthy, M.A. & Possingham, H.P. (2007) Active adaptive management for conservation. *Conservation Biology*, 21, 956–963.

McCarthy, M.A., Armstrong, D.P. & Runge, M.C. (2011) Adaptive management of reintroductions. In *Reintroduction Biology: integrating science and management*, eds J.G. Ewen, D.P. Armstrong, K.A. Parker & P.J. Seddon, Chapter 8. Wiley-Blackwell, Oxford, UK.

McClintock, B.T., Nichols, J.D., Bailey, L.L. *et al.* (2010) Seeking a second opinion: uncertainty in wildlife disease ecology. *Ecology Letters*, 13, 659–674.

Mowat, G. & Strobeck, C. (2000) Estimating population size of grizzly bears using hair capture, DNA profiling, and mark–recapture analysis. *Journal of Wildlife Management*, 64, 183–193.

Nichols, J.D. (2001) Using models in the conduct of science and management of natural resources. In *Modelling in Natural Resource Management: development, interpretation and application*, eds T.M. Shenk & A.B. Franklin, pp. 11–34. Island Press, Washington, USA.

Nichols, J.D. (2005) Capture–recapture models for open populations: recent developments. In *Handbook of Capture–Recapture Methods*, eds S.C. Amstrup, T.L. McDonald & B.F.J. Manly, pp. 88–123. Princeton University Press, Princeton, New Jersey, USA.

Nichols, J.D. & Pollock, K.H. (1990) Estimation of recruitment from immigration versus in situ reproduction using Pollock's robust design. *Ecology*, 71, 21–26.

Nichols, J.D. & Williams, B.K. (2006) Monitoring for conservation. *Trends in Ecology and Evolution*, 21, 668–673.

Nichols, J.D., Hines, J.E., MacKenzie, D.I. *et al.* (2007) Occupancy estimation with multiple states and state uncertainty. *Ecology*, 88, 1395–1400.

Nichols, J.D., Hines, J.E., Pollock, K.H. *et al.* (1994) Estimating breeding proportions and testing hypotheses about costs of reproduction with capture–recapture data. *Ecology*, 75, 2052–2065.

Nichols, J.D., Hines, J.E., Lebreton, J.-D. *et al.* (2000a) Estimation of contributions to population growth: a reverse-time capture–recapture approach. *Ecology*, 81, 3362–3376.

Nichols, J.D., Hines, J.E., Sauer, J.R. *et al.* (2000b) A double-observer approach for estimating detection probability and abundance from point counts. *Auk* 117: 393–408.

Nichols, J.D., Thomas, L. & Conn, P.B. (2008) Inferences about landbird abundance from count data: recent advances and future directions. In *Modelling Demographic Processes in Marked Populations*, eds D.L. Thomson, E.G. Cooch & M.J. Conroy, pp. 201–235. Springer, New York, USA.

Osborne, P.E. & Seddon, P.J. (2011) Selecting suitable habitats for reintroductions: variation, change and the role of species distribution modelling. In *Reintroduction Biology: integrating science and management*, eds J.G. Ewen, D.P. Armstrong, K.A. Parker, & P.J. Seddon, Chapter 3. Wiley-Blackwell, Oxford, UK.

Otis, D.L., Burnham, K.P., White, G.C. *et al.* (1978) Statistical inference from capture data on closed animal populations. *Wildlife Monographs*, 62, 1–135.

Pollock, K.H., Nichols, J.D., Brownie, C. *et al.* (1990) Statistical inference for capture–recapture experiments. *Wildlife Monographs*, 107, 1–97.

Pollock, K.H., Nichols, J.D., Simons, T.R. *et al.* (2002) Large scale wildlife monitoring studies: statistical methods for design and analysis. *Environmetrics*, 13, 1–15.

Pradel, R. (1996) Utilization of capture–mark–recapture for the study of recruitment and population growth rate. *Biometrics*, 52, 703–709.

Raiffa, H. & Schlaifer, R.O. (1961) *Applied Statistical Decision Theory*. Graduate School of Business Administration, Harvard University, Boston, Massachusetts, USA.

Richard, Y. & Armstrong, D.P. (2010) Cost distance modelling of landscape connectivity and gap-crossing ability using radio-tracking data. *Journal of Applied Ecology*, 47, 603–610.

Rotella, J.J., Dinsmore, S.J. & Shaffer, T.L. (2004) Modelling nest-survival data: a comparison of recently developed methods that can be implemented in MARK and SAS. *Animal Biodiversity and Conservation*, 27, 187–205.

Rout, T.M., Hauser, C.E. & Possingham, H.P. (2009) Optimal adaptive management for the translocation of a threatened species. *Ecological Applications*, 19, 515–526.

Royle, J.A. (2004a) N-mixture models for estimating population size from spatially replicated counts. *Biometrics*, 60, 108–115.

Royle, J.A. (2004b) Modelling abundance index data from anuran calling surveys. *Conservation Biology*, 18, 1378–1385.

Royle, J.A. & Dorazio, R.M. (2008) *Hierarchical Modelling and Inference in Ecology*. Academic Press, New York, USA.

Royle, J.A. & Link, W.A. (2005) A general class of multinomial mixture models for anuran calling survey data. *Ecology*, 86, 2505–2512.

Royle, J.A. & Nichols, J.D. (2003) Estimating abundance from repeated presence absence data or point counts. *Ecology*, 84, 777–790.

Samuel, M.D., Garton, E.O., Schlegel, M.W. *et al.* (1987) Visibility bias during aerial surveys of elk in north central Idaho. *Journal of Wildlife Management*, 51, 622–630.

Sarrazin, F. & Legendre, S. (2000) Demographic approach to releasing adults versus young in reintroductions. *Conservation Biology*, 14, 488–500.

Schwarz, C.J. & Arnason, A.N. (1996) A general methodology for the analysis of capture–recapture experiments in open populations. *Biometrics*, 52, 860–873.

Seber, G.A.F. (1965) A note on the multiple recapture census. *Biometrika*, 52, 249–259.

Seber, G.A.F. (1982) *The Estimation of Animal Abundance and Related Parameters*, 2nd edition. Macmillan Publishing Company, New York, USA.

Shaffer, T.L. (2004) A unifed approach to analyzing nest success. *Auk*, 121, 526–540.

Spendelow, J.A., Nichols, J.D., Nisbet, I.C.T. *et al.* (1995) Estimating annual survival and movement rates within a metapopulation of roseate terns. *Ecology*, 76, 2415–2428.

Sutherland, W.J., Armstrong, D., Butchart, S.H.M. *et al.* (2010) Standards for documenting and monitoring bird reintroduction projects. *Conservation Letters*, 3, 229–235.

Tavecchia, G., Covadonga, V., Martinez-Abrain, A. *et al.* (2009) Maximizing reintroduction success: assessing the immediate cost of release in a threatened waterfowl. *Biological Conservation*, 142, 3005–3012.

Thompson, S.K. (2002) *Sampling*. John Wiley & Sons, Inc., New York, USA.

Udevitz, M.S. & Pollock, K.H. (1991) Change-in-ratio estimators for populations with more than two subclasses. *Biometrics*, 47, 1531–1546.

Udevitz, M.S. & Pollock, K.H. (1992) Change-in-ratio methods for estimating population size. In *Wildlife 2001: populations*, eds D.R. McCullough & R.H. Barrett, pp. 90–101. Elsevier, New York, USA.

Usher, M.B. (1979) Markovian approaches to ecological succession. *Journal of Animal Ecology*, 48, 413–426.

Waggoner, P.E. & Stephens, G.R. (1970) Transition probabilities for a forest. *Nature*, 255, 1160–1161.

Walters, C.J. (1986) *Adaptive Management of Renewable Resources.* Macmillan Publishing Company, New York, USA.

White, G.C., Anderson, D.R., Burnham, K.P. *et al.* (1982) *Capture–Recapture and Removal Methods for Sampling Closed Populations.* Los Alamos National Laboratory Publication LA-8787-NERP, Los Alamos, New Mexico, USA.

Williams, B.K. (1997) Approaches to the management of waterfowl under uncertainty. *Wildlife Society Bulletin*, 25, 714–720.

Williams, B.K., Nichols, J.D. & Conroy, M.J. (2002) *Analysis and Management of Animal Populations.* Academic Press, San Diego, California, USA.

Williams, B.K., Szaro, R.C. & Shapiro, C.D. (2007) *Adaptive Management: the US Department of the Interior Technical Guide.* US Department of the Interior, Washington, DC, USA.

Woods, J.G., Paetkau, D., Lewis, D. *et al.* (1999) Genetic tagging of freeranging black and brown bears. *Wildlife Society Bulletin*, 27, 616–627.

Yoccoz, N.G., Nichols, J.D. & Boulinier, T. (2001) Monitoring of biological diversity in space and time. *Trends in Ecology and Evolution*, 16, 446–453.

8

Adaptive Management of Reintroduction

Michael A. McCarthy[1], Doug P. Armstrong[2]
and Michael C. Runge[3]

[1]School of Botany, University of Melbourne, Melbourne, Australia
[2]Massey University, Palmerston North, New Zealand
[3]Patuxent Wildlife Research Center, U.S. Geological Survey,
Laurel, Maryland, United States of America

*'Thinking about reintroductions from the perspective of adaptive management will
help to focus managers on clearly specifying the management objective, noting
where uncertainty clouds the management decision, identifying which aspects can
be usefully clarified by monitoring and exploiting the ability of monitoring results
to change management.'*

Page 284

Introduction

Uncertainty is prevalent in reintroduction programs (Armstrong & Seddon,
2008). The magnitude of current threats at a site relative to threats faced
previously or in other parts of a species range will usually be uncertain, as will
knowledge about how well reintroduced animals will survive and reproduce,
and whether captive-bred or wild animals are the best alternatives for sourcing
animals for reintroduction. If captive breeding is used, the best techniques
for captive husbandry to maximize performance of individuals on release and

Reintroduction Biology: Integrating Science and Management. First Edition.
Edited by John G. Ewen, Doug P. Armstrong, Kevin A. Parker and Philip J. Seddon.
© 2012 Blackwell Publishing Ltd. Published 2012 by Blackwell Publishing Ltd.

the best methods of release are also often uncertain, especially for species that have little history of captive breeding. Adaptive management (AM) can be helpful in the face of such uncertainty, by identifying aspects where suitable monitoring, and possible experiments, can help resolve the uncertainty to the benefit of management (Holling, 1978; Walters & Hilborn, 1978; Walters, 1986; McCarthy & Possingham, 2007).

The uncertain choices between management options can sometimes be represented as choices between hypotheses (Williams, 1996). The hypotheses might reflect the consequences of management options, such as how they increase the post-release survival and breeding success of released animals and, if so, by how much. Other hypotheses might reflect the key questions in reintroduction biology posed by Armstrong & Seddon (2008) (see Figure 6.3 in this volume). For example, the choice between types of release methods (e.g. providing supplementary food and/or holding animals in a pen at the release site versus immediate release with no supplementary food) can be represented as a choice between competing hypotheses; one of the release methods is most cost-effective for achieving a successful reintroduction. Some questions will be more amenable to AM than others, and this will depend on their importance for designing a management strategy, the degree of uncertainty, the ability of monitoring to help answer the questions and the ability of managers to change management in response to what is learnt.

A standard scientific response to uncertainty about the validity of competing hypotheses is to experiment and people have been urged to incorporate experiments into reintroduction programmes (Armstrong *et al.*, 1995; Sarrazin & Barbault, 1996). Data will indicate the relative performance of different management options and the evidence in favour of one option over another will become substantial with a sufficient number of samples. Traditional experimental design provides well-established methods to measure the ability of a single experiment to discriminate between hypotheses, determine the sample sizes required to achieve a particular level of certainty and design experiments within a particular budget to maximize the ability to discriminate between the hypotheses, i.e. maximize statistical power (Cohen, 1988). However, traditional experimental design does not guide how much experimentation and monitoring is warranted when improved management is the specific aim (Lee, 1999).

When faced with a seemingly novel reintroduction programme, for example captive breeding and release of a species for which little information exists, the number of uncertainties is likely to be substantial. Determining which of these uncertainties is most important for effective management of the species is beyond the scope of traditional experimental design. Nor does traditional experimental design answer some other relevant questions. For example, managers might ask how much certainty is required from experiments. At what point should experiments be stopped when one method appears superior to the alternatives? Requiring experiments to have arbitrary levels of certainty, for example based on satisfying a type-I error rate of 0.05, is unlikely to provide outcomes that maximize the performance of the reintroduction programme. This occurs because any experimental use of more than one strategy means that an inferior strategy must be used to some extent. How does one determine whether the information gained from an experiment will be sufficient to justify the use of an inferior method in an experiment?

Typical experimental design also does not account for how to maximize power in a series of experiments. A series of experiments may be usual in natural resource management generally, and is likely to be true in reintroduction biology specifically because few experiments will have a sufficient sample size and scope to provide conclusive results by themselves. In these circumstances, decisions about what experiment to use will depend on the current information that is available, such as the relative support for the different hypotheses and the predictions under each of those hypotheses.

Managers may be reluctant to do experiments because sample sizes may be too small to provide confident results or the risks of some experimental methods may appear too large. In these circumstances, how should experiments be designed to maximize the performance of management rather than maximizing statistical power? And what criteria should a manager use to change management decisions? AM can answer these questions.

AM can help managers balance the benefits of improved information against the immediate goals of management. This chapter reviews the literature on AM, with a particular focus on its application to reintroduction. A key feature of AM is re-evaluation of parameter estimates, or hypotheses more generally, as information becomes available. Bayesian inference is a natural approach for such updating hypotheses, so this chapter also describes the role of Bayesian methods in these analyses. Monitoring management outcomes is a key aspect of AM. Here we focus on the role of monitoring in AM, largely ignoring other roles and benefits of monitoring that are covered by Nichols & Armstrong (this volume, Chapter 7).

Adaptive management

Adaptive management is a structured, iterative process of decision making in the face of uncertainty (Holling, 1978; Walters, 1986). It aims to reduce uncertainty by monitoring outcomes of management actions and inaction. In this way, decision making aims to achieve particular management objectives while gaining information that improves future management. These objectives would include attaining levels of success while balancing the need to lower total costs. AM aims to help determine not just what management and monitoring is necessary but also what management and monitoring is unnecessary.

Trial and error is not AM. Simply changing management in response to changed circumstances (state-dependent decisions) is also not necessarily AM. While AM will tend to lead to changes in management, a key feature of AM, in the sense proposed by Holling (1978) and Walters (1986), is an a priori plan to change management in response to targeted monitoring. The monitoring is planned to help inform the strength of support for the different hypotheses about the managed system. If a manager says, 'We'll adapt our management in response to what happens', but does not have a plan about what that response will be for each possible management outcome, then this is not AM; they are really just flying by the seat of their pants. AM specifies how management outcomes will be monitored and how management is expected to change in response to particular monitoring data. These data will tend to support some hypotheses over others, with possible subsequent changes in management.

Define the problem

AM is often characterized as a continual cycle of 'learning by doing' through a number of steps (Figure 8.1). The step of defining the problem requires explicit statements of measurable management objectives and identifies the range of management options that might be considered. When the management objective focuses on the long term, measures to assess progress towards the objective might be required. The management objectives need to be sufficiently clear that the success in meeting the goals at agreed times in the future can be evaluated through monitoring. Achieving a minimally acceptable annual average growth rate for the population within the constraints of the projected annual budget is an example of a management objective.

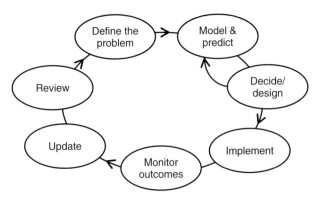

Figure 8.1 A seven-step adaptive management cycle in which managers and stakeholders (i) define the problem, (ii) develop alternative models to predict responses to management options and identify critical uncertainties, (iii) decide on management options and design monitoring strategies based on predicted responses to management actions and inactions, (iv) implement the management strategies, (v) monitor the outcomes of management actions and inactions, (vi) analyse the monitoring data to update knowledge about the effects of different management strategies and (vii) review and re-evaluate the management questions in the light of the updated knowledge. This diagram is modified from Whelan (2004) and Cawson & Muir (2008) with an explicit modelling and prediction step added. Decisions about implementing management and monitoring options often require iterative modelling and prediction, so an extra loop between these two steps is identified. These seven steps are closely related to the 10 steps for modelling reintroduced populations suggested by Armstrong & Reynolds (this volume, Chapter 6).

Model and predict

The modelling and prediction step requires clear descriptions of how the populations being managed are expected to respond to management, and the range of uncertainties in this response. Typically, these descriptions will be models of the system being managed, defining, for example, the number and characteristics of individuals in captivity and the wild, and how these are expected to change under different management scenarios. These models represent different ideas about the dynamics of the system being managed. Each model might have different assumptions, for example about the form of density dependence, the relative impacts and benefits of reintroduction strategies and the causes of population decline. Uncertainty arises because the model parameters (e.g. survival and reproduction rates) are uncertain and because the choice of the model in the first place is uncertain.

Uncertainty about the most appropriate model and its parameters can create uncertainty about the best management option to choose. The choice of the best management option might be uncertain, for example, when deciding whether to release juvenile or adult animals. Here the management objective might be to maximize the population size at a particular time in the future. While adults generally contribute more to population growth in reintroduced populations (Burgman *et al.*, 1995; McCarthy, 1995), differences in post-release survival of the different age classes makes the optimal management decision uncertain (Sarrazin & Legendre, 2000). Sarrazin & Legendre (2000) used a population model to integrate the relative benefits of releases of juvenile and adult griffon vultures (*Gyps fulvus*) in southern France. The population model was used to define the states of the system being modelled, the abundances of individuals in the different age classes, the rates of transition among these age classes and how the different management actions influenced these states. Ultimately, the effects of the management options were measured by their effects on the objective, which in this case was the population growth rate as predicted by the model. Management objectives can be developed that integrate impacts on both the source and reintroduced population (e.g. see McCarthy, 1995). Within the framework developed by Sarrazin & Legendre (2000), uncertainty about the release costs for juveniles versus adults would lead to uncertainty about which age cohort to release.

The modelling and prediction step requires appropriate data. In the case of deciding on the merits of releasing adults or juveniles, relevant data would include information on the post-release survival of individuals and on their subsequent reproductive success. Data to help address relevant management questions may extend beyond the site or even species being studied (Armstrong & Reynolds, this volume, Chapter 6). Using meta-analysis, generalities across studies and taxa can be established. For example, data compiled in a review of carnivorous mammal reintroductions (Jule *et al.*, 2008) can be used to place bounds on the relative post-release survival of captive-raised and wild-born individuals (Box 8.1; see Figure 8.2). This shows that across five families of mammals, the post-release survival of captive-raised carnivores is approximately 38% (95% CI: 23–77%) of wild-born individuals. Such data and analysis could be very useful in any decision about the best source of animals for reintroduction, and also in evaluating the success of a reintroduction.

Box 8.1 **Meta-analysis of post-release survival of mammalian carnivores**

Jule *et al.* (2008) reviewed the post-release survival of captive-born versus wild-caught mammalian carnivores. Post-release survival was measured from reintroduction over periods of around one year (6–18 months) and is expected to include 'post-release effects' (see Box 6.1) associated with the stress of translocation. The review found evidence to support the contention that wild-caught animals are more likely to survive this post-release period than captive-born animals. Here, the data that Jule *et al.* (2008) compiled on 45 case studies (17 species across five families) are analysed in a meta-analysis. The data were the number of individuals released and the number surviving over the 6–18 month post-release period.

Our meta-analysis modelled the probability of survival as a generalized linear model with the case study and species as random effects and the source of animals (wild versus captive) as an explanatory variable. While survival increases with body mass in wild populations (McCarthy *et al.*, 2008), there was no evidence of this in the reintroduction data so body mass was not included as an explanatory variable. Survival is constrained to be in the interval zero to one, which was accommodated by using a complementary log–log link function.

The meta-analysis supported the conclusion of Jule *et al.* (2008) that post-release survival of wild-caught animals was greater than that of captive-born animals (Figure 8.2). An advantage of this type of quantitative meta-analysis is that it can predict what will happen in new situations while accounting for uncertainty. For example, post-release survival of a new (random) case study involving a species not included in the dataset was predicted both for captive-born and wild-born individuals. The ratio of these two survival probabilities was also predicted. The post-release survival of captive-born animals is expected to be between 23% and 77% (based on the 95% credible interval) of that of wild-born animals, with a mean expectation for this ratio of 38% across body masses.

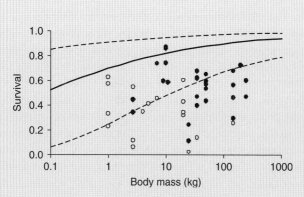

Figure 8.2   Post-release survival of reintroduced carnivorous mammals plotted versus body mass, based on the data compiled in Jule *et al.* (2008). Post-release survival was measured over periods of around one year (6–18 months) after release. Open circles are estimates of survival for individuals raised in captivity for each project and closed circles are estimates for wild-born individuals. Uncertainty in each estimate is not shown for the sake of clarity. These estimates are compared with the annual survival for non-translocated wild adult carnivores from McCarthy *et al.* 2008, represented by the median (solid lines) and 95% prediction interval (dashed lines).

The post-release survival of reintroduced mammalian carnivores is generally lower than that of non-translocated adults in wild populations (Figure 8.2). This is expected due to stresses associated with translocation (Box 6.1), but could also partially reflect cases where species were reintroduced to a poor habitat. Post-release survival does not increase with body mass, as occurs for non-translocated wild adults, suggesting that the extra mortality associated with reintroduction is typically greater for large species (Figure 8.2).

Design/decide

The design step identifies the management actions and monitoring that is required in the current cycle. This step requires that the models are analysed to determine the importance of the uncertainties, so this step often loops back iteratively to the modelling and prediction step. Each model can be analysed to determine the strategy that best meets the management objective. By varying the parameters within possible ranges and by analysing the alternative hypotheses as models, the uncertainties that are important for determining the best management strategy can be identified. This is a form of sensitivity analysis. However, rather than focusing on how uncertainty in the parameters influences the predicted responses (such as population size), it focuses on how uncertainty in the parameters influences the choice of management actions. This helps to reveal which aspects might be a useful focus for investigation with AM (Runge *et al.*, 2011). These aspects might not be those subject to the greatest uncertainty, nor subject to the greatest purely scientific interest.

For example, if deciding between whether to source reintroduced mammalian carnivores from captivity or the wild, one factor in the decision will be the relative survival of animals from these two groups. A meta-analysis of the data from previous relevant studies (e.g. Box 8.1; see Figure 8.2) would help to identify the range over which to vary parameters. If the optimal management action changed over this range of possibilities, then designing an AM programme to explore survival of captive- and wild-born individuals might be warranted.

Comparing predictions of different models and identifying sensitivities helps to determine the monitoring priorities by identifying attributes that would be desirable to monitor. The attributes to be monitored might be parameters that require estimation or outcomes that help distinguish between one model and another. The design step identifies the actions that should be undertaken and the monitoring that will contribute to learning. For example, this step might define the number of individuals to be released and how they are managed. The monitoring component might identify that the survival of reintroduced individuals and their rate of breeding under different management options should be monitored and the amount of monitoring that should be conducted.

Implement

The implementation step is the nuts and bolts of management, with on-the-ground actions undertaken. For a reintroduction this includes translocation of individuals from the source population (either captive or wild) to the release site, which is often a complex process involving a great deal of planning (Parker *et al.*, this volume, Chapter 4). Aspects of the release strategy, such as holding conditions, transport, timing, and size and composition of the release group, may all affect subsequent population performance. However, key aspects of implementation are often the restoration and ongoing management of the release site (IUCN, 1998). This could include eradication of exotics (especially predators), revegetation or restoration of hydrology before reintroduction, ongoing predator control, or provision of resources such as food, water and shelters.

Monitor outcomes

The monitoring step collects data on the response to the management actions and possibly on the response to management inaction. This contrast of management action and inaction can often be necessary to distinguish the effects of management in the face of natural variability (Nichols & Armstrong, this volume, Chapter 7). For example, although a reintroduced population might be assumed to be unable to grow without control of an exotic predator thought to have caused the original extirpation, it is possible that (a) the reintroduced species can now co-exist with the predator due to other habitat changes or (b) the reintroduced species might be able to co-exist with the predator most years but be extirpated in particular conditions.

Update and review

Following collation of the monitoring data, uncertainty about the influences of the different management strategies is updated (see the Bayesian inference section below). This updating should lead to a review where the management is re-evaluated, drawing on the results of the monitoring and also other collateral information (e.g. from different but related management programmes). In many ways, this review stage merges with the stage of defining the problem in which the available information is collated. We represent these stages separately to recognize explicitly the collation of the new information obtained in one cycle of the AM cycle (Figure 8.1) with the previously available information. Conceivably, the management objectives could be modified and the management problem redefined. This has been referred to as double-loop learning (Williams *et al.*, 2007).

Learning within AM

One of the key insights of research on AM is that the value of learning through experimentation can be expressed in terms of the expected benefit to the management objective (Walters, 1986). It is possible, at least in theory, to assess how resources should be allocated between learning about the effectiveness of management and actually managing the system. The concept of valuing learning in terms of improved management is well understood (Walters, 1986), although it can be mathematically difficult to assess the trade-off between allocating resources to conduct well-designed scientific studies and using resources that maximize the expected conservation outcome.

AM does not necessarily ensure that the best management option will be chosen. Poor decisions can occur if the range of models and management options being considered is not sufficiently broad (Peterson *et al.*, 2003), so a premium is placed on diversity of opinion and critical analysis of the alternatives. However, even when using a broad range of models, poor predictive capacity can lead to undesirable management outcomes. AM simply aims to reduce the occurrence of undesirable outcomes now and into the future. Perhaps one of the biggest advantages of AM is that it provides a framework for thinking rigorously and in a focused manner about the system being managed, the benefits of management and what needs to be monitored.

AM emphasizes that reducing uncertainty in some aspects of management will be more important than others, so monitoring should not focus on all areas of uncertainty. Monitoring as part of AM should address topics that influence management and where the monitoring is likely to help resolve the management uncertainty. In some cases, the statistical power will not be sufficient to remove the uncertainty about which management option is best. In other cases, there may be little scope to change management due to political, social or other constraints. In these cases, monitoring outcomes as part of AM will be less beneficial, although measuring the benefits of management compared with inaction would help determine management effectiveness. For example, answering the question about the order in which species are reintroduced will only be relevant when sufficient individuals of the different species are available. AM needs to consider these issues before decisions are made about appropriate monitoring strategies.

Passive and active adaptive management

Adaptive management can be described by the degree to which management is modified in an effort to improve learning. Active AM explicitly anticipates the value of learning about the effectiveness of management. Management actions might be modified and the outcomes monitored, with the intention to improve knowledge that can contribute to better management in the future. It contrasts with passive AM in which monitoring of management effectiveness and consequent learning still occur, but management actions are not chosen in anticipation of their effect on learning.

At least two possible approaches to active AM exist. In one, which could be called sequential or two-phase AM, the explicit aim of the first phase is to compare the effectiveness of actions and inaction to gain knowledge that will improve future management in the second phase. Here, the immediate goal of the first phase is to fill a knowledge gap that has been identified as a barrier to a good decision. This form of AM is very similar to traditional experimental design in that it aims to reduce uncertainty. However, the chosen management treatments aim to fill a critical knowledge gap that clouds a particular management decision, rather than simply focusing on an interesting knowledge gap.

A second form of active AM, which could be termed simultaneous AM, keeps the ultimate decision, and therefore the objective of management, in mind at all times. The management and monitoring are modified to reduce uncertainty only to the extent that this reduction in uncertainty improves management. The management and monitoring programmes are designed with a view to optimize the management objective (Williams, 1996). This is accomplished by including the information state, along with the usual system state variables, in the objective function.

In passive AM, the management options are not modified with a view to improving knowledge. Instead, learning occurs somewhat serendipitously in passive AM, depending on the suite of actions being taken and the environment in which they are implemented (Box 8.2). Any knowledge is then incorporated into management plans (Parma *et al.*, 1998; Shea *et al.*, 1998, 2002). Passive AM can be distinguished from what might be usual management practices because it formally considers how monitoring can reduce uncertainty to assist management in the future. It does this by identifying the uncertainties that cloud the optimal management action and designing monitoring programmes to reduce those uncertainties. The key difference between passive and active AM is that active AM explicitly modifies management to help improve learning. Passive AM implements what is thought to be best practice at the time, rather than deviating from what appears to be best in the short-term to help learn the best management option.

Box 8.2 Passive versus active adaptive management

Assume that a management agency is planning to release captive-raised animals to establish a new population. Two options are available. The first involves releasing animals that have been raised in a standard breeding facility. The second and more expensive option involves raising animals in a specialist facility that is constructed at the actual release site, with the hope that this will foster greater site fidelity after release. Under the first option, the managers expect that their budget is sufficient to release 20 animals per year, but they only expect two of these (10%) to remain at the site and survive to the breeding season. Under the second option, the budget is sufficient to release only 10 animals per year, but five (50%) are expected to remain at the site and survive to the breeding season. In analysing this as a structured decision problem (Nichols & Armstrong, this volume, Chapter 7), we ignore set-up costs (for the sake of a simple illustration) and assume that the goal is to maximize the expected number of released animals that breed at least once.

Of course, the survival probabilities under each option will be uncertain. In modelling the problem, the managers would have considered the available data to determine the expected survival probabilities described above, but they would also have represented the uncertainty in these parameters. AM would be identified as potentially useful if the uncertainty about survival to the breeding season under both strategies is sufficiently large that the optimal management decision is unclear. The competing models being resolved by AM would represent the different possible survival rates under the two management options.

Under passive AM, the managers would initially implement the second strategy because the expected success rate is higher. The results of this strategy would be monitored and this strategy would continue while the estimated number of successful breeders from each release exceeded two, which is the number expected under the first release strategy. After each release, the new data would sequentially reduce the uncertainty in the survival probability and the expected value of further data. The managers would switch to the first strategy if monitoring revealed that the expected number of successes per year from the second strategy was less than two. After switching, monitoring of the first strategy would continue to check whether its success rate bettered that of the second and managers would not switch back to the second strategy unless the first strategy performed worse. Under passive AM, the apparently best strategy is implemented and the results monitored, updating the managers' expectations and switching strategies when an alternative has apparently better outcomes.

Under active AM, the managers could potentially split their budget between the two options and implement them concurrently as an experiment, or alternate among years. The outcomes of the two strategies would be monitored, with more emphasis placed on one method over the other in subsequent decisions, depending on the outcomes and how well the experiment can discriminate between the two options. Implementing this style of AM is illustrated later in the chapter using hypothetical reintroduction problems (see the section Illustrative applications of adaptive management for reintroduction).

There are several examples of passive and active AM in the ecological literature (e.g. Johnson *et al.*, 1993; Varley & Boyce, 2006; McCarthy & Possingham, 2007; Nichols *et al.*, 2007), but neither is applied very often. Further, active AM presents major conceptual and theoretical challenges. Active AM involves designing conservation measures in such a way that managers can learn efficiently about the system for which they are responsible so that future management is improved, bearing in mind the needs of managing the system in the present. It is recognized that experimentation is useful in reintroduction biology (Armstrong *et al.*, 1995; Sarrazin & Barbault, 1996) and environmental management generally (Ferraro & Pattanayak, 2006), but it is not clear how conservation resources should be split between learning through experimentation and management based on what is known currently (McDonald-Madden *et al.*, 2010). Mathematical analysis of AM problems can help resolve these questions (e.g. see McCarthy & Possingham, 2007). Optimizing active AM programmes is computationally difficult for all but the simplest problems, but implementing AM does not rely on optimization (Box 8.3).

Box 8.3 **Adaptive management of habitat quality after reintroduction**

Habitat quality is probably the main factor determining the success of reintroduction programmes. In some cases, habitat quality will depend on long-term processes such as vegetation maturation, making it impossible to manage habitat quality adaptively after reintroduction. However, AM is possible if factors likely to affect habitat quality can be rapidly manipulated. When the hihi (*Notiomystis cincta*), an endangered New Zealand forest bird, was reintroduced to Mokoia Island in 1994, it was hypothesized that reintroductions at other sites had failed due to food shortage. Because food availability could be manipulated with feeders (Figure 8.3), Armstrong *et al.* (2007) conducted an AM programme to estimate growth of the Mokoia hihi population under alternative management actions. Alternatives differed in timing (e.g. breeding versus non-breeding), quality (sugar water versus full food supplement) and distribution of food provided, and also in whether nest mites were treated. These manipulations showed that provision of sugar water during the breeding season greatly improved population growth, but that quality and distribution of supplementary food had little effect and that provision outside the breeding season also had little effect. Mean population projections suggested that the population would grow under sustained supplementary feeding and mite control, but there was great uncertainty even after eight years of AM with the possibility of population decline (Figure 8.4). Due to the risk of decline under continuing management, it was decided in 2002 to discontinue this management on Mokoia and concentrate investment in other reintroduction sites where the management protocols developed on Mokoia had resulted in greater population growth. The remaining hihi on Mokoia were therefore translocated to one of those sites.

The AM programme on Mokoia followed the six steps shown in Figure 8.1 from 1994 to 2002, with each cycle taking one year.

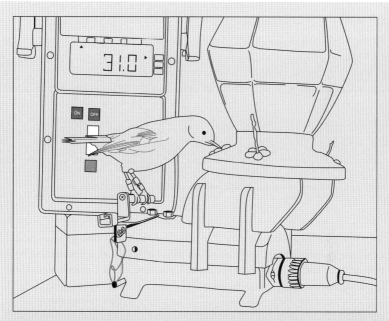

Figure 8.3   A female hihi at a feeding station on Mokoia Island, New Zealand. These stations were used in an on–off food supplementation experiment at the start of the AM programme for the reintroduced population. The stations had weigh scales and automatic video recording systems, making it possible to measure drops in weight when feeders were temporarily removed. The aim was to identify times of year when birds would starve if no food were provided, but to avoid them dying. It turned out that feeders had no effect on the condition or survival of adult birds, but had a strong effect on reproduction, which accounts for the shift in growth rate shown in Figure 8.4.

Monitoring was undertaken each year to estimate adult survival, juvenile survival and reproduction (fledglings per female) as a function of management. These estimates were combined into a population model to predict population growth, accounting for demographic stochasticicy and uncertainty in parameter estimates and the choice of model (Armstrong *et al.*, 2007). These were presented to the hihi recovery group each year and management actions for the following year were

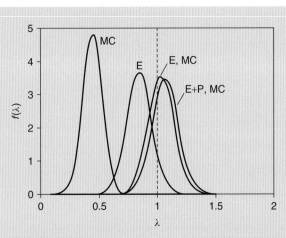

Figure 8.4   **Probability density functions for the finite rate of increase (λ) of the reintroduced hihi population on Mokoia Island as a function of four different combinations of management actions: energy supplementation from sugar water (E); energy plus protein from full food supplement (E + P); and mite control (MC). The dashed line indicates a growth rate that leads to a stable population size, with population growth to the right of the line and decline to the left. (Reproduced from Armstrong *et al.*, 2007).**

decided at this stage. The aim was to select actions that would provide information needed while simultaneously avoiding undue risk to the population. However, these decisions were made intuitively rather than following an optimization procedure. This intuitive approach facilitated direct involvement of the recovery group in the decision process, but it would be interesting to assess the improvement in efficiency that could have been gained through optimization. Optimization would have been quite complex in this scenario, as the range of management actions evolved throughout the programme, requiring a 'double loop learning' approach (Nichols & Armstrong, this volume, Chapter 7).

Active AM intentionally seeks to compare multiple management actions. However, it will recommend experiments and monitoring that are very different, at least in some cases, from those recommended by traditional experimental design. For example, conclusively answering the question 'How heavily should source populations be harvested?' (Figure 6.3) using a traditional experiment might require harvest rates that have impacts on source populations that range from negligible to unacceptably high. Intentionally harvesting a threatened population at an unsustainable rate may have unacceptable costs. Active AM that seeks to optimize the management objective, rather than treating improved knowledge as a separate explicit objective, might be beneficial in these circumstances to judge appropriate levels of harvest.

Passive AM might lead to similar management as active AM when examining appropriate levels of harvest that aim to ensure that the costs (monetary costs, but also the costs to the viability of the source population) are outweighed by the benefits derived from the harvested individuals. In contrast, active and passive AM might recommend very different management strategies when answering the question 'How is the probability of establishment affected by the size and composition of the release group?' because these properties might be readily manipulated, the outcomes might conceivably be monitored (e.g. by measuring post-release survival and breeding success) and the costs of failure might be relatively small before results of monitoring are available. When experimentation is feasible and cost-effective, active AM might be expected to recommend different management and monitoring strategies compared with passive AM.

Bayesian inference

AM uses a cycle of problem formulation, decisions, implementation, monitoring, analysis and review (Figure 8.1). Within each loop of this cycle, the new data arising from the monitoring needs to be combined with existing information. This existing information will include data from previous cycles, but could also include information from other projects or even other species. Bayesian analyses are able to integrate theoretical predictions, data from multiple sources and any initial judgements about the performance of different management options.

Box 8.2 gives a hypothetical example where managers have initial expectations about the performance of the two management options. The first strategy is expected to have a 10% success rate (probability of a released animal surviving to the breeding season) and the second strategy a 50% success rate. However, these success rates will be uncertain, with the degree of uncertainty depending on the amount of prior data available. Uncertainty in the success rate can be represented by expressing it as a probability distribution (McCarthy, 2007). As data accumulate, there will be a decline in uncertainty represented by the probability distribution narrowing. This increase in precision with accumulation of information can be formalized with Bayesian analysis, which defines how an estimate should change as data accumulate (McCarthy, 2007).

Bayesian analyses are being used increasingly in environmental science (Clark, 2005; McCarthy, 2007; Royle & Dorazio, 2008; Kery, 2010). These analyses have four main components (McCarthy, 2007):

(1) a prior distribution (also known simply as 'the prior'), which represents the state of knowledge and degree of uncertainty about parameters prior to collecting the new set of data;
(2) the new data;
(3) a model that relates the new data to the parameters; and
(4) a posterior distribution (also known simply as 'the posterior'), which represents the state of knowledge, including the degree of uncertainty about parameters, after considering both the new data and the prior.

The posterior distribution is obtained using Bayes' rule (McCarthy, 2007). Bayesian analyses provide a formal mechanism for updating information as it accumulates over time, thereby being a natural framework for data analysis in AM.

Bayesian analyses provide a formal mechanism for incorporating information from multiple sources. Results of previous monitoring, results from other sites or studies, and results from other species can be used to generate the prior distribution. For example, existing data on post-release survival can be analysed and used to generate a prior distribution (Box 8.1). The probability of a mammalian carnivore raised in captivity surviving the first 6–18 months after release is expected to be 0.27 with a 95% CI of [0.04, 0.77] (Box 8.1). This prior distribution means that a reintroduction of captive stock that relied on survival greater than approximately 0.75 over this period would be unlikely to succeed, unless the programme was fundamentally different from those in the dataset of Jule et al. (2008).

If the reintroduction went ahead, new data could be used to update the prior. The resulting posterior distribution would form the basis of the prior for a further translocation, which might involve releasing more animals at the same site or for another reintroduction programme. Such updating can be

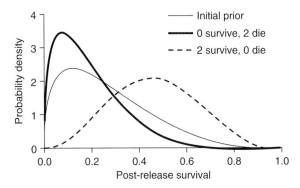

Figure 8.5   Probability density functions for the post-release survival of mammalian carnivores raised in captivity. The initial prior (thinner line) is derived from a meta-analysis of data compiled by Jule *et al.* (2008) as described in Box 8.1. The two posterior distributions are based on this prior and alternatively assuming that two individuals are released and both die (thicker solid line) or survive (dashed line) a period of one year after release.

illustrated by considering how data on the fates of two individuals released from captivity would update the prior described above. The distribution shifts towards smaller values if the two individuals die within the defined post-release time period. In contrast, if both individuals survive the interval, the estimated survival rate would increase (Figure 8.5).

Optimal AM

It can be difficult to determine the trade-offs associated with switching between management strategies, investing in learning about management effectiveness and dealing with the uncertainties that are inherent in reintroduction biology. Reliably determining these trade-offs requires an assessment of the costs and benefits of undertaking different management options and learning about their performance. The costs can be monetary, but they can also be opportunity costs, such as the forgone production of captive animals when individuals are released to the wild, or vice versa. The structure of these costs and benefits can be complex, occurring over multiple time periods and potentially in different units that may not be easily exchanged (e.g. money versus numbers of individuals). Further, the outcomes of management are uncertain, with inference from monitoring often limited by small sample sizes and case-specific circumstances, although, with sufficient data, each case can be treated as a random effect (Box 8.1). Intuitive, subjective judgements of these factors can be difficult and subject to frailties of human judgement (Walters, 1986; Burgman, 2005).

Mathematical analysis can help assess these trade-offs reliably. The simplest problems can be assessed using analytical solutions in which the various costs and benefits in different time periods are calculated and the optimal management option is assessed (Box 8.4). Numerical methods are required for more complex problems. Stochastic dynamic programming determines the optimal combination of management options over time, accounting for stochasticity (randomness) in the dynamics of the system being monitored. For example, Rout *et al.* (2009) used stochastic dynamic programming to determine the number of individuals to move from one population, in which the survival rate was known, to a second, where the survival rate was unknown. The dynamics of the populations were uncertain, due to the uncertain reproductive success and survival of individuals (demographic stochasticity). Stochastic dynamic programming was used to find the optimal number of individuals to move in each period, which depended on the number of individuals in each population, the success of previous reintroductions to the second site and the remaining time horizon for management.

Box 8.4 **Active adaptive management of release and monitoring strategies**

Consider a case where a manager is aiming to release animals using two different methods, the success of each being uncertain. The manager proposes a trial of the two methods and will then implement the method with the highest apparent success rate. The success rate is the probability of surviving a specified period post-release. The cost of implementing each option could be expressed in staff time or dollars, but here the focus of the costs will be on the reduced survival rate due to marking. For example, return rates of toe-clipped frogs decline with the number of toes removed, suggesting possible adverse impacts on survival (McCarthy & Parris, 2004). This is currently a topical issue because there can be heated debate about whether the information from toe clipping justifies possible impacts on survival. Therefore, the manager needs to assess the difficult trade-off between identifying the best release method, learning about which relies on individual recognition of marked animals, and the possible harm that marking causes.

McCarthy & Parris (2008) show how this trade-off can be assessed mathematically. The analysis integrates over the a priori probability of achieving particular success rates for the two different methods and how those estimated success rates influence the subsequent release decision. The optimal proportion of individuals to mark in the trial depends on the number of individuals released during and after the trial, and the mortality caused by the marking. For even low levels of mortality ($<1\%$) caused by marking, the optimal proportion of individuals to mark can be less than one, even ignoring other costs. Costs of marking could be monetary, but also include ethical considerations, such as the harm to individuals, beyond the harm to the population caused by the mortality (McCarthy & Parris, 2008; Parris *et al.*, 2010).

A management decision in the presence of uncertainty can often be analysed as a Markov decision process, which is a mathematical framework where outcomes depend on both randomness and management actions. In the context of reintroductions, the Markovian component might refer to the probabilistic population dynamics in which the future state of the population, conditional on the present state, is independent of its past state. Stochastic dynamic programming is used to solve Markov decision processes (e.g. Rout *et al.*, 2009). The generated solution typically varies depending on the current state of the system being managed, such as population size in the study by Rout *et al.* (2009). However, the current state might not be known with certainty, obscuring the optimal decision. For example, the number of individuals in each population is usually only estimated. Uncertainty about the state of the system means that the Markov decision process is only partially observable. Methods for solving partially observable Markov decision processes are available and have been applied to environmental management. For example, Chadès *et al.* (2008) incorporated imperfect knowledge about the persistence of the population into the management decision, with investment in monitoring being part of that decision. Similar analyses could be applied to reintroduction decisions, particularly if the fate or viability of the reintroduced population were uncertain.

Computational constraints can often limit the mathematical analysis of the costs and benefits of different management strategies. Algorithms and approximations for these types of analyses continue to be developed and applied in environmental management (e.g. Nicol & Chades, 2011). However, these methods can require specialist technical skills that might not be available to managers of reintroduction programmes. Difficulties of applying these mathematical methods to find optimal solutions do not preclude the application of AM to reintroduction. Indeed, using a structured approach of thinking about management problems is often beneficial in its own right without the extra benefit of finding a (nominally) optimal decision (Brook et al., 2002). Thinking about reintroductions from the perspective of AM will help to focus managers on clearly specifying the management objective, noting where uncertainty clouds the management decision, identifying which aspects can be usefully clarified by monitoring and exploiting the ability of monitoring results to change management. Examining these factors will help define appropriate management and monitoring, and therefore contribute to better AM, even in the absence of formal optimization of the trade-offs involved. Box 8.3 describes such AM of a reintroduced population of hihi, a threatened New Zealand bird.

It is an open question whether it is possible to judge appropriate levels and foci of monitoring in AM reliably. The optimal amount of experimentation and the best level of monitoring in AM problems can be relatively small, even when the benefit of the information is large (McCarthy & Possingham, 2007; McCarthy & Parris, 2008; Moore & McCarthy, 2010). It is possible that ecologists would be tempted to be more experimental rather than conforming to the recommended optimal strategy. However, some factors mitigate this possibility. First, ecologists, and possibly also managers of reintroductions, can overestimate the information content of their data when making subjective judgements (Burgman, 2005). Therefore, experiments and monitoring might not be larger than necessary. Second, managers might be unwilling to trial novel management strategies when current strategies are 'tried and true'. Further, risks of over-experimentation might be weighed against novel information and insights that were not foreshadowed in the optimisation (Wintle et al., 2010). Thus, formal methods of optimizing AM problems may underestimate the value derived from monitoring. Research into the ability of managers to use AM reliably without conducting formal optimization seems warranted. Developing strategies to assist such choices without complex optimizations, such as measuring the value of information (Runge et al., 2011), would be valuable.

Illustrative applications of adaptive management for reintroduction

This section describes two hypothetical examples to illustrate how optimal AM can assist reintroduction decisions when post-release mortality is uncertain. The first example accounts for uncertainty in post-release survival of juveniles relative to adults. This relative survival can determine whether it is better to release adults or juveniles. Increased information about survival can improve the success of the reintroduction programme by improving the decision, but does the expected gain offset the costs of obtaining the information? Here we explore this question for reintroduction of the griffon vulture by describing an AM framework for this problem developed by Runge (in preparation).

Runge (in preparation) developed an active AM framework to model the release of griffon vultures (*Gyps fulvus*) and account for the value of information. The population model was based on that developed by Sarrazin & Legendre (2000) to assess the relative merits of releasing juveniles or adults. The management objective was to maximize the expected population size in 50 years time. The survival of adults for a period of a year from reintroduction was uncertain, with the variance of the estimate given by $s(1-s)/c$, where s is the estimated survival probability and c defines the effective sample size on which the estimate of survival is based. A larger value of c represents a more precise estimate of survival. Other parameters in the model, including the post-release survival of juveniles and the subsequent annual survival and reproduction rate of released and wild-born birds, were considered fixed in this hypothetical scenario. The decision each year is whether to release juveniles, adults or a mixture of both, with a total of 20 individuals released annually. Differential costs of releasing juveniles and adults could be incorporated, for example, if releasing adults had greater cost to the source population.

In each time step, the post-release survival of adults is monitored and the estimate of this probability is modified using Bayesian inference. The influence of the data collected each year on the estimate (i.e. the posterior distribution) will be progressively less (i.e. as c increases). The optimal decision strategy is determined using stochastic dynamic programming.

In the scenario modelled, the optimal release strategy in the first year is to release only adults when the prior mean for post-release survival of adults is greater than approximately 0.76. At lower prior mean survival probabilities, the optimal decision depends on the uncertainty in the estimate. When the precision of the estimate is low (low values of *c*), it is still optimal to release some adults, even when the expected survival is approximately 0.5. Because the key uncertainty concerns the adult survival rate, uncertainty can only be reduced if adults are released; thus, the recommended release of adults helps to reduce the uncertainty associated with this key parameter for the purpose of making better decisions in the future. The parameter range over which both adults and juveniles are released becomes narrower as the estimate of the survival probability becomes more certain (Figure 8.6). This occurs because new data have less ability to change the parameter estimate in the face of larger amounts of previous data.

The value of planned learning to management outcomes can be illustrated by comparing passive and active AM. Rout *et al.* (2009) optimised both passive and active AM decisions about whether to release bridled nailtail wallabies (*Onychogalea fraenata*) to a current site where the survival rate is known or to a new site where survival is uncertain. They did not consider post-release effects (see Box 6.1 in Armstrong & Reynolds, this volume, Chapter 6), so annual survival probabilities were assumed to be constant after release. The assumed management objective was to maximize the number of individuals, summed

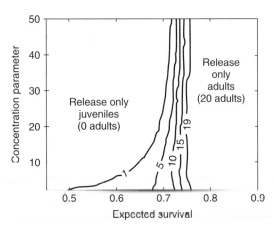

Figure 8.6 **A contour plot of the optimal number of adult griffon vultures to release in the first year as a function of the expected post-release survival probability of adults and the concentration parameter (c, the effective sample size on which the expected survival is based). The scenario assumes that 20 individuals are released each year, so the balance of the 20 released individuals are juveniles. (Reproduced from Runge, in preparation).**

across both populations. The managers were further assumed to have two animals that could be released each year, with neither, one or both individuals released to the new site and any remaining animals released to the original site. The managers can gain knowledge about the uncertain survival at the new site by releasing wallabies there, but the information gain is limited with only two individuals available each year. Rout *et al.* (2009) determined the optimal AM design using a model of population dynamics, with Bayesian updating of the mortality at the new site as data accumulate. They used stochastic dynamic programming to find the optimal solution under both active and passive AM.

The passive AM solution is to release individuals to the site that, based only on the current expected survival, will result in the greatest expected number of individuals in the future. This solution might change in the future if results of monitoring suggest it should. However, it differs from active AM where the management decision is modified to specifically provide strategies that improve learning. The difference between active and passive AM is illustrated in Figure 8.7. The active AM solution has a wider range of survival estimates where individuals are released to the new site, particularly when few data exist for survival at the new site. This greater propensity to release to the new site occurs because these releases also contribute to helping better management decisions in the future by contributing to learning.

The results are qualitatively similar to those for the griffon vulture (Figure 8.6); i.e. in both cases there is a greater propensity to explore when management outcomes are more uncertain. Although these hypothetical scenarios are simpler than problems that are typically faced in reintroductions, they illustrate the potential to apply AM to reintroductions. The assumptions of these scenarios could be relaxed to address real decisions, although optimizing these modified problems might be more complex.

Impediments to adaptive management in reintroduction

Adaptive management should help managers decide how best to allocate scarce management resources and determine how most effectively to monitor the outcomes. However, there are relatively few examples where it has been applied to pressing conservation and environmental management problems (Keith *et al.*, 2011). At present, the Mokoia hihi example (Box 8.3) is the only reintroduction programme we know of that has used AM to assess competing

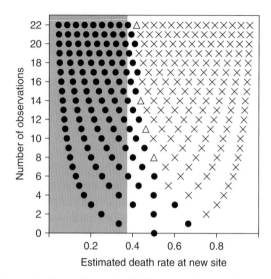

Figure 8.7   The optimal number of wallabies to release each year to a new site. Symbols represent the optimal reintroduction decision under active AM: release two wallabies per year to the new site (circles); one individual to each site (triangles); and two individuals to the original site (crosses). The optimal strategy depends on the expected survival in the new site and the number of animals that have been monitored (number of observations). The shading distinguishes survival estimates where the optimal decision under passive AM is to release both individuals to the new site (grey) or the original site (white). It is assumed that individuals at the original site have an annual survival of 0.7 (death rate of 0.3), that both sites have a carrying capacity of 30 individuals and that there were initially 20 individuals at the original site (reproduced from Rout et al., 2009).

hypotheses, although it did not optimize the design of the trials. Several impediments to the successful application of AM have been identified (e.g. Jacobson et al., 2006; Duncan & Wintle; 2008, Allen & Gunderson, 2011; Keith et al., 2011), and many of these apply to reintroduction biology. Some of these impediments are discussed here, noting how they might be overcome.

One impediment is misunderstanding of the approach. For example, applications under the rubric of 'adaptive management' sometimes examine a single management strategy at a time, changing strategy only when it appears to have failed (Duncan & Wintle, 2008). However, this is not AM as we have defined it here, which requires a priori identification of alternative possibilities.

While experts are important for identifying the alternative models that are necessary for AM, experts can tend to limit the choice of possible models due to various human foibles (Burgman, 2005). Overconfidence, willingness to conform and unwillingness to shift from established positions all contribute to narrowing the set of possible models being considered. There can be a reluctance to embrace uncertainty and attempt management options that are not regarded as best practice (Keith *et al.*, 2011; van Wilgen & Biggs, 2011). Further, institutional inertia can reduce capacity and inspiration to trial different approaches. These impediments to imagination can be overcome by, for example, embracing uncertainty and differences of opinion, managing social interactions during model development to encourage fruitful rather than counterproductive differences of opinion and calibrating elicited opinion (Burgman, 2005).

When resources are plentiful, traditional experimental design may well suffice to identify appropriate management practices. However, the limited time, money and number of animals typically available for reintroductions will restrict the scope for traditional experiments (Armstrong *et al.*, 1995). These aspects will also limit the ability of managers to learn from AM programmes. However, they are not impediments to the application of AM. Indeed, AM is specifically designed to address the difficult trade-offs involved when the scope for experimentation is limited. AM is designed to optimize performance given limited resources and to identify when aspects of experimentation are warranted and also when they are not (e.g. Figures 8.6 and 8.7). Formal meta-analysis (e.g. Box 8.1) helps to integrate results from multiple studies, thereby lending support to an individual study where data might be limited. Greater use of meta-analysis in reintroduction biology is warranted, as recommended by Seddon *et al.* (2007).

Ethical considerations might also impose constraints on experimentation. However, as is the case with other constraints to AM, costs to individuals that are released (e.g. effects of marking on survival; see Box 8.4) can also be considered in the design. AM provides an objective basis for assessing such ethical issues. As with budgets, impacts on individuals being released might influence the optimal AM solution, but they do not impede the application of AM. In fact, ethical issues are a motivation for designing a reintroduction programme using AM. For example, in the AM of hihi on Mokoia (Armstrong *et al.*, 2007; see Box 6.3), the influence of food limitation was initially assessed by measuring weight loss and comparing reproduction of birds close and far from feeders. Food supplementation was removed to understand food limitation better only after a few years, rather than implementing this action immediately.

AM requires that an explicit measurable objective is established. In even some of the simplest management scenarios, the choice of objective might not be obvious and the optimal management strategy can depend on this objective. The choice of objective might be further clouded in more complex management environments with multiple stakeholders and possibly fragmented management responsibility. Initiating AM programmes in these cases can be hindered at the first stage by difficulty in establishing agreed objectives. Achieving consensus might be possible, perhaps through explicitly weighting the different objectives (Linkov *et al.*, 2006). However, consensus might be an unrealistic and perhaps undesirable goal in many cases of marked disagreement and approaches based on game theory that retain differences might be more suitable (Colyvan *et al.*, 2011). Any informed and purposeful management will be hampered by difficulties of setting objectives, but this is not restricted to AM.

AM requires a range of technical skills and knowledge. Full implementation of AM requires skills in modelling, complex data analysis, optimization, expert elicitation, managing differences of opinion, field research and monitoring. Ecological and technical knowledge of the system and species being managed in a reintroduction programme is clearly important. Although individuals will typically have some of these particular skills, interdisciplinary teams are required to undertake all the required tasks. The necessary investment in developing such teams can be an impediment to AM.

While efficient computational techniques exist to solve smaller problems, computational capacity can limit the application of these methods to optimization of more complex AM problems. Continuing development of efficient algorithms and approximations (e.g. Nicol & Chadès, 2011) can help to overcome these restrictions. However, rules of thumb and other heuristics (e.g. Runge *et al.*, 2011) to help guide AM are likely to become increasingly important because managers will not necessarily have continual access to the scientists with the requisite skills to implement and update AM.

AM of reintroductions would rely on developing models that encompass the range of possibilities about the dynamics of the source and reintroduced populations under different management options. Development of these models requires sufficient knowledge that the scope of possibilities can be conceived by the managers. When confronted with a novel reintroduction, imagining the scope of possibilities might be difficult, which might restrict the application of AM. However, wider applications of meta-analysis would help to inform the range of possible responses in novel reintroduction programmes, which would assist the application of AM. Nevertheless, the range of models being considered is likely to expand over the course of a reintroduction programme.

Despite the potential benefits of AM in reintroductions of animals, there have been more hypothetical than real examples, but few examples regardless. AM provides the necessary framework to implement experiments within reintroduction programmes, which would satisfy calls for wider use of experiments (Armstrong *et al.*, 1995; Sarrazin & Barbault, 1996). We see great scope for increasing applications of AM to help improve the success of reintroduction programmes.

Acknowledgements

The authors are grateful to Jim Nichols, Tracy Rout and Libby Rumpff for their insightful comments on previous versions of this chapter and to Liz Grant for the drawing that Figure 8.3 is based on. This chapter was supported by an Australian Research Council Discovery Grant to Michael McCarthy and Peter Vesk (DP0985600).

References

Allen, C.R. & Gunderson, L.H. (2011) Pathology and failure in the design and implementation of adaptive management. *Journal of Environmental Management*, 92, 1379–1384.

Armstrong, D.P. & Reynolds, M.H. (2011) Modelling reintroduced populations: the state of the art and future directions. In *Reintroduction biology: integrating science and management*, eds J.G. Ewen, D.P. Armstrong, K.A. Parker & P.J. Seddon, Chapter 6. Wiley-Blackwell, Oxford, UK.

Armstrong, D.P. & Seddon, P.J. (2008) Directions in reintroduction biology. *Trends in Ecology and Evolution*, 23, 20–25.

Armstrong, D.P., Castro, I. & Griffiths, R. (2007) Using adaptive management to determine requirements of reintroduced populations: the case of the New Zealand hihi. *Journal of Applied Ecology*, 44, 953–962.

Armstrong, D.P., Soderquist, T. & Southgate, R. (1995) Designing experimental reintroductions as experiments. In *Reintroduction Biology of Australian and New Zealand Fauna*, ed. M. Serena, pp. 27–29. Surrey Beatty & Sons, Chipping Norton, Australia.

Brook, B.W., Burgman, M.A., Akçakaya, H.R. *et al.* (2002) Critiques of PVA ask the wrong questions: throwing the heuristic baby out with the numerical bath water. *Conservation Biology*, 16, 262–263.

Burgman, M.A. (2005) *Risks and Decisions for Conservation and Environmental Management*. Cambridge University Press, Cambridge, UK.

Burgman, M., Ferson, S. & Lindenmayer, D. (1995) The effect of the initial age–class distribution on extinction risks: implication for the reintroduction of Leadbeater's possum. In *Reintroduction Biology of Australian and New Zealand Fauna*, ed. M. Serena, pp. 15–19. Surrey Beatty & Sons, Chipping Norton, Australia.

Cawson, J. & Muir, A. (2008) Flora monitoring protocols for planned burning: a user's guide. Fire and Adaptive Management Report 74, Department of Sustainability and Environment, Melbourne.

Chadès, I., McDonald-Madden, E., McCarthy, M.A. *et al.* (2008) When to stop managing or monitoring cryptic threatened species. *Proceedings of the National Academy of Sciences*, 105, 13936–13940.

Clark, J.S. (2005) Why environmental scientists are becoming Bayesians. *Ecology Letters*, 8, 2–15.

Cohen, J. (1988) *Statistical Power Analysis for the Behavioral Sciences*, 2nd edition. Erlbaum Associates, Hillsdale, New Jersey, USA.

Colyvan, M., Justis, J. & Regan, H.M. (2011) The conservation game. *Biological Conservation*, 144, 1246–1253.

Duncan, D.H. & Wintle, B.A. (2008) Towards adaptive management of native vegetation in regional landscapes. In *Landscape Analysis and Visualisation. Spatial models for natural resource management and planning*, eds C. Pettit, I. Bishop, W. Cartwright, D. Duncan, K. Lowell & D. Pullar, pp. 159–182. Springer-Verlag GmbH, Berlin, Germany.

Ferraro, P.J. & Pattanayak, S.K. (2006) Money for nothing? A call for empirical evaluation of biodiversity conservation investments. *PLoS Biology*, 4, e105.

Holling, C.S. (1978) *Adaptive Environmental Assessment and Management*. Blackburn Press, Caldwell, New Jersey, USA.

IUCN (1998) *IUCN Guidelines for Re-introductions*. IUCN, Gland, Switzerland.

Jacobson, S.K., Morris, J.K., Sanders, J.S. *et al.* (2006) Understanding barriers to implementation of an adaptive land management program. *Conservation Biology*, 20, 1516–1527.

Johnson, F.A., Williams, B.K., Nichols, J.D. *et al.* (1993) Developing an adaptive management strategy for harvesting waterfowl in North America. *Transactions of the North American Wildlife and Natural Resources Conference*, 58, 565–583.

Jule, K.R., Leaver, L.A. & Lea, S.E.G. (2008) The effects of captive experience on reintroduction survival in carnivores: a review and analysis. *Biological Conservation*, 141, 355–363.

Keith, D.A., Martin, T.G., McDonald-Madden, E. *et al.* (2011) Uncertainty and adaptive management for biodiversity conservation. *Biological Conservation*, 144, 1175–1178.

Kery, M. (2010) *Introduction to WinBUGS for Ecologists: Bayesian approach to regression, ANOVA, mixed models and related analyses.* Academic Press, Burlington, Massachusetts, USA.

Lee, K.N. (1999) Appraising adaptive management. *Conservation Ecology*, 3(2), 3. [online] URL: http://www.consecol.org/vol3/iss2/art3.

Linkov, I., Satterstrom, F.K., Kiker, G. *et al.* (2006) From comparative risk assessment to multi-criteria decision analysis and adaptive management: recent developments and applications. *Environment International*, 32, 1072–1093.

McCarthy, M.A. (1995) Population viability analysis of the helmeted honeyeater: risk assessment of captive management and reintroduction. In *Reintroduction Biology of Australian and New Zealand fauna*, ed. M. Serena, pp. 21–25. Surrey Beatty & Sons, Chipping Norton, Australia.

McCarthy, M.A. (2007) *Bayesian Methods for Ecology.* Cambridge University Press, Cambridge, UK.

McCarthy, M.A. & Parris, K.M. (2004) Clarifying the effect of toe clipping on frogs with Bayesian statistics. *Journal of Applied Ecology*, 41, 780–786.

McCarthy, M.A. & Parris, K.M. (2008) Optimal marking of threatened species to balance benefits of information with impacts of marking. *Conservation Biology*, 22, 1506–1512.

McCarthy, M.A. & Possingham, H.P. (2007) Active adaptive management for conservation. *Conservation Biology*, 21, 956–963.

McCarthy, M.A., Citroen, R. & McCall, S.C. (2008) Allometric scaling and Bayesian priors for annual survival of birds and mammals. *American Naturalist*, 172, 216–222.

McDonald-Madden, E., Probert, E., Hauser, C. (2010) Active adaptive conservation of threatened species in the face of uncertainty. *Ecological Applications*, 20, 1476–1489.

Moore, A.L. & McCarthy, M.A. (2010) On valuing information in adaptive management models. *Conservation Biology*, 24, 984–993.

Nicol, S. & Chadès, I. (2011) Beyond stochastic dynamic programming: a heuristic sampling method for optimizing conservation decisions in very large state spaces. *Methods in Ecology and Evolution*, 2 (2), 221–228, April 2011.

Nichols, J.D. & Armstrong, D.P. (2011) Monitoring for reintroductions. In *Reintroduction biology: integrating science and management*, eds J.G. Ewen, D.P. Armstrong, K.A. Parker & P.J. Seddon, Chapter 7. Wiley-Blackwell, Oxford, UK.

Nichols, J.D., Runge, M.C., Johnson, F.A. *et al.* (2007) Adaptive harvest management of North American waterfowl populations: a brief history and future prospects. *Journal of Ornithology*, 148 (Suppl. 2), S343–S349.

Parker, K.A., Dickens, M.J., Clarke, R.H. *et al.* (2011). The theory and practice of catching, holding, moving and releasing animals. In *Reintroduction Biology: integrating science and management*, eds J.G. Ewen, D.P. Armstrong, K.A. Parker & P.J. Seddon, Chapter 4. Wiley-Blackwell, Oxford, UK.

Parma, A., Amarasekare, P., Kareiva, P. *et al.* (1998) What can adaptive management do for our fish, food, forests and biodiversity? *Integrative Biology*, 1, 16–26.

Parris, K.M., McCall, S.C., McCarthy, M.A. *et al.* (2010) Assessing ethical trade-offs in ecological field studies. *Journal of Applied Ecology*, 47, 227–234.

Peterson, G.D., Carpenter, S.R. & Brock, W.A. (2003) Uncertainty and the management of multistate ecosystems: an apparently rational route to collapse. *Ecology*, 84, 1403–1411.

Rout, T.M., Hauser, C.E. & Possingham, H.P. (2009) Optimal adaptive management for the translocation of a threatened species. *Ecological Applications*, 19, 515–526.

Royle, J.A. & Dorazio, R.M. (2008) *Hierarchical Modeling and Inference in Ecology: the analysis of data from populations, metapopulations and communities*. Academic Press, London, UK.

Runge, M.C. (in preparation) Adaptive management of reintroductions to re-establish an animal population.

Runge, M.C., Converse, S.J. & Lyons, J.E. (2011) Which uncertainty? Using expert elicitation and expected value of information to design an adaptive program. *Biological Conservation*, 144, 1214–1223.

Sarrazin, F. & Barbault, R. (1996) Reintroduction: challenges and lessons of basic ecology. *Trends in Ecology and Evolution*, 11, 474–478.

Sarrazin, F. & Legendre S. (2000) Demographic approach to releasing adults versus young in reintroductions. *Conservation Biology*, 14, 488–500.

Seddon, P.J., Armstrong, D.P. & Maloney, R.F. (2007) Developing the science of reintroduction biology. *Conservation Biology*, 21, 303–312.

Shea, K., Amarasekare, P., Kareiva, P. *et al.* (1998) Management of populations in conservation, harvesting and control. *Trends in Ecology and Evolution*, 13, 371–374.

Shea, K., Possingham, H.P., Murdoch, W.W. *et al.* (2002) Active adaptive management in insect pest and weed control: management with a plan for learning. *Ecological Applications*, 12, 927–936.

van Wilgen, B.W. & Biggs, H.C. (2011) A critical assessment of adaptive ecosystem management in a large savanna protected area in South Africa. *Biological Conservation*, 144, 1179–1187.

Varley, N. & Boyce, M.S. (2006) Adaptive management for reintroductions: updating a wolf recovery model for Yellowstone National Park. *Ecological Modelling*, 193, 315–339.

Walters, C.J. (1986) *Adaptive Management of Renewable Resources*. Blackburn Press, Caldwell, New Jersey, USA.

Walters, C.J. & Hilborn, R. (1978) Ecological optimization and adaptive management. *Annual Review of Ecology and Systematics*, 9, 157–188.

Whelan, R.J. (2004) Adaptive management: what does it mean and how can it be used in fire management? In *Bushfire: managing the risk*, ed. S. Halse. New South Wales Nature Conservation Council, Sydney, Australia.

Williams, B.K. (1996) Adaptive optimization and the harvest of biological systems. *Mathematical Biosciences*, 136, 1–20.

Williams, B.K., Szaro, R.C & Shapiro, C.D. (2007) *Adaptive Management: the US Department of the Interior Technical Guide*. US Department of the Interior, Washington, DC, USA.

Wintle, B.A., Runge, M.C. & Bekessy, S.A. (2010) Allocating monitoring effort in the face of unknown unknowns. *Ecology Letters*, 13, 1325–1337.

9

Empirical Consideration of Parasites and Health in Reintroduction

John G. Ewen[1], Karina Acevedo-Whitehouse[1,2], Maurice R. Alley[3], Claudia Carraro[4], Anthony W. Sainsbury[1], Kirsty Swinnerton[5] and Rosie Woodroffe[1]

[1]Institute of Zoology, Zoological Society of London, United Kingdom
[2]Facultad de Ciencias Naturales, Universidad Autonoma de Queretaro, Queretaro, Mexico
[3]Institute of Veterinary, Animal and Biomedical Sciences, Massey University, New Zealand
[4]Facoltà di Medicina Veterinaria, Università degli Studi di Padova, Italia
[5]Island Conservation, Center for Ocean Health, Santa Cruz, United States of America

'Our review includes cases where disease introduction has occurred and where translocations have failed because of disease, providing some response to Griffith et al.'s (1993) comment that "it is surprising that disease introduction or disease caused translocation failure is not more common, given the large number of translocations conducted annually." Reported cases of either parasite introduction or disease-caused translocation failure, however, remain uncommon.'

Page 320

Reintroduction Biology: Integrating Science and Management. First Edition.
Edited by John G. Ewen, Doug P. Armstrong, Kevin A. Parker and Philip J. Seddon.
© 2012 Blackwell Publishing Ltd. Published 2012 by Blackwell Publishing Ltd.

Introduction

Reintroduction as a form of conservation management is increasing in frequency and taxonomic breadth (Seddon *et al.*, 2005, 2007). Reintroduction fits within the translocation spectrum (Seddon, 2010; Seddon *et al.*, this volume, Chapter 1) and all translocations involve moving organisms from one environment to another and hence will necessarily affect both source and release locations (either captive or wild). Some effects may be negative to the species being translocated or to the wider ecosystem, and it is the responsibility of managers to identify and reduce or prevent negative aspects of these conservation initiatives. One important risk involved with moving organisms is that they will carry pathogens (disease-causing parasites; see Box 9.1).

Box 9.1 **Glossary of key terms**

Definitive host	The host in which the parasite can reproduce.
Density-dependent transmission	Where monoxenic parasites are transmitted via direct contact, the rate of transmission is dependent on host population density.
Endemic parasite	We use the epidemiological definition that the parasite can be maintained in the population without need for repeated external inputs.
Frequency-dependent transmission	Where the transmission of heteroxenic parasites is dependent on the proportion of the host population that is infected, which in turn affects the frequency of interaction between infected intermediate hosts or vectors and the final hosts. Monoxenic parasites that transmit by host sexual contact are also frequency dependent.
Heteroxenic parasites	Parasites that are transmitted via indirect transmission. Transmission occurs by one of two mutually exclusive strategies, via a vector or an intermediate host.

Host specificity (description)	Parasites lie on a spectrum of host-specificity, from highly host specific parasites that infect only a single host species to generalist parasites that can infect a range of species. Typically, parasites that are transmitted sexually, vertically or by close contact tend to have higher degrees of host specificity while parasites transmitted via the environment and heteroxenic parasites tend to have lower degrees of host specificity (Pedersen *et al.*, 2005).
Monoxenic parasites	Parasites that are transmitted directly including those producing infective stages that infect another host via sexual contact, vertical transmission or via environmental contamination (such as food or water).
Opportunistic pathogen	A parasite that does not usually cause disease but can do so under certain conditions.
Parasite	An organism that obtains energy for reproduction (or replication in the case of viruses) directly from the host and where the host suffers a cost. Here we include any viral, bacterial, fungal, protozoan and metazoan parasite.
Pathogen	Parasites that are the causative agents of infectious disease.
Prevalence	Measured as the proportion of instances of disease (or infection) in a given population over a given sampling period.
Reservoir host	A host species that acts as a source of infection for other species. This normally requires three conditions: (1) cases where the reservoir host population shows persistent infection, (2) cases of disease (or infection) occurring in reservoir hosts in the absence of cases in other species and (3) cases where outbreaks in other species should follow cases in the reservoir host (Cleaveland & Dye, 1995).

| *Seroprevalence* | Proportion of individuals in a given population that have detectable antibodies against a given pathogen. |
| *Virulence* | Measured as the proportion of infected hosts that become diseased. |

Parasites (Box 9.1) are being increasingly cited as major threats in conservation (Daszak *et al.*, 2000; Cleaveland *et al.*, 2002; Tompkins *et al.*, 2003; Tompkins & Poulin, 2006). Some now infamous case studies focus on the spread of exotic infectious disease in often naive and susceptible hosts, often following the introduction of exotic reservoir species (as an example of pathogen pollution, see Cunningham *et al.*, 2003). Examples include the co-introduction of the infectious viral disease rinderpest with domestic cattle in East Africa, causing mass mortality in wild ruminant species (Plowright, 1982), and exposure of United Kingdom red squirrels, *Sciurus vulgaris*, to squirrelpox virus co-introduced with the more resistant North American grey squirrels, *Sciurus carolinensis* (Tompkins *et al.*, 2002). These cases have generated repeated calls for managing high-risk pathogens during reintroduction programmes (e.g. Viggers *et al.*, 1993; Cunningham, 1996; IUCN, 1998; Woodford, 2001; Mathews *et al.*, 2006; Parker *et al.*, 2006; Breed *et al.*, 2009). Some characteristics of high-risk pathogens are obvious (for a recent review in mammals, see Breed *et al.*, 2009). Pathogens that are host generalist and are maintained in common (often domestic) and less susceptible reservoir hosts and/or those where the transmission rate is frequency-dependent can be more risky (and are often more obvious). The disastrous consequences for endemic Hawaiian land birds of a disease caused by *Plasmodium relictum* (carried by exotic avian hosts) following the establishment of the exotic mosquito vector, *Culex quinquefasciatus* (Van Riper *et al.*, 1986; Atkinson & van Riper, 1991; Scott *et al.*, 2001), provides a clear example. Prior to these exotic introductions it is thought that no *Plasmodium* parasites were present in the terrestrial resident avifauna of Hawaii because there was no competent vector (Beadell *et al.*, 2006) and hence the complete naivety of the Hawaiian land birds to these parasites. Pathogens that have previously caused problems are often flagged as potentially posing a high level of risk in subsequent management, and the *Plasmodium relictum* case provides a good example with concerns that this parasite will be present and moved in current reintroduction programmes in New Zealand.

There is no empirical evidence for any global host extinction that has resulted directly from disease (de Castro & Bolker, 2005; Smith *et al.*, 2006). A lack of evidence may be because (i) theoretical modelling of extinction is more advanced than the required detailed empirical data of global host extinction, (ii) the absence of evidence is not evidence of the absence of global host extinction and (iii) cases of successful intervention have prevented global host extinction. Successful intervention has occurred with black-footed ferrets, *Mustela nigripes*, when 16 individuals were captured in an emergency operation to avoid complete species extirpation due to dual epizootics of native canine distemper virus (Thorne & Williams, 1988) and exotic plague caused by the bacterium *Yersinia pestis* (Lockhart *et al.*, 2006). In addition to being a direct threat, parasites can also be detrimental if they suppress the survival or reproduction of individuals and thus the resilience of populations, making individuals and populations vulnerable to other threatening processes (Cleveland *et al.*, 2002). Parasites often have deleterious effects that act in combination with other extrinsic (e.g. habitat degradation or predation) and intrinsic (e.g. low genetic diversity, see Box 9.2, or physiological stress, see Parker *et al.*, this volume, Chapter 4) factors commonly affecting endangered host species. This complexity and interplay of extrinsic and intrinsic factors will increase the range of parasites that may pose a risk to reintroduction programmes (e.g. adding those that are host specific and density regulated).

Box 9.2 **Genetic diversity and disease**

In the past decade, increasing numbers of studies have reported correlations between aspects of disease resistance and individual levels of heterozygosity (an association known as heterozygosity–fitness correlation, HFC), measured using a panel of presumed neutral markers. Disease HFCs have been documented for a number of natural populations, with effects ranging from increased susceptibility to infection (Acevedo-Whitehouse *et al.*, 2005, 2006), higher parasite burdens (Coltman *et al.*, 1999), to more severe presentations (Acevedo-Whitehouse *et al.*, 2003, 2005) and a higher risk of death (Valsecchi *et al.*, 2004).

One explanation for HFCs is inbreeding, as incestuous mating increases the proportion of homozygous alleles of the offspring, leading to the increased likelihood of reduced fitness through inbreeding depression (see Keller *et al.*, this volume, Chapter 11). There are various mechanisms by which inbreeding might affect the occurrence of disease. First, disease could occur by harmful recessive alleles in immune-related genes being expressed more frequently in inbred individuals (partial dominance hypothesis; see Charlesworth & Charlesworth, 1987; Houle, 1994), thus suppressing or altering key immune functions. A second mechanism might be that inbred individuals would expectedly have a lower frequency of heterozygote loci experiencing balancing selection, and thus they would have lower chances of expressing overdominance (Crow, 1948; Charlesworth & Charlesworth, 1987) and would consequently have decreased immune fitness. Alternatively, effects could be due to higher numbers of harmful alleles that have evaded selective pressures (Lynch *et al.*, 1995). Many loci related to the immune system, particularly those involved with antigen recognition, depend on high levels of allelic variation (Zinkernagel *et al.*, 1985; Kurtz *et al.*, 2004) or haplotypic complexity (Yoder & Litman, 2011). Incestuous mating might influence disease resistance by increasing homozygosity in these genetic regions, leading individuals to be less successful at recognizing pathogens (O'Brien & Evermann, 1998; Potts & Wakeland, 1990).

More recently, theoretical (DeWoody & DeWoody, 2005), simulation-based (Balloux *et al.*, 2004) and empirical studies (Ferreira & Amos, 2006; Slate *et al.*, 2004) have shown that the primary mechanism for HFCs is likely to reflect associative overdominance, i.e. linkage between one or more of the markers used and a gene or genes under balancing selection. In most genomes, balanced polymorphisms are sufficiently common for the probability of at least 10% of the markers being located close enough to a gene under selection to show a correlation with fitness by association (Amos & Acevedo-Whitehouse, 2009), an expectation that has gathered empirical support for various aspects of fitness (Bierne *et al.*, 1998; Acevedo-Whitehouse *et al.*, 2005; Ferreira & Amos, 2006; Hansson *et al.*, 2001). However, obtaining such empirical evidence is difficult for disease-based studies in natural populations, mainly because acquiring accurate, detailed, data on disease and health status for free-ranging individuals is fraught with complications.

A case study: the California sea lion

The California sea lion, *Zalophus californianus* (Figure 9.1), has proven to be an ideal model for exploring the effect of inbreeding on disease susceptibility and elucidating the role that pathogens may play in maintaining genetic variation by selection against inbred individuals (Acevedo-Whitehouse *et al.*, 2003). The species is sexually dimorphic, highly polygynous and displays strong site fidelity (Riedman, 1990), traits that increase the odds that consanguineous matings occur (Acevedo-Whitehouse *et al.*, 2006) even though the population is by no means small or fragmented. Importantly, the diseases that affect this pinniped have been studied in depth for the past 40 years, and detailed tissue archives and datasets on cause of death and clinical illness exist due to the many stranding networks and rehabilitation centres located along its distribution.

Figure 9.1   **California sea lion pup (photo: Karina Acevedo-Whitehouse).**

Using this non-traditional model, it was shown that relatively less heterozygous California sea lions were more likely to develop acute and severe infectious diseases, such as leptospirosis and herpesvirus-related cancer, than more heterozygous individuals (Acevedo-Whitehouse *et al.*, 2003), and that this pattern held even when accounting for population structure. Interestingly, even tolerance to domoic acid, an algal-produced neurotoxin, was influenced by levels of heterozygosity (Acevedo-Whitehouse *et al.*, 2003), as was rehabilitation tenure once accounting for disease class and treatment (Acevedo-Whitehouse, 2004). A study that examined the possibility of these results being explained by associative overdominance found that inbreeding depression is a likely explanatory mechanism (Balloux *et al.*, 2004). However, further studies have shown that sea lion tolerance and susceptibility to some pathogens and diseases are also explained by polymorphisms at some of the loci used to estimate individual levels of inbreeding. For instance, while relatively inbred sea lions were more likely to die due to *Uninaria* spp. (a haematophagus hookworm) infections, polymorphisms at a single microsatellite determined whether infected sea lions developed anaemia (Acevedo-Whitehouse *et al.*, 2006). Interestingly, this 'single locus' effect was mirrored in a different species, the New Zealand sea lion *Phocarctos hookeri*, host to a seemingly less-virulent *Uncinaria* (Castinel *et al.*, 2007), suggesting that aspects of blood clotting or platelet synthesis may be under pathogen-driven selection. However, individual levels of inbreeding were not significantly related to pathogen resistance (Acevedo-Whitehouse *et al.*, 2009). This was surprising as the New Zealand sea lion population is extremely small and has a restricted distribution (Campbell *et al.*, 2006), so that inbreeding depression would be expected.

Taken together, these studies show that similar diseases can be influenced in markedly different ways by host genetics, possibly owing to differences in pathogen virulence, evolutionary relationships between the host and pathogen, and parallel pressures acting on a given population. In this light, drawing general conclusions regarding the importance of high levels of genetic diversity for disease resistance per se are difficult (see Boyce *et al.*, 2011) and one should consider these findings cautiously when designing management plans for wildlife reintroduction and translocation programmes.

How have the risks of disease been incorporated into reintroduction planning? It is helpful here to consider a general framework of reintroduction within which parasite management should fit. Armstrong & Seddon (2008) proposed 10 key questions for reintroduction biology, four of which (questions 2, 3, 8 and 9) have direct relevance to parasites. When made specific to parasite management, the four questions are: (i) how is post-release survival affected by parasite management, (ii) how do parasites affect suitability of release sites, (iii) are parasites native to the ecosystem and (iv) how will the ecosystem be affected by the parasites? These questions move along a continuum from individual, to population, and wider ecosystem impacts. At the individual level there is a focus on the general health of individuals selected for translocation in addition to any specific attention to parasites. The aims of this chapter are therefore: (i) to provide a summary of empirical considerations of health and parasites in reintroduction programmes and (ii) to identify current gaps in empirical studies in order to direct future research. To do this we will first summarize the published literature and then provide a more detailed review of empirical considerations during each stage of the reintroduction process (including long-term captivity, pre-translocation screening of wild source populations, short-term captivity and quarantine, and post-release). We will then provide a brief summary of all empirical research under each of the four related questions of Armstrong & Seddon (2008).

Early reviews and the empirical literature

We are aware of four early literature reviews that highlight disease risks in reintroduction programmes (Griffith et al., 1993; Viggers et al., 1993; Cunningham, 1996; Davidson & Nettles, 1992). All have similar general conclusions. First, parasites and health evaluation had been largely neglected in reintroduction programmes. Second, the lack of empirical evidence for disease effects suggests that the effects of parasites and health status on reintroduction outcomes are negligible. In the most quantitative survey, Griffith et al. (1993) summarized data from Australia, New Zealand, Canada and the United States (the last including game animal translocations) totalling 400 translocations between 1973 and 1986. This dataset reveals some critical points; first, less than a quarter of translocations collected sufficient data to calculate the proportion of animals that were lost to infectious disease and, second, in about a quarter of translocations there was no direct examination of animals for parasites or disease by either a veterinarian or biologist prior to release. Given that few detailed data have been collected on causes of mortality it is hard to conclude whether disease effects are few or whether we have been simply ignorant of the degree to which diseases have had an impact on reintroduction programmes (Cunningham, 1996).

For our own literature review we entered the key phrases *reintroduction, translocation, parasite* and *disease* within two online search engines, Web of Science and Google Scholar (searches conducted July 2010). We primarily compiled papers published since those used in the previous reviews (since the early 1990s) and supplemented those papers that appeared relevant with our own bibliographies and references cited within sourced articles. This list is not exhaustive, but is assumed to be representative of the available literature on the topic. Eighty-two case studies involving some component of reintroduction are included in this review. These cases include a broad host taxonomic base, including birds (40%), mammals (46%), fish (2%), reptiles (10%), amphibians (1%) and invertebrates (1%). The high proportion of studies on birds and mammals probably reflects the fact that most reintroduction projects involve these groups (Seddon *et al.*, 2005). Not surprisingly, when key parasites are targeted in these studies, there is a strong focus on host generalist parasites (90%; 57 of 63 studies). In addition, far more emphasis has been on parasites with monoxenic transmission (78%; 49 of 63 studies) as opposed to heteroxenic transmission (see Box 9.1 for definitions). A large proportion of studies focused on health and disease in captive populations. We provide a summary of studies from captive breeding programmes and the sometimes unique conditions they face.

Long-term captivity and vaccination

It has been predicted that over 3000 vertebrate species and subspecies will require captive breeding programmes to aid their conservation (Seal, 1991). This presents unique challenges for managing pathogens in captivity and preventing unintentional establishment of novel pathogens at release sites during host reintroduction. Common problems arising due to long-term captivity (multigeneration) are well known and include: (i) inbreeding and loss of immunogenetic variation, (ii) adaptation to captivity, (iii) lack of exposure to native parasites and (iv) exposure to exotic parasites. Furthermore, captivity can (i) induce stress that compromises immune capacity and/or (ii) cause health conditions unique to captive environments. The above problems are well known and have been dealt with in detail elsewhere (Frankham & Loebel, 1992; Viggers *et al.*, 1993; Snyder *et al.*, 1996; Allendorf & Luikart, 2007). Here we accept the necessity for captive breeding and focus on important lessons from reintroduction case studies or cases where reintroduction was about to occur.

Beyond the problem of losing captive individuals to disease are the wider implications for reintroduction. For example, mortality in captive wartbiter crickets, *Decticus verrucivorus*, due to mycotic infection resulted in postponement of a reintroduction attempt and a shift of release sites to areas with no extant wartbiter cricket population (Cunningham *et al.*, 1997). Avoiding wild wartbiter cricket populations reduced the risk of introducing an infectious disease to a remnant wild population. It is not known whether *Verticillium* sp., the fungal pathogen suspected to have caused disease, was also present in wild populations. Similar concerns also resulted in release sites being constrained to areas without conspecifics in the case of golden lion tamarins, *Leontopithecus rosalia* (Ballou, 1993). Although this is one way of preventing infection of remnant populations, it prohibits re-inforcement of free-living populations and fails to account for what may happen if subsequent population increases result in merging between remnant wild and establishing populations derived from captivity.

Accidental introduction of infectious disease agents has occurred through translocations, justifying caution in releasing captive animals. For example, Walker *et al.* (2008) traced the introduction of *Batrachochytrium dendrobatidis*, the pathogen causing the disease chytridiomycosis, in the Mallorcan midwife toad, *Alytes muletensis*, from a captive-breeding facility that had been used for reintroducing this species to the wild (Figure 9.2). At the time of reintroduction this pathogen was unknown, so preventative measures via pathogen screening were not possible. In other cases, a lack of thorough pathogen management has resulted in accidental co-introduction of infectious agents, e.g. when a pet orangutan, *Pongo pygmaeus*, was released despite having tested positive for tuberculosis (Bonner, 1995).

In an effort to reduce chances of pathogen co-introduction there is also growing attention to identifying health problems in captive animals destined for release (reviewed in mammals by Breed *et al.*, 2009). An important component of this involves undertaking and reporting causes of mortality through necropsy examination (e.g. Cunningham *et al.*, 1997; Acton *et al.*, 2000) and conducting health and parasite evaluations (e.g. Alberts *et al.*, 1998; Kilbourn *et al.*, 2003; Britt *et al.*, 2004; Lisboa *et al.*, 2004; Munson *et al.*, 2004;

Figure 9.2   **Mallorcan midwife toad (photo: Richard Griffiths).**

Frölich *et al.*, 2005; Gartrell & Hare, 2005; Gartrell *et al.*, 2006; Mul *et al.*, 2007; Martel *et al.*, 2009). Numerous studies have recommended the screening of wild populations at release sites prior to reintroduction to determine whether parasites identified in captive animals are also present in the wild, and vice versa (Ballou, 1993; Williams *et al.*, 1994; Alberts *et al.*, 1998; Kirkwood & Cunningham, 2006; Mul *et al.*, 2007; Breed *et al.*, 2009). For example, Rüegg *et al.* (2002) screened feral horses (*Equus caballus*) at a reintroduction site for the Przewalski horse (*Equus przewalskii*) in Mongolia for the presence of equine piroplasmosis, a disease caused by either *Babesia caballi* or *Theileria equi*, both of which are transmitted by ticks and are possible causes of mortality in the Przewalski horse.

Where made, comparisons between the parasites of captive and wild populations do provide a better insight for risk assessment in reintroduction (e.g. lesser kestrel, *Falco naumanni*, Lierz *et al.*, 2008; ring-tailed lemur, *Lemur catta*, Villers *et al.*, 2008; Arabian oryx, *Oryx leucoryx*, Frölich *et al.*, 2005). In some cases parasite fauna is similar (e.g. between captive and wild Sumatran orangutans, *Pongo abelii*, Mul *et al.*, 2007), whereas in others the captive population has a greater diversity or higher prevalence of parasites (e.g. intestinal coccidia in New Zealand hihi, *Notiomystis cincta*, Twentyman, 2001; Helminth species in New Zealand kakī, *Himantopus novaezelandiae*, Robertson, 2009), and vice versa (e.g. intestinal coccidia in Cuban iguana, *Cyclura nubile*, Alberts *et al.*, 1998; ticks on red wolves, *Canis rufus*, Phillips & Scheck, 1991). The best justification for management decisions constraining releases occurs when screening identifies an apparent lack of infectious agent or disease in either the captive or the remnant wild populations when an infectious disease exists in the other. For example, a neoplastic disease, reticuloendotheliosis, caused by the reticuloendotheliosis virus in captive Attwater's prairie chickens, *Tympanuchus cupido attwateri*, has not yet been detected in remnant wild populations of this endangered species leading to restrictions on release of captive birds (Drew *et al.*, 1998; Zavala *et al.*, 2006). Also, canine distemper virus and *Yersinia pestis* are now widespread in the historic range of black-footed ferrets, resulting in careful release site selection to avoid areas of high pathogen prevalence in alternative hosts (Williams *et al.*, 1994; Antolin *et al.*, 2002; Gober, 2008).

Two important factors increase the risk of a captive population becoming infected with a novel parasite: (i) captive facilities being located in exotic environments and (ii) captive facilities holding multispecies, multiorigin animals, especially closely related species that are normally allopatric. For example, the mycotic infection of wartbiter crickets described above may have resulted from the fungus already being present at the captive location (Cunningham *et al.*, 1997) and, similarly, the *Batrachochytrium dendrobatidis* traced from captivity to wild Mallorcan midwife toads may have originated from infected *Xenopus gilli*, endangered frogs endemic to South Africa that were housed in the same room (Walker *et al.*, 2008). Other examples appear in the literature; for example *Mycobacterium avium* infection (the causal agent of avian tuberculosis) of captive New Zealand kiwi, *Apteryx mantelli*, was likely transmitted from game birds (pheasants, *Phasianus* sp.) held in the same enclosure (Davis *et al.*, 1984). The zoonotic pathogen *Trypanosoma cruzi*, which causes chagas disease, has recently been isolated in nine primate species across two genera, *Leontopithecus* and *Saguinus*, held at a multispecies/multiorigin captive primate unit in Brazil (Lisboa *et al.*, 2004). Arabian oryx (*Oryx leucoryx*) acquired capripox virus from infected sheep that grazed along the perimeter fence of the breeding centre (Greth *et al.*, 1992).

While centres in exotic locations that house multispecies and multiorigin animals are often necessary and perform important conservation roles, their use in reintroduction programmes requires stringent parasite screening and quarantine periods. Even with screening and quarantine in place, the risk of co-introducing a novel and potentially damaging pathogen is not removed because the required diagnostic tests may not be available, or may not have the required sensitivity to detect the parasite, or may not be included in screening (Kirkwood & Sainsbury, 1997). Consequently, captive programmes with a realistic goal of reintroduction should be single species units ideally located within the historic range of the species (Wilson et al., 1994; Cunningham, 1996; Walker et al., 2008; Breed et al., 2009), as recently conducted in the cirl bunting, *Emberiza cirlus*, reintroduction in Cornwall, UK (McGill et al., 2010) and in the island fox, *Urocyon littoralis*, reintroduction in California, USA (US Fish and Wildlife Service, in review). This may be logistically onerous but in many cases will be a feasible strategy, especially during the planning phases of removing wild individuals into captivity to initiate captive breeding. While there are merits of multiple isolated captive populations, e.g. to safeguard against catastrophe within a single locality, these must be balanced against the above-mentioned parasite risks. For instance, recommendations to initiate captive breeding populations of Puerto Rican parrots, *Amazona vittata*, at various locations, including zoos, outside Puerto Rico (Lacy et al., 1989) were rejected based in large part on disease risks (Wilson et al., 1994).

The idea of maintaining captive populations as 'parasite-free' has also come under reconsideration. There is a growing realization that parasites are important for host evolution and that conservationists must not forget that parasites themselves deserve conservation as important components of biodiversity (Gompper & Williams, 1998; Pérez et al., 2006; Pizzi, 2009). In addition, reintroducing 'parasite-naive' individuals may lead to establishment failure if they encounter a pathogen to which they have no resistance. How often this may occur is hard to judge given a lack of detailed studies. However, complete reintroduction failure due to disease has been documented, e.g. in a population of black-footed ferret when a plague outbreak occurred immediately after their release (Antolin et al., 2002; Gober, 2008) and perhaps in Hawaiian 'Alala due to *Toxoplasma gondii* infections (Work et al., 2000). Furthermore, it may be unrealistic to remove parasites that will be commonly encountered once hosts are reintroduced and when these parasites cause insignificant mortality in the captive environment. For example, an experimental withholding of a standard anthelmintic treatment that was provided to captive-bred New Zealand kakī allowed infection with gastrointestinal helminths (Robertson, 2009). Following release, birds that were not treated had no change in helminth load whereas those that were treated became parasitized once exposed to common reservoir species (such as pied stilts, *Himantopus himantopus leucocephalus*) at the release site (Robertson, 2009). Importantly, there was no difference in survival between these groups, suggesting little utility in treatment (Robertson, 2009) and reiterating the cautions of attempting to remove parasites from translocated hosts when they might be exposed to the parasites again at the release site (De Leo & Dobson, 2005).

The idea that maintaining natural host–parasite relationships is beneficial has recently been elegantly tested within an experimental study using a model species, the guppy *Poecillia reticulate* (Faria *et al.*, 2010). In their study, the authors tested two common arguments about long-term captivity: first, that captive animals are naive to natural parasites they will encounter on release and, second, that immunogenetic variation will decline due to a loss of natural host–parasite co-evolution. Both of these hypothesized manifestations of captivity are expected to increase risks to individuals on release. By pre-exposing guppies to their most common ectoparasite, the trematode *Gyrodactylus turnbulli*, prior to release, Faria *et al.* (2010) showed that parasites in populations with pre-exposed guppies were more likely to go extinct and that parasite load was lower in guppies within pre-exposed treatments. These patterns were even stronger if susceptibility (assumed to be linked to immunogenetic competence) was taken into account and the most susceptible individuals were not reintroduced. This study helps justify maintaining *natural* host–parasite relationships in captivity when possible. In another case study, careful and minimal treatment of captive-reared cirl buntings with the anti-coccidial drug toltrazuril allowed control of the disease, isosporosis, without eliminating the *Isospora normanlevinei* parasite from the hosts (McGill *et al.*, 2010). Subsequent screening of released individuals detected *I. normanlevinei*, indicating that this host–parasite relationship had been conserved (McGill *et al.*, 2010).

If release sites are deemed to be appropriate (IUCN, 1998) then preventative measures such as vaccination can also be a powerful tool, with noted precautions (see Box 9.3). Examples where vaccination has occurred prior to the release of either captive or wild-caught individuals include: whooping cranes, *Grus americana*, to Eastern equine encephalitis (Langenberg *et al.*, 2002); pink pigeon, *Columba mayeri*, to avipox (Swinnerton, unpublished data); wild rabbits, *Oryctolagus cuniculus*, to myxomatosis and rabbit haemorrhagic disease (Moreno *et al.*, 2004; Calvete *et al.*, 2005); kakapo, *Strigops habroptilus*, to erysipelas (Gartrell *et al.*, 2005); black-footed ferret to canine distemper (Marinari & Kreeger, 2006); and golden lion tamarins to tetanus and rabies

(Montali *et al.*, 1995). Vaccination has also been recommended for Laysan duck, *Anas laysanensis*, against avian botulism (Work *et al.*, 2010), for future wild beaver, *Castor fiber*, translocations to the Netherlands to protect against pseudotuberculosis (*Yersinia pseudotuberculosis* infection) and leptospirosis (Nolet *et al.*, 1997), and is being developed for protection of black-footed ferrets against plague (Rocke *et al.*, 2006). There are warnings against vaccine use because of the dangers of inadvertent development of disease (a particular risk for modified live vaccines, which may revert to virulence) and because of the disruption vaccines can cause to natural host–parasite relationships (e.g. in cheetah, *Acinonyx jubatus*; see Munson *et al.*, 2004). Inadvertent development of disease has been noted as a risk pathway for disease introduction to wild populations. Walzer *et al.* (2000), for example, recommended against the use of modified live vaccines in Przewalski horse reintroductions.

Box 9.3 **Advantages and disadvantages of vaccinating threatened wildlife**

The risks of population extinction due to infectious disease are highest in small populations (Berger, 1990; De Castro & Bolker, 2005; Woodroffe, 1999). On this basis alone, populations being restored through reintroduction are expected to be especially vulnerable to disease. It may be appealing, therefore, to vaccinate animals intended for release against pathogens they may encounter post-release, and perhaps to continue such vaccination within the reintroduced population. In deciding whether, and how, to conduct such management, it is important to balance the advantages and disadvantages of vaccination.

The potential advantages of vaccination are straightforward: as long as the vaccine and its delivery method are safe and effective, vaccination should reduce the impacts of specific pathogens on hosts' survival and/or reproduction. Such beneficial effects may improve the chances of reintroduction leading to successful population recovery. However, vaccination of threatened wildlife also has three potential disadvantages.

First, vaccines may occasionally have negative side effects. Most of such problems were associated with older modified-live vaccines, many of which have now been replaced by safer alternatives. For example, when black-footed ferrets, *Mustela nigripes*, were first taken into captivity for breeding in 1971, all were given a modified-live vaccine against canine distemper virus (CDV) which killed them (Carpenter *et al.*, 1976). Fortunately, modern recombinant CDV vaccine is far safer (Wimsatt *et al.*, 2003). In another high-profile case, vaccination of African wild dogs (*Lycaon pictus*) against rabies virus was blamed for the species' local extinction in the Serengeti (Burrows, 1992), although evidence was scanty (Woodroffe, 2001), and subsequent studies suggest that such vaccination can reduce rabies mortality in this species (Hofmeyr *et al.*, 2004). Nevertheless, the perception that vaccination had such negative consequences for wild dogs influenced decisions not to vaccinate other threatened species (Laurenson *et al.*, 1997). Clearly, safety trials are vital to inform decisions about whether to proceed with vaccination.

Second, apart from the effects of vaccines themselves, vaccine delivery methods may risk injury or stress to the animals; this was one reason why vaccination of harbour seals (*Phoca vitulina*) against phocine distemper virus was not undertaken (Hall & Harwood, 1990). Ill effects of handling can be avoided by using oral vaccines where available; however, such vaccines present their own safety challenges. Because oral vaccines may potentially be consumed by multiple species, safety testing must involve non-target species as well as the threatened species of concern. As an example of the risks, one strain of modified-live rabies vaccine being considered for use in jackals was found to induce rabies in baboons (*Papio ursinus*; see Bingham *et al.*, 1992), though a more modern strain caused no such disease (Bingham *et al.*, 1997). Recently developed recombinant vaccines are expected to be safer than modified-live vaccines, but still need to be safety tested. For such vaccines, safety trials may need to focus on non-target species likely to be affected by the vector virus. For example, the recombinant vaccine against CDV (a virus of mammalian carnivores) is a genetically modified strain of canarypox (a virus of birds), and so safety trials for oral use might need to assess potential risks to birds.

Third, in addition to these safety concerns for individuals, vaccination has the potential to erode natural disease resistance in host populations. Where hosts and pathogens co-exist, some proportion of hosts will usually have acquired immunity to the pathogen and thus be protected in the event of an outbreak. However, if vaccination leads to local eradication of the pathogen, within a generation the host population will be immunologically naive and potentially vulnerable to epidemics if vaccination is halted. If maintained for longer, vaccination could potentially impede natural selection for pathogen resistance. For these reasons, reintroduction programmes may need to balance the benefits of vaccination in the early stages of recovery (when every individual may be important for population viability) against possible costs later on (when occasional deaths may be a worthwhile price to pay for population immunity).

A case study: reintroduction of island foxes

In recent years, vaccination has been an important component of rein-troduction strategies for the critically endangered island fox (*Urocyon littoralis*). This species, endemic to six islands off the California coast, suffered major population crashes on four islands in the 1990s, with two island subspecies becoming extinct in the wild (Roemer *et al.*, 2001; Timm *et al.*, 2009). Although only one of these crashes was caused by a pathogen (likely CDV; see Timm *et al.*, 2009), disease was identified as a risk to recovery on all of the affected islands (US Fish and Wildlife Service, in review).

Disease management strategies for island foxes were influenced by the recognition that population extinction risks can be reduced by vaccinating a subset of animals, which would be expected to survive major epidemics and act as founders for post-epidemic recovery (Hay-don *et al.*, 2002, 2006). Not only does this approach minimise the number of individuals to be vaccinated, it also leaves unvaccinated animals to continue interacting with the pathogen, hopefully helping to maintain natural resistance. In the island fox programme, a wild strain of CDV was known to be circulating on the islands with extant fox populations (Clifford *et al.*, 2006), and maintaining this virus was considered important for long-term fox conservation (US Fish and Wildlife Service, in review). Simulation modelling revealed that disease risks could not be effectively managed by responding to epidemics: by the time elevated mortality had been recognized, causes identified and appropriate responses mounted, a CDV epidemic could already have devastated such small isolated populations (Doak *et al.*, submitted). A prophylactic vaccination strategy was therefore preferred (US Fish and Wildlife Service, in review).

The closely-related grey fox (*Urocyon cineroargentaeus*) was known to be highly sensitive to modified-live CDV vaccines (Henke, 1997). Fortunately, captive trials revealed that recombinant CDV vaccine could be used safely on island foxes (Timm *et al.*, 2002). As wild populations have been re-established on all islands, partly by reintroduction, the strategy has been to vaccinate all captive animals (and hence all animals destined for release; see Island Fox Working Group, 2010). In small populations in the early phases of recovery, all captured foxes are vaccinated against CDV, with the intention of transitioning to maintaining a 'vaccinated subset' of specified size as the populations grow (Island Fox Working Group, 2010).

Pre-translocation parasite screening of wild populations and risk assessment

Despite the increasing awareness of infectious disease risks, a surprisingly small number of studies have been conducted on wild populations prior to reintroduction (either at source or release locations). Rather, most studies appear post hoc and document the current extent of parasites across wild (including re-established) populations (e.g. Hopkins *et al.*, 1990; Tocidlowski *et al.*, 1997; Gartrell *et al.*, 2007; Ortiz-Catedral *et al.*, 2009a, 2009b). For example, surveys of infectious disease in wild turkey, *Meleagridis gallopavo silvestris*, sampled during translocation programmes identified a range of pathogens that may be co-introduced (including *Pasteurella multocida, Bordetella avium* and Newcastle disease virus; see Hopkins *et al.*, 1990). Similarly, there is an extensive database of haematology and plasma biochemistry parameters accrued through sampling of North American river otters, *Lontra canadensis*, during translocation (Tocidlowski *et al.*, 2000; Kimber and Kollias, 2005). Serologic surveys of river otters during these translocations have revealed substantial exposure to *Toxoplasma gondii* (Tocidlowski *et al.*, 1997) and various viral agents, including canine distemper virus, canine herpesvirus-1 and canine parvovirus-2 (Kimber *et al.*, 2000). Although these data were not collected to inform a priori disease risk assessments, the information obtained is valuable for understanding of disease threats in these and other host species (wild, domestic and human) and hence in providing wildlife agencies protection against liability for disease outbreaks (Hopkins *et al.*, 1990; Tocidlowski *et al.*, 1997; Kimber *et al.*, 2000).

Some studies have, however, been conducted prior to reintroduction (e.g. Berry & Christopher, 2001; Calle *et al.*, 2001; Peterson *et al.*, 2002; Work *et al.*, 2004; Middleton *et al.*, 2010). For example, Laysan ducks within their remnant population are susceptible to the nematode *Echinuria uncinata* during periods of low rainfall and poor food availability. Work *et al.* (2004) recommended that translocations of Laysan duck involve screening to avoid co-translocation of this parasite, given its known pathogenicity in this endangered host and the lack of recorded pathogenic infection of other waterfowl species at release sites. Similarly, Middleton *et al.* (2010) screened 18 lizard species across eight geographically isolated islands in New Zealand for *Salmonella enteric* serovars with an a priori aim to determine spatial distributions of this potential pathogenic agent and to inform disease risk assessments for an increasing number of lizard translocations. Perhaps the biggest distinction of studies in wild versus captive populations is that: (i) typically, wild populations should have a more *natural* and endemic (see Box 9.1 for epidemiological definition) parasite fauna, but that (ii) far less is known about parasitism in free-living non-domestic and non-captive animals. However, just because a specific translocation uses free-living wild animals as the source does not mean that parasite risk is removed. Remnant populations are often highly fragmented and exist in highly modified environments. Free-living source populations may be host to novel infectious agents and/or the selected release locations may contain infectious agents not present in source locations. Small populations may also lose parasites due to the effects of disrupted transmission (particularly when transmission is density dependent) (Swinton *et al.*, 2002). The four important questions of Armstrong & Seddon (2008) still apply.

A recurring statement in disease-related reintroduction literature is risk assessment. This simple term is the backbone to any work relating to parasites and reintroduction. It is also contentious, with disagreement between different stakeholders (Wilson *et al.*, 1994). For example, pre-translocation disease screening in New Zealand hihi detected a low prevalence infection of an unidentified *Plasmodium* parasite within a free-living population, and the coordinating personnel were divided whether or not to allow infected individuals to be translocated to a new site (Ewen *et al.*, in press). Some argued against translocation because no information was available on the presence of this host generalist parasite at the release location (where other threatened potentially naive bird species occur), whereas others cited the known low pathogenicity and high prevalence of this parasite in other species at the source location as indicating that this *Plasmodium* sp. was probably a low threat.

Risk assessment can take many forms, from epidemiologically informed yet qualitative calls of judgement to more sophisticated quantitative modelling of known or estimated effects and parasite distributions. When risk is assessed, or when practitioners state that risk assessments have been undertaken, it is beneficial to understand the level at which any analysis has been made and exactly what data have been used. Parasite risk assessments are now a required component of reintroduction permit applications in some countries, e.g. New Zealand under Department of Conservation and United Kingdom under Natural England, where in the United Kingdom risk assessments must be made publically available under the Freedom of Information Act.

By far the most common form of risk assessment is the qualitative disease risk assessment. In most cases our knowledge is limited; we often know little about parasites in source or release locations and have little information of the pathogenicity or regulatory roles any given parasite plays on the target species being translocated or the wider ecosystem. Hence many studies screen for a parasite with traits that identify it as high risk and, if found, recommend caution, preventative treatment, elimination of the parasite if it is presumed alien or rejection of the animal from translocation. This may include situations when the parasites' detection was as an apparently non-pathogenic infection, e.g. where a suspected alien cestode parasite was eliminated from dormice (*Muscardinus avellanarius*) prior to reintroduction to prevent the possibility of disease in naive conspecifics at release locations (Sainsbury *et al.*, 2004). Other studies draw on previously collected necropsy datasets that go a step further in identifying which parasites are causative agents of disease and mortality in the target species to be translocated (or in their environments), and often similarly recommend treatment or rejection of an animal from translocation (e.g. New Zealand hihi; see Cork *et al.*, 1999; Ewen *et al.*, in press).

One way these qualitative risk assessments may be formalized is by following standardized steps of identifying and categorizing risk from different parasites at different stages throughout the reintroduction pathway (e.g. elk *Cervus elaphus*; see Corn & Nettles, 2001; Davidson & Nettles, 1992). Qualitative risk assessments follow formats adapted from disease risk assessments for movements of domesticated animals (Murray *et al.*, 2004; Armstrong *et al.*, 2003). These early documents continue to be refined as wildlife disease specialists grapple with the unique challenges faced when trying to predict and reduce disease risks in often poorly understood systems (Vaughan & Sainsbury, submitted; Sainsbury *et al.*, this volume, Chapter 10). The benefits of this risk assessment are that: (i) risk assessment is structured through the different stages of the reintroduction pathway, (ii) the information used to make an assessment is stated at each stage of the process and (iii) there is transparency in the process so recommendations can be weighed against data quality.

Despite the benefits of qualitative risk assessment, they rely at some level on informed judgement calls. Informed judgement occurs across many components of reintroduction and is essential to allow expert knowledge to direct strategy. However, this level of assessment must be continually refined to allow evidence-based approaches (Gusset *et al.*, 2008). We present a recent qualitative disease risk assessment format in the following chapter (Sainsbury *et al.*, this volume, Chapter 10). Quantitative risk assessments have also been advocated but are rarely seen in published literature. In an early effort to provide a quantitative framework, Ballou (1993) outlined the utility of population viability analysis (PVA) to examine the potential effects of disease on establishing reintroduced populations. Ballou (1993) modelled the effects of co-introducing callitrichid hepatitis (an arena virus) on the remnant golden lion tamarin population in Brazil. Using detailed demographic data from the wild population and virulence data from captivity (about 75% mortality in infected captive colonies) this study was able to predict the fate of the wild population if the disease was introduced and outbreaks occurred at various frequencies (extinction risk was substantially increased) (Ballou, 1993). A problem with this form of PVA is that detailed primary data are required to make meaningful predictions, and even in the above example it is difficult to know whether virulence data from captivity are representative of wild populations.

Disease risks are also posed to domestic animals from reintroduced and infected wild hosts, or even uninfected wild hosts, which may facilitate transmission of commercially damaging pathogens. For example, plans to reintroduce wild boar, *Sus scrofa*, in Denmark are complicated because wild boar are potential reservoirs of classical swine fever virus, which causes a viral disease of economic concern to domestic pig farming. It is thought, for example, that 60% of classical swine fever outbreaks in domestic pig farms in Germany are linked to cycling of the virus in wild boar populations (Fritzemeier *et al.*, 2000). Fernández *et al.* (2006) combined known wild boar population dynamics from other European countries with spatial modelling of suitable habitat and location of domestic pig farms (ranked according to the potential of contact between wild boar and domestic pig) in Denmark. Combining predictions of population growth in establishing wild boar populations with known locations of high-risk farms identified suitable release locations and more generally have provided an important example of reconciling species recovery with the affected health and economic interests of the pig-farming sector (Fernández *et al.*, 2006).

Short-term captivity, quarantine and stress

In wild-to-wild translocations of animals there is usually a period of captivity, either during transfer, or because quarantine is imposed, or because release is delayed for some other purpose. Quarantine is common when disease and health assessments are being undertaken and is considered essential to prevent contact between the animals and any external sources of parasites while the animals are prepared for reintroduction. The length of quarantine may be: (i) determined by the time needed for diagnostic results from tests made at capture or (ii) designed to be longer than the incubation time of targeted diseases (Viggers *et al.*, 1993; Wobeser, 2002; Calvete *et al.*, 2005). Quarantine being longer than the incubation time of disease has, for example, been implemented to prevent transmission of *Yersinia pestis* between black-footed ferrets (Williams *et al.*, 1994) and for preventing oral Panagrolaimidae nematode infections in Theraphosidae, so-called tarantula spiders (Pizzi, 2004). Animals held in quarantine could suffer various stressors related to confinement, crowding and close contact with humans, and these may further compromise an individual's immune capacity. Stress-induced disease during captivity and translocation has been hypothesized in the literature (Viggers *et al.*, 1993; McCallum & Dobson, 1995; Nolet *et al.*, 1997; Mathews *et al.*, 2006). Recently, stress management in its own right has been the focus of reviews (Teixeira *et al.*, 2007; Dickens *et al.*, 2010) and experimental evaluation of reintroduction programmes indicates that it is an important factor contributing to individual health and survival (Groombridge *et al.*, 2004; Calvete *et al.*, 2005; Hartup *et al.*,

2005; Dickens *et al.*, 2009; Pinter-Wollman *et al.*, 2009) (see Parker *et al.*, this volume, Chapter 4). One study concluded that a major factor causing increases in stress hormone levels was restraint for health screening (whooping cranes; see Langenberg *et al.*, 2002). Obviously stress requires management through appropriate husbandry and handling and, importantly, should always be considered within any disease risk assessment.

Another problem of confinement is the risk of parasite transmission and disease outbreak, complicating the purpose of quarantine. For example, coccidia are often targeted for parasite screening during bird translocations (Forrester *et al.*, 1978; McGill *et al.*, 2010; Ewen *et al.*, in press; Vaughan & Sainsbury, submitted). Coccidian infections are common in captivity where their oral–faecal transmission is facilitated by confinement and increased density around feeding and roosting sites. Furthermore, coccidian occysts are resistant and remain viable in substrate for long periods (Marquardt *et al.*, 2000). In small reserves in New Zealand, which act as temporary homes for juvenile kiwi, *Apteryx mantelli*, a possible increased environmental load of coccidia oocysts is causing disease expression in birds temporarily held there (K. Morgan, unpublished data). To reduce environmental loading of parasites like coccidia, permanent aviaries used for holding hihi at source locations during quarantine have a fresh layer of natural leaf litter spread over the floor to trap faecal material for removal once birds have been translocated (Ewen *et al.*, in press). In some cases it is possible to prevent contact between individuals during transfer, thus eliminating any parasite transmission (e.g. kakapo; see Gartrell *et al.*, 2005).

Preventative medication is also often used to reduce parasite density and minimize parasite transmission. Frequently, broad-spectrum antiparasitic treatment is provided as a standard medication to individuals held in captivity (Phillips & Scheck, 1991; Britt *et al.*, 2004; Moreno *et al.*, 2004). In hihi, for example, a mix of toltrazuril (Baycox®, Bayer Animal Health) for the reduction of the intensity of coccidial infection, itraconazole (Sporanox®, Jannsen-Cilag) as prophylaxis for aspergillosis and a pyrantel pamoate (Combantrim®, Johnson & Johnson) and praziquantel (Droncit®, Bayor Animal Health) combination to reduce the population of intestinal helminths has been administered during quarantine (Ewen *et al.*, in press). The choice of treatment reflects known health problems in this species based on necropsy results from captive and wild populations (Ewen *et al.*, in press). Another commonly administered medication in translocations is the broad-spectrum parasiticide ivermectin (e.g. in Laysan ducks, Work *et al.*, 2010; red wolves, Phillips &

Scheck, 1991; Przewalski horse, Walzer *et al.*, 2000; and Sumatran orang-utans, *Pongo abelii*, Mul *et al.*, 2007). Medication use can be an important and efficient way of controlling parasites during captive stages of translocation and may also help offset any stress-induced reduction in immune capacity. However, there are disadvantages to broad-spectrum treatments. For example, treatment can remove or disrupt other commensal parasites that may perform beneficial roles. One suggested benefit of maintaining a natural host–commensal parasite community is that this affords a natural resistance (Cunningham, 1996). Also, where conspecifics are absent from a reintroduction site, maintaining the host–commensal parasite community should ensure that parasites are conserved in future generations, and these generations will therefore have immunity if subsequently exposed to conspecifics with the same parasites. Furthermore, in some species there is a risk of injury during capture and treatment; e.g. administering anthelmintics in kaki has caused injuries such as broken bills and legs, and capture myopathy (Maurice Alley, personal observation). Critically, inadvertent loss of parasites through broad-spectrum treatments can also result in endangered parasite extinction, such as likely occurred with a louse species, *Colpocephalum californici*, when its Californian condor, *Gymnogyps californianus*, hosts were treated during the founding of the captive population (Pizzi, 2009).

Disease risk assessments for wild-to-wild translocations are often made with little prior information. It is not surprising then that more general assessments of health are often also used to select or reject individuals for translocation (e.g. Ewen *et al.*, in press). As with parasite screening, however, *normal* ranges, e.g. haematology parameters, are rarely known, albeit with a growing number of exceptions. Examples include a survey of haematology and serum biochemistry parameters in North American river otters (Tocidlowski *et al.*, 2000) and a survey of weights and vital signs, haematology and serum biochemistry parameters in red ruffed lemurs, *Varecia rubra* (Dutton *et al.*, 2008). One objective of health screening is to select generally healthy individuals that may have an increased chance of post-release survival (Tocidlowski *et al.*, 2000; Mathews *et al.*, 2006; Kwon *et al.*, in press). Some health metrics may also reveal an underlying disease agent that is not identified by direct parasite screening and unusual results can be used to reject individuals as potential founders. Common health parameters that are measured in animals to be translocated include weight and body condition, various haematology and serum biochemistry parameters (see the above references) and a condition index might be assigned (Bell *et al.*, 2004). As with all methods of parasite management, the choice of which and how many tests to undertake needs consideration within a transparent risk assessment.

In an earlier section of this chapter, we summarized the risks to translocations from captive facilities in exotic locations housing multiple species from multiple origins. Similar risks apply to temporary captivity during wild-to-wild translocation. Ex situ facilities present risks of disease transmission, and ideally captive-holding enclosures should be single species and situated at source locations. Recent examples of temporarily holding wild-caught animals in ex situ enclosures include North Island saddleback, *Philesturnus rufusater*, being held at a multispecies from multiorigin zoo during disease diagnosis in translocations from Cuvier Island to Boundary Stream (http://rsg-oceania.squarespace.com/nz-saddleback/; Armstrong & Davidson, 2006) and from Mokoia Island to Bushy Park (Thorne, 2007).

Post-release survival and ongoing parasite management

Everything that happens up until the point of releasing individuals should be done to maximize the chances that these individuals will survive and establish or re-inforce a wild population without damage to either source or release ecosystems. Despite the multitude of factors considered (parasites and health only being a small part) it is surprisingly rare that the fates of reintroductions are monitored with sufficient intensity to allow estimation of the effectiveness of management or to identify what alternative management may be required. Furthermore, we often know little about parasite (and other) risks at the release site. Given the low success rates of reintroductions (Griffith *et al.*, 1989; Fischer & Lindenmayer, 2000) it is crucial that appropriate post-release monitoring is used to evaluate outcomes (Ewen & Armstrong, 2007; Sutherland *et al.*, 2010; Chapters 6 to 8 in this volume).

The most common post-release data published about parasites is documentation of parasites, disease and causes of death, e.g. recording metazoan parasites in Atlantic salmon, *Salmo salar*, reintroduced to the Czech Republic, Moravec, in 2003. Such work provides important information about possible disease threats facing reintroduced populations. Work *et al.* (2010), for example, recommended that systematic monitoring of translocated populations should always occur so that any new or additional disease risks are identified. Examples of disease causing post-release mortality are most often associated with parasites already present at release sites and include: the clinical and pathological findings from hihi necropsies indicating aspergillosis as a significant cause of mortality (Alley *et al.*, 1999); the documentation of combined avian-pox virus and *Plasmodium elongatum* infections in South Island saddleback, *Philesturnus carunculatus* (Alley *et al.*, 2010); erysipelas causing mortality in kakapo (Gartrell *et al.*, 2005); *Toxoplasma gondii* infections causing mortality in 'Alala (Work *et al.*, 2000); documenting plague infections as a mortality factor in reintroduced Canadian lynx, *Lynx Canadensis* (Wild *et al.*, 2006) and in black-footed ferrets (Antolin *et al.*, 2002; Gober, 2008); rabies in African wild dogs, (Hofmeyr *et al.*, 2000; Scheepers & Venzke, 1995); yersiniosis and leptospirosis causing mortality in reintroduced beavers (Nolet *et al.*, 1997); and exposure to infectious bursal disease virus in whooping cranes released in the USA (Candelora *et al.*, 2010).

If mortality events are associated with infectious disease outbreaks then monitoring for disease occurrence can provide an early warning mechanism and allow mitigation. For example, an outbreak of avian botulism, caused by toxins released by *Clostridium botulinum*, led to high mortality in Laysan ducks in a translocated population on Midway Atoll (Work *et al.*, 2010). Disease management strategies are restricted in such remote sites, but could include a level of monitoring that can detect if another outbreak occurs, the removal of carcasses, especially during outbreaks, and, if possible, the draining or flooding of affected wetlands (Work *et al.*, 2010). With this critically endangered duck there are also suggestions that demographically or genetically important individuals be vaccinated against type C botulism toxin.

However, more data are often required in order to investigate whether a pathogen is causing disease at unsustainably high levels such that population establishment is compromised (either as a direct or multifactor effect on mortality or fecundity). Some case studies attempt this by estimating the demographic or population level impact on hosts from a recognized pathogen. For example, population monitoring during a recent *Salmonella typhimurium* outbreak in hihi on Tiritiri Matangi Island showed about 76% of the adult population died due to salmonellosis (Ewen *et al.*, 2007), but that this had only minor impact on the populations continued growth since reintroduction (Armstrong & Ewen, 2011). Similarly, *Leucocytozoon marchouxi* infections in pink pigeons, *Columba mayeri*, occur at a high prevalence and have been noted as a mortality factor, but screening blood smears showed no difference in survival between infected and uninfected individuals (Swinnerton *et al.*, 2005b). The authors suggested that *L. marchouxi* was probably introduced to Mauritius with exotic pigeons and that the pink pigeon has acquired sufficient immunity now to be a maintenance host (Swinnerton *et al.*, 2005b). Demographic data can help re-evaluate pathogen risk as populations establish and, in some cases, disease due to pathogens can be treated as one sustainable component of mortality affecting wild populations.

We have seen in the above examples that parasites already present at release sites can cause mortality within establishing populations. Exotic pathogen pollution may have contributed to the original extirpation of native species or arrived subsequently. In either scenario the release habitat may require substantial management or rehabilitation before it becomes suitable. Parasites already present at release sites may also be native to those sites but able to cause unsustainable mortality due to anthropogenic environmental changes leading to novel interactions between the parasite and endangered host. For example, evidence suggests that a macroparasite roundworm, *Baylisascaris procyonis*, is reducing population viability within a reintroduced intermediate (and endangered) host, the Allegheny woodrat, *Neotoma magister* (LoGiudice, 2003). LoGiudice (2003) suggested that suburban development led to increased densities of the parasites' definitive host, the racoon *Procyon lotor*, and has created parasite environments unsuitable for the highly susceptible intermediate host. Intrinsic factors such as stress (Parker *et al.*, this volume, Chapter 4) and low genetic diversity (e.g. Acevedo-Whitehouse *et al.*, 2003; see Box 9.2) may also contribute to novel host–parasite interaction.

Where parasites have sufficient demographic impacts to reduce a reintroduced population's viability, but where available release locations cannot avoid them, then in situ parasite management is required. This should be treated alongside other factors that may be affecting population vital rates (survival and reproduction). Predation (and the potential interaction between parasitism and vulnerability to predation), for example, is another common problem for reintroduced populations and one where complete removal of predators is not always appropriate. Modelling mortality rates under different parasite pressures may reveal a level of sustainable mortality that enables population persistence and can be a target for management (such as early worked examples of PVA by Ballou, 1993). A powerful approach is modelling population dynamics within an adaptive management framework, as this can evaluate the effectiveness of different management systems and allow the testing of alternative methods. For example, *Ornithonyssus bursa* ectoparasites in hihi nests can cause reduced chick quality and also nest failure (Ewen *et al.*,

2009). Using an adaptive management approach and modelling population vital rates Armstrong *et al.* (2007) showed that management of *O. bursa* in nests of a reintroduced hihi population on Mokoia Island significantly improved population growth rate, but that population persistence remained uncertain due to other factors, possibly because of the disease aspergillosis. In another example, some reintroduced pink pigeon subpopulations are thought to be constrained by trichomoniasis, where monitoring has shown that up to 66% of squabs may die of the disease in a given breeding season (Bunbury, 2006). The parasite, *Trichomonas gallinae*, is a presumed exotic but now endemic parasite of Mauritius that was probably co-introduced with exotic pigeons (Bunbury *et al.*, 2007). The severity of this disease and its prevalence has resulted in treatment of individuals and attempts to treat the Ile aux Aigrettes subpopulation via medicating drinking water (Swinnerton *et al.*, 2005a). While individual treatment of squabs substantially increased survival, the population level treatments were less successful as adults did not utilize medicated drinking sites frequently enough (Swinnerton *et al.*, 2005a). The vital rates of reintroduced pink pigeon populations are now being modelled to assess better how important trichomoniasis is, and will be used to evaluate different management strategies (C.G. Jones, A.G. Greenwood & K.J. Swinnerton, unpublished).

Importantly, our literature review has highlighted that very few studies have evaluated how well the general health measures or medical treatments given to individuals earlier in the reintroduction pathway can predict post-release survival. This would seem a particularly relevant objective especially considering the requirement to hold and manipulate animals (risking stress or injury) and the monetary cost of tests and treatments (Ewen *et al.*, in press). Obtaining such data can provide the context to general metrics of health and perhaps identify where management can be streamlined. The few studies that have done this should be applauded (e.g. Mathews *et al.*, 2006; Robertson, 2009). Mathews *et al.* (2006), for example, provide detailed insights to the fates of reintroduced water voles, *Arvicola terrestris*, using data collected during health and parasite screening before release into a restored wetland in London, United Kingdom. The mass of water voles was a significant predictor of initial survival but not longevity, and this association was dependent on the timing of release (Mathews *et al.*, 2006). Furthermore, erythrocyte concentrations were positively associated with both initial survival and longevity, while both haematocrit and haemoglobin values were positively associated with longevity (Mathews *et al.*, 2006).

Revisiting Armstrong & Seddon's (2008) four questions (a conclusion and future direction)

(i) *How is post-release survival affected by parasite management?* Most past study has focused on the health and parasite screening of individuals to be reintroduced. These studies frequently claim that this is to enhance post-release survival as well as to prevent the introduction of alien parasites to the release site and to avoid zoonoses. Very rarely is this followed through with adequate post-release monitoring. Insufficient monitoring is not restricted to health and parasite considerations in reintroduction biology, but it is none-the-less concerning that our conclusion is necessarily similar to that made about two decades ago (Griffith *et al.*, 1993).

(ii) *How do parasites affect suitability of release sites?* Very little attention is paid to assessing release sites for endemic disease risk prior to reintroduction. Release site suitability is often considered in ad hoc investigations following mortality events. The majority of post-release mortality and health problems reviewed here *suggest* that it is parasites that are already present at release sites that more commonly present problems. Given the fact that endemic parasites at release sites are frequently noted as problems raises a challenge to reintroduction biologists to assess release site suitability better before reintroduction takes place. Determining suitable release areas for translocations remains a difficult problem in reintroduction biology.

(iii) *Are parasites native to the ecosystem?* This can sometimes be an easy question. If the parasite is host specific and the host is range restricted, it is therefore by nature native to the ecosystem. In such cases this means that it should be co-introduced from a restoration perspective, although it may require some management such that the parasite does not compromise host population establishment. This question becomes more complicated when (i) the host has a wide range and there is geographic structuring in host-parasite relationships and (ii) with host-generalist parasites. While there is good demonstration of screening captive individuals for reintroduction (especially where captive centres are in exotic locations and are multispecies from multiorigin units), most studies have not assessed release sites in enough detail to know if host generalist parasites are already present. Common conclusions either suggest that release site screening is important prior to reintroduction and/or a cautious approach to reintroduction is recommended, such as avoiding areas with extant populations of the same or other endangered species.

(iv) *How will the ecosystem be affected by the parasites?* Very little research has been devoted to parasite impacts on the wider ecosystem at release sites beyond documenting disease in released individuals resulting from co-introduced parasites in some studies. Wider ecosystem impacts may often go unnoticed because attention is focused on the reintroduced host. Most awareness of the risks of exotic pathogen introduction draws on studies of pathogen co-introduction with exotic hosts. Here, the number of case studies is substantial, making it surprising that more information is not available from native species reintroductions. The similarity between the risks from reintroduction of native hosts and from the introduction of exotic hosts deserves attention.

Early literature reviews suggested that parasites and health status had been largely neglected in the planning of reintroductions and that the difficulties in (and lack of) monitoring parasites and health perhaps falsely gave an impression of low risk and minimal impact of parasites (Griffith et al., 1993; Viggers et al., 1993; Cunningham, 1996). It has been about 18 years since these early empirical summaries and even longer since the seminal paper of Thorne & Williams (1988), so it is worth judging how parasite management has progressed. We feel that over this time substantial attention has been paid to publishing case studies of parasite consideration in reintroductions. Our review includes cases where disease introduction has occurred and where translocations have failed because of disease, providing some response to Griffith et al.'s (1993) comment that 'it is surprising that disease introduction or disease caused translocation failure are not more common, given the large number of translocations conducted annually'. Reported cases of either parasite introduction or disease-caused translocation failure, however, remain uncommon. This may be due to the difficulties in detecting disease in wild populations (Wobeser, 2006), linked with the noted poor post-release monitoring of most reintroduced populations (Seddon et al., 2007). Some researchers therefore still argue that parasite and health considerations in reintroduction biology remain neglected (e.g. Mathews et al., 2006; Faria et al., 2010). This may reflect a publishing bias towards clinical case studies, partly as a consequence of a clinical veterinary perspective. It is encouraging, however, to see that governing bodies are starting to request parasite risk assessments during the planning stages of reintroduction.

We conclude by highlighting four areas where we encourage focused attention into the future:

1. The majority of studies are post hoc, conducted in reaction to parasite or disease occurrence, with far too few published results from well-planned and a priori considered studies.
2. Risk assessments need further development including the building of stronger ties with population ecology. Most parasite and health considerations have been the focus of veterinary practitioners with a strong clinical viewpoint. This is not unusual as conservation scientists with different expertise often work in isolation and publish in specialist journals.
3. More attention is needed at the ecosystem level in order to determine the wider implications of parasite co-introduction. This is equally relevant in both reintroduction biology and in invasion biology.
4. There is a current lack of detailed study post-release to provide a complete understanding of parasites and health screening in relation to management. This is essential to justify imposed management. Gusset *et al.* (2008) best summarize this point by arguing a need for reintroduction biology in general to move from a scientific discipline often based on subjective beliefs to one based on critical appraisal, or evidence-based conservation. In another recent study, Mathews *et al.* (2006) suggested that there is no consensus on the policies or protocols that health and parasite screening should take. We note here that the frequent recommendations to screen for high-risk parasites in reintroduction programmes very rarely state the accuracy of current test procedures and therefore how reliable screening results are. Furthermore, high-risk parasites may often be unknown because in their native environments they can be constrained by host immunity and will cause significant disease only in new ecosystems.

There is unlikely to be a single health and pathogen screening protocol for any given species, as wildlife disease is a dynamic process with new and emergent diseases hard to predict, and risk assessments are influenced by the characteristics of different source and release sites. Furthermore, test procedures continue to be refined, both improving test validity and increasing the number of different tests (e.g. Artois et al., 2009). This makes parasite and health management in reintroduction also a dynamic process and one that offers considerable challenges and rewards to scientists and reintroduction practitioners.

References

Acevedo-Whitehouse, K. (2004) Influence of inbreeding on disease susceptibility in natural populations. PhD Thesis, Department of Zoology, University of Cambridge, Cambridge, UK.

Acevedo-Whitehouse, K., Gulland, F., Greig, D. et al. (2003) Inbreeding. Disease susceptibility in California sea lions. Nature, 422, 3.

Acevedo-Whitehouse, K., Petetti, L., Duignan, P. et al. (2009) Hookworm infection, anaemia and genetic variability of the New Zealand sea lion. Proceedings of the Royal Society B, 276, 3523–3529.

Acevedo-Whitehouse, K., Spraker, T. R., Lyons, E. et al. (2006) Contrasting effects of heterozygosity and survival and hookworm resistance in California sea lion pups. Molecular Ecology, 15, 1973–1982.

Acevedo-Whitehouse, K., Vicente, J., Höfle, U. et al. (2005) Genetic resistance to bovine tuberculosis in wild boars. Molecular Ecology, 14, 3209–3217.

Acton, A. E., Munson, L. & Waddell, W. T. (2000) Survey of necropsy results in captive red wolves (Canis rufus), 1992–1996. Journal of Zoo and Wildlife Medicine, 31, 2–8.

Alberts, A. C., Oliva, M. L., Worley, M. B. et al. (1998) The need for pre-release health screening in animal translocations: a case study of the Cuban iguana (Cyclura nubile). Animal Conservation, 1, 165–172.

Allendorf, F. W. & Luikart, G. (2007) Conservation and the Genetics of Populations. Blackwell, Oxford, UK.

Alley, M. R., Castro, I. & Hunter, J. E. B. (1999) Aspergillosis in hihi (Notiomystis cincta) on Mokoia Island. New Zealand Veterinary Journal, 47, 88–91.

Alley, M. R., Hale, K. A., Cash, W. et al. (2010) Concurrent avian malaria and avipox virus infection in translocated South Island saddlebacks (Philesturnus carunculatus carunculatus). New Zealand Veterinary Journal, 58, 218–223.

Amos, W. & Acevedo-Whitehouse, K. (2009) A new test for genotype–fitness associations reveals a single microsatellite allele that strongly predicts the nature of tuberculosis infections in wild boar. Molecular Ecology Research, 9, 1102–1111.

Antolin, M. F., Gober, P., Luce, B. et al. (2002) The influence of sylvatic plague on North American wildlife at the landscape level, with special emphasis on black-footed ferret and prairie dog conservation. In Transactions of the 67th North American Wildlife and Natural Resources Conference, ed. J. Rahm, Washington, DC.

Armstrong, D. P. & Davidson, R. S. (2006) Developing population models for guiding reintroductions of extirpated bird species back to the New Zealand mainland. *New Zealand Journal of Ecology*, 30, 73–85.

Armstrong, D. P. & Ewen, J. G. (2011) Consistency, continuity and creativity: long-term studies of population dynamics on Tiritiri Matangi Island. *New Zealand Journal of Ecology* (in press).

Armstrong, D. P. & Seddon, P. J. (2008) Directions in reintroduction biology. *Trends in Ecology and Evolution*, 23, 20–25.

Armstrong, D. P., Castro, I. & Griffiths, R. (2007) Using adaptive management to determine requirements of reintroduced populations: the case of the New Zealand hihi. *Journal of Applied Ecology*, 44, 953–962.

Armstrong, D., Jakob-Hoff, R. & Seal, U. S. (eds) (2003) *Animal Movements and Disease Risk: A Workbook*. Conservation Breeding Specialist Group (SSC/IUCN), Apple Valley, Minnesota, USA.

Artois, M., Bengis, R., Delahay, R. J. *et al.* (2009) Wildlife disease surveillance and monitoring. In *Management of Disease in Wild Mammals*, eds J. J. Delahay, G. C. Smith & M. R. Hutchings. Springer, New York, USA.

Atkinson, C. T. & van Riper, C. III. (1991) Pathogenicity and epizootiology of avian haematozoa: *Plasmodium, Leucocytozoon*, and *Haemoproteus*. In *Bird–Parasite Interactions: ecology, evolution and behaviour*, eds J. E. Loye & M. Zuk, pp. 19–48. Oxford University Press, Oxford, UK.

Ballou, J. D. (1993) Assessing the risks of infectious disease in captive breeding and reintroduction programs. *Journal of Zoo and Wildlife Medicine*, 24, 327–335.

Balloux, F., Amos, W. & Coulson, T. (2004) Does heterozygosity estimate inbreeding in real populations? *Molecular Ecology*, 13, 3021–3031.

Beadell, J. S., Ishtiaq, F., Covas, R. *et al.* (2006) Global phylogeographic limits of Hawaii's avian malaria. *Proceedings of the Royal Society of London B*, 273, 2935–2944.

Bell, B. D., Pledger, S. & Dewhurst, P. L. (2004) The fate of a population of the endemic frog *Leiopelma pakeka* (Anura: Leiopelmatidae) translocated to restored habitat on Maud Island, New Zealand. *New Zealand Journal of Zoology*, 31, 123–131.

Berger, J. (1990) Persistence of different-sized populations: an empirical assessment of rapid extinctions in bighorn sheep. *Conservation Biology*, 4, 91–98.

Berry, K. H. & Christopher, M. M. (2001) Guidelines for the field evaluation of desert tortoise health and disease. *Journal of Wildlife Diseases*, 37, 427–450.

Bierne, N., Launey, S., Naciri-Graven, Y. *et al.* (1998) Early effect of inbreeding as revealed by microsatellite analyses on *Ostrea edulis* larvae. *Genetics*, 148, 1893–1906.

Bingham, J., Foggin, C. M., Gerber, H. *et al.* (1992) Pathogenicity of SAD rabies vaccine given orally in chacma baboons (*Papio ursinus*). *Veterinary Record*, 131, 55–56.

Bingham, J., Schumacher, C. L., Aubert, M. F. A. *et al.* (1997) Innocuity studies of SAG-2 oral rabies vaccine in various Zimbabwean wild non-target species. *Vaccine*, 15, 937–943.

Bonner, J. (1995) Orang-utan release sparks off TB row. *New Scientist*, 148, 9.

Boyce, W. M., Weisenberger, M. E., Penedo, M. C. *et al.* (2011) Wildlife translocation: the conservation implications of pathogen exposure and genetic heterozygosity. *BMC Ecology*, 11, 5.

Breed, A. C., Plowright, R. K., Hayman, D. T. S. *et al.* (2009) Disease management in endangered mammals. In *Management of Disease in Wild Mammals* eds R. J. Delahar, G. C. Smith & M. R. Hutchings. Springer, New York, USA.

Britt, A., Welch, C., Katz, A. *et al.* (2004) The re-stocking of captive-bred ruffed lemurs (*Varecia variegata variegata*) into the Betampona Reserve, Madagascar: methodology and recommendations. *Biodiversity and Conservation*, 13, 635–657.

Bunbury, N. (2006) *Parasitic disease in the endangered Mauritian pink pigeon* Columba mayeri. PhD Thesis, University of East Anglia, Norwich, UK.

Bunbury, N., Jones, C. G., Greenwood, A. G. *et al.* (2007) *Trichomonas gallinae* in Mauritian columbids: implications for an endangered endemic. *Journal of Wildlife Diseases*, 43, 399–407.

Burrows, R. (1992) Rabies in wild dogs. *Nature*, 359, 277.

Calle, P. P., Rivas, J., Muñoz, M. *et al.* (2001) Infectious disease serologic survey in free-ranging Venezuelan anacondas (*Eunectes murinus*). *Journal of Zoo and Wildlife Medicine*, 32, 320–323.

Calvete, C., Angulo, E., Estrada, R. *et al.* (2005) Quarantine length and survival of translocated European wild rabbits. *Journal of Wildlife Management*, 69, 1063–1072.

Campbell, R. A., Chilvers, B. L., Childerhouse, S. *et al.* (2006) Conservation management issues and status of the New Zealand (*Phocarctos hookeri*) and Australian (*Neophoca cinerea*) sea lions. In *Sea Lions of the World*, eds A. W. Trites, D. P. DeMaster, L. W. Fritz, L. D. Gelatt, L. D. Rea & K. M. Wynne, pp. 455–471. Lowell Wake Weld Fisheries Symposuim, Alaska.

Candelora, K. L., Spalding, M. G. & Sellers, H. S. (2010) Survey for antibodies to infectious bursal disease virus serotype 2 in wild turkeys and sandhill cranes of Florida, USA. *Journal of Wildlife Diseases*, 46, 742–752.

Carpenter, J. W., Appel, M. J. G., Erickson, R. C. *et al.* (1976) Fatal vaccine-induced canine distemper virus infection in black-footed ferrets. *Journal of the American Veterinary Medical Association*, 169, 961–964.

Castinel, A., Duignan, P. J., Pomroy, W. E. *et al.* (2007) Neonatal mortality in New Zealand sea lions (*Phocarctos hookeri*) at Sandy Bay, Enderby Island, Auckland Islands from 1998 to 2005. *Journal of Wildlife Disease*, 43, 461–474.

Charlesworth, D. & Charlesworth, B. (1987) Inbreeding depression and its evolutionary consequences. *Annual Review of Ecology and Systematics*, 18, 237–268.

Cleaveland, S. & Dye, C. (1995) Maintenance of a microparasite infecting several host species: rabies in the Serengeti. *Parasitology*, 111, S33–S47.

Cleaveland, G. R., Hess, G. R., Dobson, A. P. *et al.* (2002). The role of pathogens in biological conservation. In *The Ecology of Wildlife Diseases*, eds P. J. Hudson, A. Rizzoli, B. T. Grenfell, H. Heesterbeek & A. P. Dobson. Oxford University Press, Oxford, UK.

Clifford, D. A., Mazet, J. K., Dubovi, E. J. *et al.* (2006) Pathogen exposure in endangered island fox (*Urocyon littoralis*) populations: implications for conservation management. *Biological Conservation*, 131, 230–243.

Coltman, D. W., Pilkington, J. G., Smith, J. A. *et al.* (1999) Parasite-mediated selection against inbred Soay sheep in a free-living, island population. *Evolution*, 53, 1259–1267.

Cork, S. C., Alley, M. R., Johnstone, A. C. *et al.* (1999) Aspergillosis and other causes of mortality in the stitchbird in New Zealand. *Journal of Wildlife Diseases*, 35, 481–486.

Corn, J. L. & Nettles, V. F. (2001) Health protocol for translocation of free-ranging elk. *Journal of Wildlife Diseases*, 37, 413–426.

Crow, J. F. (1948) Alternative hypotheses of hybrid vigor. *Genetics*, 33, 477–487.

Cunningham, A. A. (1996) Disease risks of wildlife translocations. *Conservation Biology*, 10, 349–353.

Cunningham, A. A., Daszak, P. & Rodriguez, J. P. (2003) Pathogen pollution: defining a parasitological threat to biodiversity conservation. *Journal of Parasitology*, 89, 78–83.

Cunningham, A. A., Frank, J. M., Croft, P. *et al.* (1997) Mortality of captive British wartbiter crickets: implications for reintroduction programs. *Journal of Wildlife Diseases*, 33, 673–676.

Daszak, P., Cunningham, A. & Hyatt, A. D. (2000) Emerging infectious diseases of wildlife – threats to biodiversity and human health. *Science*, 287, 443–449.

Davidson, W. R. & Nettles, V. F. (1992) Relocation of wildlife: identifying and evaluating disease risks. *Transactions of the North American Wildlife and Natural Resources Conference*, 57, 466–473.

Davis, G. B., Watson, P. R. & Billing, A. E. (1984) Tuberculosis in a kiwi, *Apteryx mantelli*. *New Zealand Veterinary Journal*, 32, 30.

de Castro, F. & Bolker, B. (2005) Mechanisms of disease-induced extinction. *Ecology Letters*, 8, 117–126.

De Leo, G. & Dobson, A. (2005) Virulence management in wildlife populations. In *Adaptive Dynamics of Infectious Diseases: in pursuit of virulence management*, U. Dieckman, J. A. J. Metz, M. W. Sabelis & K. Sigmun, pp. 26–38. Cambridge University Press, Cambridge, UK.

DeWoody, Y. D. & DeWoody, J. A. (2005) On the estimation of genome-wide heterozygosity using molecular markers. *Journal of Heredity*, 96, 85–88.

Dickens, M. J., Delehanty, D. J. & Romero, L. M. (2009) Stress and translocation: alterations in the stress physiology of translocated birds. *Proceedings of the Royal Society London B*, 276, 2051–2056.

Dickens, M. J., Delehanty, D. J. & Romero, L. M. (2010) Stress: an inevitable component of animal translocation. *Biological Conservation*, 143, 1329–1341.

Doak, D. F., Bakker, V. J. & Vickers, W. (submitted) Assessing alternative strategies to minimize disease threats to an endangered carnivore using population viability criteria.

Drew, M. L., Wigle, W. L., Graham, D. L. *et al.* (1998) Reticuloendotheliosis in captive greater and Attwater's prairie chickens. *Journal of Wildlife Disease*, 34, 783–791.

Dutton, C. J., Junge, R. E. & Louis, E. E. (2008) Biomedical evaluation of free-ranging red ruffed lemurs (*Varecia rubra*) within the Masoala National Park, Madagascar. *Journal of Zoo and Wildlife Medicine*, 39, 76–85.

Ewen, J. G. & Armstrong, D. P. (2007) Strategic monitoring of reintroductions in ecological restoration programmes. *Ecoscience*, 14, 401–409.

Ewen, J. G., Armstrong, D. P., Empson, R. *et al.* (in press) Pathogen management in translocations: lessons from an endangered New Zealand bird. *Oryx*.

Ewen, J. G., Thorogood, R., Nicol, C. *et al.* (2007) *Salmonella typhimurium* in hihi, New Zealand. *Journal of Emerging Infectious Diseases*, 13, 788–790.

Ewen, J. G., Thorogood, R., Brekke, P. *et al.* (2009) Maternally invested carotenoids compensate costly ectoparasitism in the hihi. *Proceedings of the National Academy of Sciences USA*, 106, 12798–12802.

Faria, P. J., van Oosterhout, C. & Cable, J. (2010) Optimal release strategies for captive-bred animals in reintroduction programs: experimental infections using the guppy as a model organism. *Biological Conservation*, 143, 35–41.

Fernández, N., Kramer-Schadt, S. & Thulke H. (2006) Viability and risk assessment in species restoration: planning reintroductions for the wild boar, a potential disease reservoir. *Ecology and Society*, 11, 6.

Ferreira, A. G. A. & Amos, W. (2006) Inbreeding depression and multiple regions showing heterozygote advantage in *Drosophila melanogaster* exposed to stress. *Molecular Ecology*, 15, 3885–3893.

Fischer, J. & Lindenmayer, D. B. (2000) An assessment of the published results of animal relocations. *Biological Conservation*, 96, 1–11.

Forrester, D. J., Carpenter, J. W. & Blankinship, D. R. (1978) Coccidia of whooping cranes. *Journal of Wildlife Diseases*, 14, 24–27.

Frankham, R. & Loebel, D. A. (1992) Modeling problems in conservation genetics using captive *Drosophila* populations – rapid genetic adaptation to captivity. *Zoo Biology*, 11, 333–342.

Fritzemeier, J., Teuffert, J., Greiser-Wilke, I. *et al.* (2000) Epidemiology of classical swine fever in Germany in the 1990s. *Veterinary Microbiology*, 77, 29–41.

Frölich, K., Hamblin, C., Jung, S. *et al.* (2005) Serologic surveillance for selected viral agents in captive and free-ranging populations of Arabian oryx (*Oryx leucoryx*) from Saudi Arabia and the United Arab Emirates. *Journal of Wildlife Diseases*, 41, 67–79.

Gartrell, B. D. & Hare, K. M. (2005) Mycotic dermatitis with digital gangrene and osteomyelitis, and protozoal intestinal parasitism in Marlborough green geckos (*Naultinus manukanus*). *New Zealand Veterinary Journal*, 53, 363–367.

Gartrell, B. D., Alley, M. R., Mack, H. *et al.* (2005) Erysipelas in the critically endangered kakapo (*Strigops habroptilus*). *Avian Pathology*, 34, 383–387.

Gartrell, B. D., Jillings, E., Adlington, B. A. *et al.* (2006) Health screening for a translocation of captive-reared tuatara (*Sphenodon punctatus*) to an island refuge. *New Zealand Veterinary Journal*, 54, 344–349.

Gartrell, B. D., Youl, J. M., King, C. M. *et al.* (2007) Failure to detect *Salmonella* species in a population of wild tuatara (*Sphenodon punctatus*). *New Zealand Veterinary Journal*, 55, 134–136.

Gober, J. (2008) *US Fish and Wildlife Service 5-Year Review of the Black-Footed Ferret*. US Fish and Wildlife Service.

Gompper, M. E. & Williams, E. S. (1998) Parasite conservation and the black-footed ferret recovery program. *Conservation Biology*, 12, 730.

Greth, A., Gourreau, J. M., Vassart, M. *et al.* (1992) Capripoxvirus disease in an Arabian oryx (*Oryx leucoryx*) from Saudia Arabia. *Journal of Wildlife Diseases*, 28, 295–300.

Griffith, B., Scott, J. M. Carpenter, J. W. *et al.* (1989) Translocation as a species conservation tool: status and strategy. *Science*, 245, 477–480.

Griffith, B., Scott, M. J., Carpenter, J. W. *et al.* (1993) Animal translocations and potential disease transmission. *Journal of Zoo and Wildlife Medicine*, 24, 231–236.

Groombridge, J. J., Massey, J. G., Bruch, J. C. *et al.* (2004) Evaluating stress in a Hawaiian honeycreeper, *Paroreomyza montana*, following translocation. *Journal of Field Ornithology*, 75, 183–187.

Gusset, M., Ryan, S. J., Hofmeyr, M. *et al.* (2008) Efforts going to the dogs? Evaluating attempts to re-introduce endangered wild dogs in South Africa. *Journal of Applied Ecology*, 45, 100–118.

Hall, A. & Harwood, J. (1990) *The Intervet Guidelines to Vaccinating Wildlife*. Sea Mammal Research Unit, Cambridge, UK.

Hansson, B., Bensch, S., Hasselquist, D. *et al.* (2001) Microsatellite diversity predicts recruitment of sibling great reed warblers. *Proceedings of the Royal Society of London B*, 268, 1287–1291.

Hartup, B. K., Olsen, G. H. & Czekala, N. M. (2005) Fecal corticoid monitoring in whooping cranes (*Grus americana*) undergoing reintroduction. *Zoo Biology*, 24, 15–28.

Haydon, D. T., Laurenson, M. K. & Sillero-Zubiri, C. (2002) Integrating epidemiology into population viability analysis: managing the risk posed by rabies and canine distemper to the Ethiopian wolf. *Conservation Biology*, 16, 1372–1385.

Haydon, D. T., Randall, D. A., Matthews, L. *et al.* (2006) Low coverage vaccination strategies for the conservation of endangered species. *Nature*, 443, 692–695.

Henke, S. E. (1997) Effects of modified live-virus canine distemper vaccines in gray foxes. *Journal of Wildlife Rehabilitation*, 20, 3–7.

Hofmeyr, M., Bingham, J., Lane, E. P. *et al.* (2000) Rabies in African wild dogs (*Lycaon pictus*) in the Madikwe Game Reserve, South Africa. *Veterinary Record*, 146, 50–52.

Hofmeyr, M., Hofmeyr, D., Nel, L. *et al.* (2004) A second outbreak of rabies in African wild dogs (*Lycaon pictus*) in Madikwe Game Reserve, South Africa, demonstrating the efficacy of vaccination against natural rabies challenge. *Animal Conservation*, 7, 193–198.

Hopkins, B. A., Skeeles, J. K., Houghten, G. E. *et al.* (1990) A survey of infectious diseases in wild turkeys (*Meleagridis gallopavo silvestris*) from Arkansas. *Journal of Wildlife Diseases*, 26, 468–472.

Houle, D. (1994) Adaptive distance and the genetic basis of heterosis. *Evolution*, 48, 1410–1417.

Island Fox Working Group (2010) *Report of 12th Annual Meeting.* National Park Service, Ventura, California.

IUCN (World Conservation Union) (1998) *Guidelines for Re-introductions.* IUCN/SSC Re-introduction Specialist Group, IUCN, Gland, Switzerland and Cambridge, UK.

Keller, L. F., Biebach, I., Ewing, S. R. *et al.* (2011) The genetics of reintroductions: inbreeding and genetic drift. In *Reintroduction Biology: integrating science and management*, eds J. G. Ewen, D. P. Armstrong, K. A. Parker & P. J. Seddon, Chapter 11. Wiley-Blackwell, Oxford, UK.

Kilbourn, A. M., Karesh, W. B., Wolfe, N. D. *et al.* (2003) Health evaluation of free-ranging and semi-captive orangutans (*Pongo pygmaeus pygmaeus*) in Sabah, Malaysia. *Journal of Wildlife Diseases*, 39, 73–83.

Kimber, K. & Kollias, G. V. (2005) Evaluation of injury severity and hematologic and plasma biochemistry values for recently captured North American river otters (*Lontra canadensis*). *Journal of Zoo and Wildlife Medicine*, 36, 371–384.

Kimber, K., Kollias, G. V. & Dubovi, E. J. (2000) Serologic survey of selected viral agents in recently captured wild North American river otters (*Lontra canadensis*). *Journal of Zoo and Wildlife Medicine*, 31, 168–175.

Kirkwood, J. K. & Sainsbury, A. W. (1997) Diseases and other considerations in wildlife translocations and releases. In *Proceedings of the World Association of Wildlife Veterinarians Symposium on Veterinary Involvement with Wildlife Reintroduction and Rehabilitation*, Ballygawley, UK, WAWV, pp. 12–16.

Kirkwood, J. K. & Cunningham, A. A. (2006) Portrait of prion diseases in zoo animals. In *Prions, Humans and Animals*, eds B. Hornlimann, D. Riesner & H. Kretzschmar, pp. 250–256. Walter de Gruyter, Berlin and New York.

Kurtz, J., Kalbe, M., Aeschlimann, P. B. *et al.* (2004) Major histocompatibility complex diversity influences parasite resistance and innate immunity in sticklebacks. *Proceedings of the Royal Society London B*, 271, 197–204.

Lacy, R. C., Flesness, N. R. & Seal, U. S. (1989) Puerto Rican parrot (*Amazona vittata*) population viability analysis and recommendations. Captive Breeding Specialist Group, Apple Valley, Minnesota.

Langenberg, J., Hartup, B., Olsen, G. *et al.* (2002) Health management for the reintroduction of whooping cranes (*Grusa americana*) using ultra-light guided migration. In *Proceedings of the American Association for Zoo Veterinarians*, pp. 246–248.

Laurenson, K., Shiferaw, F. & Sillero-Zubiri, C. (1997) Disease, domestic dogs and the Ethiopian wolf: the current situation. In *The Ethiopian Wolf: status survey and conservation action plan*, eds C. Sillero-Zubiri & D. W. Macdonald. IUCN, Gland, Switzerland.

Lierz, M., Obon, E., Schink, B. *et al.* (2008) The role of mycoplasmas in a conservation project of the lesser kestrel (*Falco naumanni*). *Avian Diseases*, 52, 641–645.

Lisboa, C. V., Mangia, R. H., Rubião, E. *et al.* (2004) *Trypanosoma cruzi* transmission in a captive primate unit, Rio de Janeiro, Brazil. *Acta Tropica*, 90, 97–106.

Lockhart, J. M., Thorne, E. T. & Gober, D. R. (2006) A historical perspective on recovery of the black-footed ferret and the biological and political challenges affecting its future. In *Recovery of the Black-Footed Ferret: progress and continuing challenges*, eds J. E. Roelle, B. J. Miller, J. L. Godbey & D. E. Biggins. US Geological Survey Scientific Investigations Report 2005–5293, Reston, Virginia, USA.

LoGiudice, K. (2003) Trophically transmitted parasites and the conservation of small populations: raccoon roundworm and the imperiled Allegheny woodrat. *Conservation Biology*, 17, 258–266.

Lynch, M., Conery, J. & Burger, R. (1995) Mutation accumulation and the extinction of small populations. *American Naturalist*, 146, 489–518.

Marinari, P. E. & Kreeger, J. S. (2006) An adaptive management approach for black-footed ferrets in captivity. In *Recovery of the Black-Footed Ferret: progress and continuing challenges*, eds J. E. Roelle, B. J. Miller, J. L. Godbey & Biggins. US Geological Survey.

Marquardt, W. C., Demaree, R. S. & Grieve, R. B. (2000) *Parasitology and Vector Biology*, 2nd edn. Harcourt/Academic Press, Massachusetts, USA.

Martel, A., Blahak, S., Vissenaekens, H. *et al.* (2009) Reintroduction of clinically healthy tortoises: the herpesvirus Trojan horse. *Journal of Wildlife Diseases*, 45, 218–220.

Mathews, F., Moro, D., Strachan, R. *et al.* (2006) Health surveillance in wildlife reintroductions. *Biological Conservation*, 131, 338–347.

McCallum, H. & Dobson, A. (1995) Detecting disease and parasite threats to endangered species and ecosystems. *Trends in Ecology and Evolution*, 10, 190–194.

McGill, I., Feltrer, Y., Jeffs, C. *et al.* (2010) Isosporoid coccidiosis in translocated cirl buntings (*Emberiza cirlus*). *Veterinary Record*, 167, 656–660.

Middleton, D. M. R. L., Minot, E. O. & Gartrell, B. D. (2010) *Salmonella enterica* serovars in lizards of New Zealand's offshore islands. *New Zealand Journal of Ecology*, 34, 247–252.

Montali, R. J., Bush, M., Hess, J. *et al.* (1995) *Ex situ* diseases and their control for reintroduction of the endangered lion tamarin species (*Leontopithecus* spp.). *Verh ber Erkrg Zootiere*, 37, 93–98.

Moravec, F. (2003) Observations on the metazoan parasites of the Atlantic salmon (*Salmo salar*) after its reintroduction into the Elbe River basin in the Czech Republic. *Folia Parasitologica*, 50, 298–304.

Moreno, S., Villafuerte, R., Cabezas, S. *et al.* (2004) Wild rabbit restocking for predator conservation in Spain. *Biological Conservation* 118: 183–193.

Mul, I. F., Paembonan, W., Singleton, I. *et al.* (2007) Intestinal parasites of free-ranging, semicaptive, and captive *Pongo abelii* in Sumatra, Indonesia. *International Journal of Primatology*, 28, 407–420.

Munson, L., Marker, L., Dubovi, E. *et al.* (2004) Serosurvey of viral infections in free-ranging Namibian cheetahs (*Acinonyx jubatus*). *Journal of Wildlife Diseases*, 40, 23–31.

Murray, N., Macdiarmid, S. C., Wooldridge, M. *et al.* (2004) *Handbook on Import Risk Analysis for Animals and Animal Products*. Office of International Epizootics (OIE).

Nolet, B. A., Broekhuizen, S., Dorrestein, G. M. *et al.* (1997) Infectious diseases as main causes of mortality to beavers *Castor fiber* after translocation to the Netherlands. *Journal of Zoology*, 241, 35–42.

O'Brien, S. J. & Evermann, J. F. (1998) Interactive influence of infectious disease and genetic diversity in natural populations. *Trends in Ecology and Evolution*, 3, 254–259.

Ortiz-Catedral, L., Ishmar, S. M. H., Baird, K. *et al.* (2009a) No evidence of *Campylobacter, Salmonella* and *Yersinia* in free living populations of the red-crowned parakeet (*Cyanoramphus novaezelandiae*). *New Zealand Journal of Zoology*, 36, 379–383.

Ortiz-Catedral, L., McInnes, K., Hauber, M. E. *et al.* (2009b) First report of psittacine beak and feather disease (PBFD) in wild red-crowned parakeets (*Cyanoramphus novaezelandiae*) in New Zealand. *Emu*, 109, 244–247.

Parker, K. A., Brunton, D. H. & Jakob-Hoff, R. (2006) Avian translocations and disease; implications for New Zealand conservation. *Pacific Conservation Biology*, 12, 155–162.

Parker, K. A., Dickens, M. J., Clarke, R. H. *et al.* (2011). The theory and practice of catching, holding, moving and releasing animals. In *Reintroduction Biology: integrating science and management*, J. G. Ewen, D. P. Armstrong, K. A. Parker & P. J. Seddon, Chapter 4. Wiley-Blackwell, Oxford, UK.

Pedersen, A. B., Altizer, A., Poss, M. *et al.* (2005) Patterns of host specificity and transmission among parasites of wild primates *International Journal of Parasitology*, 35, 647–657.

Pérez, J. M., Meneguz, P. G., Dematteis, A. *et al.* (2006) Parasites and conservation biology: the 'ibex-ecosystem'. *Biodiversity and Conservation*, 15, 2033–2047.

Peterson, M. J., Ferro, P. J., Peterson, M. N. *et al.* (2002) Infectious disease survey of lesser prairie chickens in North Texas. *Journal of Wildlife Diseases*, 38, 834–839.

Phillips, M. K. & Scheck, J. (1991) Parasitism in captive and reintroduced red wolves. *Journal of Wildlife Diseases*, 27, 498–501.

Pinter-Wollman, N., Isbell, L. A. & Hart, L. A. (2009) Assessing translocation outcome: comparing behavioral and physiological aspects of translocated and resident African elephants (*Loxodonta africana*). *Biological Conservation*, 142, 1116–1124.

Pizzi, R. (2004) Disease diagnosis and control in ex-situ terrestrial invertebrate conservation programs. Presented at European Association of Zoo and Wildlife Veterinarians 5th Scientific Meeting.

Pizzi, R. (2009) Veterinarians and taxonomic chauvinism: the dilemma of parasite conservation. *Journal of Exotic Pet Medicine*, 18, 279–282.

Plowright, W. (1982) The effects of rinderpest and rinderpest control on wildlife in Africa. *Symposium of the Zoological Society of London*, 50, 1–28.

Potts, W. K. & Wakeland, E. K. (1990) Evolution of diversity at the major histocompatibility complex. *Trends in Ecology and Evolution*, 5, 181–187.

Riedman, M. (1990) *The Pinnipeds, Seals, Sea Lions and Walruses*. University of California Press, Berkeley, California, USA.

Robertson, L. M. (2009) *Anthelmintic treatment and digestive organ morphology of captive-reared kaki* (Himantopus novaezelandiae). MSc Thesis, Massey University, New Zealand.

Rocke, T. E., Nol, P., Marinari, P. E. *et al.* (2006) Vaccination as a potential means to prevent plague in black-footed ferrets. In *Recovery of the Black-Footed Ferret: progress and continuing challenges*, eds J. E. Roelle, B. J. Miller, J. L. Godbey & D. E. Biggins. US Geological Survey.

Roemer, G. W., Coonan, T. J., Garcelon, D. K. *et al.* (2001) Feral pigs facilitate hyperpredation by golden eagles and indirectly cause the decline of the island fox. *Animal Conservation*, 4, 307–318.

Rüegg, S., Walzer, C., Robert, N. *et al.* (2002) Disease risk assessment: piroplasmosis at the reintroduction site of the Przewalski horse (*Equus przewalskii*) in the Dsungarian Gobi, Mongolia. In *Proceedings of the European Association of Zoo and Wildlife Veterinarians 4th Scientific Meeting*. Heidelberg, Germany.

Sainsbury, A. W., Armstrong, D. P. & Ewen, J. G. (2011) Methods of disease risk analysis for reintroduction programmes. In *Reintroduction Biology: integrating science and management*, eds J. G. Ewen, D. P. Armstrong, K. A. Parker & P. J. Seddon, Chapter 10. Wiley-Blackwell, Oxford, UK.

Sainsbury, A. W., Flach, E. J., Sayers, G. *et al.* (2004) Disease risk analysis for a reintroduction programme: common dormice (*Muscardinus avellanarius*). In *Proceedings of the Sixth Conference of the European Wildlife Disease Association*, 8–12 September 2004, Uppsala, Sweden.

Scheepers, J. L. & Venzke, K. A. E. (1995) Attempts to reintroduce African wild dogs *Lycaon pictus* into Etosha National Park, Namibia. *South African Journal of Wildlife Research*, 25, 138–140.

Scott, J. M., Conant, S. & van Riper III, C. (2001) Evolution, ecology, conservation, and management of Hawaiian birds: a vanishing avifauna. Studies in Avian Biology 22.

Seal, U. S. (1991) Life after extinction. In *Beyond Captive Breeding: re-introducing endangered animals to the wild*, ed. J. H. W. Gipps, pp 39–55. Zoological Society of London, UK.

Seddon, P. J. (2010) From reintroduction to assisted colonization: moving along the conservation translocation spectrum. *Restoration Ecology*, 18, 796–802.

Seddon, P. J., Armstrong, D. P. & Maloney, R. F. (2007) Developing the science of reintroduction biology. *Conservation Biology*, 21, 303–312.

Seddon, P. J., Soorae, P. S. & Launay, F. (2005) Taxonomic bias in reintroduction projects. *Animal Conservation*, 8, 51–58.

Seddon, P. J., Strauss, W. M. & Innes, J. (2011) Animal translocations: what are they and why do we do them? In *Reintroduction Biology: integrating science and management*, eds J. G. Ewen, D. P. Armstrong, K. A. Parker & P. J. Seddon, Chapter 1. Wiley-Blackwell, Oxford, UK.

Slate, J., David, P., Dodds, K. G. *et al.* (2004) Understanding the relationship between the inbreeding coefficient and multilocus heterozygosity, theoretical expectations and empirical data. *Heredity*, 93, 255–265.

Smith, K. F., Sax, D. F. & Lafferty, K. D. (2006) Evidence for the role of infectious disease in species extinction and endangerment. *Conservation Biology*, 20, 1349–1357.

Snyder, N. F. R., Derrickson, S. R., Beissinger, S. R. *et al.* (1996) Limitations of captive breeding in endangered species recovery. *Conservation Biology*, 10, 338–348.

Sutherland, W. J., Armstrong, D., Butchart, S. H. M. *et al.* (2010) Standards for documenting and monitoring bird reintroduction projects. *Conservation Letters*, 3, 229–235.

Swinnerton, K. J., Greenwood, A. G., Chapman, R. E. *et al.* (2005a) The incidence of the parasitic disease trichomoniasis and its treatment in reintroduced and wild pink pigeons *Columba mayeri. Ibis*, 147, 772–782.

Swinnerton, K. J., Peirce, M. A., Greenwood, A. *et al.* (2005b) Prevalence of *Leucocytozoon marchouxi* in the endangered pink pigeon *Columba mayeri. Ibis*, 147, 725–737.

Swinton, J., Woolhouse, M. E. J., Begon, M. E. *et al.* (2002) Microparasite transmission and persistence. In *The Ecology of Wildlife Diseases*, eds P. J. Hudson, A. Rizzoli, B. T. Grenfell, H. Hesterbeek & A. P. Dobson, pp. 83–101. Oxford University Press, Oxford, UK.

Teixeira, C. P., Schetini de Azevedo, C., Mendl, M. *et al.* (2007) Revisiting translocation and reintroduction programmes: the importance of considering stress. *Animal Behaviour*, 73, 1–13.

Thorne, E. T. & Williams, E. S. (1988) Disease and endangered species: the black-footed ferret as a recent example. *Conservation Biology*, 2, 66–74.

Thorne, J. M. (2007) *An experimental approach to the translocation of the North Island saddleback* (Philesturnus rufusater) *to Bushy Park Reserve, Wanganui*. MSc Thesis, Massey University, New Zealand.

Timm, S. F., Barker, W. D., Johnson, S. A. *et al.* (2002) *Island Fox Recovery Efforts on Santa Catalina Island, California, September 2000–October 2001*. Institute for Wildlife Studies, Arcata, California, USA.

Timm, S. F., Munson, L., Summers, B. A. *et al.* (2009) A suspected canine distemper epidemic as the cause of a catastrophic decline in Santa Catalina Island foxes (*Urocyon littoralis catalinae*). *Journal of Wildlife Diseases*, 45, 333–343.

Tocidlowski, M. E., Lappin, M. R., Sumner, P. W. *et al.* (1997) Serologic survey for toxoplasmosis in river otters. *Journal of Wildlife Diseases*, 33, 649–652.

Tocidlowski, M. E., Spelman, L. H., Sumner, P. W. *et al.* (2000) Hematology and serum biochemistry parameters of North American river otters (*Lontra canadensis*). *Journal of Zoo and Wildlife Medicine*, 31, 484–490.

Tompkins, D. M. & Poulin, R. (2006) Parasites and biological invasions. In *Biological Invasions in New Zealand*, eds R. B. Allen & W. G. Lee. *Ecological Studies*, 186, 67–84.

Tompkins, D. M., Sainsbury, A. W., Nettleton, P. *et al.* (2002) Parapoxvirus causes a deleterious disease in red squirrels associated with UK population declines. *Proceedings of the Royal Society London B*, 269, 529–533.

Tompkins D. M., White A. R. & Boots, M. (2003) Ecological replacement of native red squirrels by invasive greys driven by disease. *Ecology Letters*, 6, 189–196.

Twentyman, C. M. (2001) *A study of coccidial parasites in the hihi* (Notiomystis cincta). MSc in Veterinary Science, Massey University, Palmerston North, New Zealand.

US Fish and Wildlife Service (in review) *Recovery Plan for the Island Fox*. US Fish and Wildlife Service, Ventura, California, USA.

Valsecchi, E., Amos, W., Raga, J. A. *et al.* (2004) The effects of inbreeding on mortality during a morbillivirus outbreak in the Mediterranean striped dolphin (*Stenella coeruleoalba*). *Animal Conservation*, 7, 139–146.

van Riper III, C., van Riper, S. G., Goff, M. L. *et al.* (1986) The epizootiology and ecological significance of malaria in Hawaiian land birds. *Ecological Monographs*, 56, 327–344.

Vaughan, R. J. & Sainsbury, A. W. (submitted) Analyzing disease risks associated with translocations for conservation purposes.

Viggers, K. L., Lindenmayer, D. B. & Spratt, D. M. (1993) The importance of disease in reintroduction programmes. *Wildlife Research*, 20, 786–698.

Villers, L. M., Jang, S. S., Lent, C. L. *et al.* (2008) Survey and comparison of major intestinal flora in captive and wild ring-tailed lemur (*Lemur catta*) populations. *American Journal of Primatology*, 70, 175–184.

Walker, S. F., Bosch, J., James, T. Y. *et al.* (2008) Invasive pathogens threaten species recovery programs. *Current Biology*, 18, 853–854.

Walzer, C., Baumgartner, R., Robert, N. *et al.* (2000) Medical considerations in the reintroduction of the Przewalski horse (*Equus przewalskii*) to the Dzungarian Gobi, Mongolia. In *Proceedings of the 3rd Scientific Meeting of the European Association of Zoo and Wildlife Veterinarians*, Paris, France.

Wild, M. A., Shenk, T. M. & Spraker, T. R. (2006) Plague as a mortality factor in Canada lynx (*Lynx Canadensis*) reintroduced to Colorado. *Journal of Wildlife Diseases*, 42, 646–650.

Williams, E. S., Mills, K., Kwiatkowski, D. R. *et al.* (1994) Plague in a black-footed ferret (*Mustela nigripes*). *Journal of Wildlife Diseases*, 30, 581–585.

Wilson, M. H., Kepler, C. B., Snyder, N. F. R. *et al.* (1994) Puerto Rican parrots and potential limitations of the metapopulation approach to species conservation. *Conservation Biology*, 8, 114–123.

Wimsatt, J., Biggins, D., Innes, K. *et al.* (2003) Evaluation of oral and subcutaneous delivery of an experimental canarypox recombinant canine distemper vaccine in the Siberian polecat (*Mustela eversmanni*). *Journal of Zoo and Wildlife Medicine*, 34, 25–35.

Wobeser, G. (2002) Disease management strategies for wildlife. *Revue Scientifique et Technique – Office International des Epizooties*, 21, 159–178.

Wobeser, G. (2006) *Essentials of Disease in Wild Animals*. Blackwell Publishing, Iowa, USA.

Woodford, M. H. (2001) *Quarantine and Health Screening Protocols for Wildlife Prior to Translocation and Release into the Wild*. IUCN/SSC CBSG, Gland, Switzerland, OIE, Paris, France, Care for the Wild UK, and the European Association of Zoo and Wildlife Veterinarians, Switzerland.

Woodroffe, R. (1999) Managing disease risks to wild mammals. *Animal Conservation*, 2, 185–193.

Woodroffe, R. (2001) Assessing the risks of intervention: immobilization, radio-collaring and vaccination of African wild dogs. *Oryx*, 35, 234–244.

Work, T. M., Klavitter, J. L., Reynolds, M. H. *et al.* (2010) Avian botulism: a case study in translocated endangered Laysan ducks (*Anas laysanensis*) on Midway Atoll. *Journal of Wildlife Diseases*, 46, 499–506.

Work, T. M., Massey, J. G., Rideout, B. A. *et al.* (2000) Fatal toxoplasmosis in free-ranging endangered 'Alala from Hawaii. *Journal of Wildlife Diseases*, 36, 205–212.

Work, T. M., Meteyer, C. U. & Cole, R. A. (2004) Mortality in Laysan ducks (*Anas laysanensis*) by emaciation complicated by *Echinuria uncinata* on Laysan Island, Hawaii, 1993. *Journal of Wildlife Diseases*, 40, 110–114.

Yoder, J. A. & Litman, G. W. (2011) The phylogenetic origins of natural killer receptors and recognition: relationships, possibilities, and realities. *Immunogenetics*, 63, 123–141.

Zavala, G., Cheng, S., Barbosa, T. *et al.* (2006) Enzootic reticuloendotheliosis in the endangered Attwater's and greater prairie chickens. *Avian Diseases*, 50, 520–525.

Zinkernagel, R. M., Pfau, C. J., Hengartner, H. *et al.* (1985) Susceptibility to murine lymphocytic choriomeningitis maps to class I MHC genes – a model for MHC/disease associations. *Nature*, 316, 814–817.

10

Methods of Disease Risk Analysis for Reintroduction Programmes

Anthony W. Sainsbury[1], Doug P. Armstrong[2] and John G. Ewen[1]

[1]Institute of Zoology, Zoological Society of London, United Kingdom

[2]Massey University, Palmerston North, New Zealand

'A crucial difficulty with assessing the disease risks of free-living wild animal translocations is that the components of the biological package are not known for any free-living species in entirety because (i) the number of parasites harboured by each species is large and (ii) studies on parasite ecology and disease in wild animals are in their infancy.'

Page 344

Introduction

Concomitant with our increasing understanding of the threat of disease to translocation programmes, there has been some attempt by wildlife disease professionals over the last 20 years to develop appropriate methods to assess the risks. However, considering the scale of wild animal translocation, in terms of the number of animals moved per annum (Griffith *et al.*, 1993; Karesh *et al.*, 2005) and the interest in disease as a threat to wild animals, the literature in this area has been weak. Two approaches have been used (i) to conduct qualitative

Reintroduction Biology: Integrating Science and Management. First Edition.
Edited by John G. Ewen, Doug P. Armstrong, Kevin A. Parker and Philip J. Seddon.
© 2012 Blackwell Publishing Ltd. Published 2012 by Blackwell Publishing Ltd.

risk analysis adapted from those methods originally developed to assess health and environmental risks to humans and/or (ii) to build quantitative models of the impact of specific diseases on translocated populations. Both of these methods have been reported for reintroduction programmes.

In this chapter we look at the origins of disease risk analysis (DRA) methods and their development for domestic animal movements, explore how these methods have been extrapolated for use in wild animals, examine the reasons why the methods required for wild animal reintroduction or more general translocation programmes are necessarily different to those used for domestic animal movements, examine recent approaches to undertaking disease risk analysis for multiple types of wild animal translocation programme and discuss the monitoring and research needed to advance our understanding of the costs and benefits of a risk analysis approach to assessing the potential impact of disease on these programmes.

The evolution of qualitative disease risk analysis methods

Over the last forty years the literature in risk assessment and risk analysis has burgeoned, perhaps largely in response to a perceived need to quantify the risk to humans from various environmental and health threats as disparate as, for example, nuclear power, car accidents and the use of machinery at work. A major portion of this literature examined the assessment of the risk of various diseases, e.g. cancer, to humans (Bruckner *et al.*, 2010). Following such general advances made to analyse disease risks to humans, similar methods began to be adopted in assessing the disease risks of the importation of domestic animals by, for example, the Office International des Epizooties (OIE, currently referred to as the World Organisation for Animal Health). The history of the development of methods to assess the disease risks of importation of domestic animals has been dealt with by Bruckner *et al.* (2010) and Leighton (2002). However, a key contribution, of relevance to this chapter, in the development of DRA was made by Covello & Merkhofer (1993).

Covello & Merkhofer (1993) proposed a risk assessment and risk analysis framework using a common terminology, which cut across disciplines and could be as equally used by those assessing the risks of nuclear power to those assessing the risk posed by car accidents or cancer. In their system the term *risk analysis* was used to denote an overarching process, which encompassed

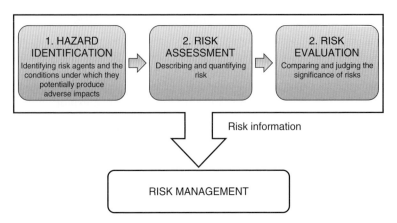

Figure 10.1 **Covello & Merkhofer's three stages of risk analysis (adapted from Covello & Merkhofer, 1993).**

three elements: hazard identification, risk assessment and risk evaluation (Figure 10.1). Risk management was a separate entity, which was fed with 'risk information' from the risk analysis, and risk management was subcategorized as 'option generation', 'option evaluation', 'option selection' and 'implementation and enforcement'. This framework has been widely adopted, including by the OIE in their animal import risk analysis (Bruckner *et al.*, 2010). Covello & Merkhofer (1993) were insistent that hazard identification should be seen to be a separate entity to disease risk assessment *and* should be completed prior to disease risk assessment, because of the importance of determining 'cause and effect, weighing the available evidence, and characterising the nature and strength of the evidence of causation' in hazard identification. Disease risk assessment was defined as 'a systematic process for describing and quantifying the risks associated with hazardous substances, processes, action or events' and Covello & Merkhofer (1993) stated that 'our definition implies (1) the possibility of adverse health or environmental consequences, and (2) uncertainty over the occurrence, magnitude, or timing of those consequences'. They suggested that quantitative methods were best used to assess the magnitude of the adverse consequences, their probability of occurrence and their timing, and that probability theory could be used to evaluate the degree of uncertainty.

Methods to assess the disease risks of domestic animal movements were developed in the 1990s (OIE, 1999) using the Covello & Merkhofer (1993) framework. The OIE expanded the fledgling methodology published in 1999 into a two volume guide to qualitative and quantitative disease risk analysis (Murray *et al.*, 2004) and updated the 'qualitative guide' in 2010 (Bruckner *et al.*, 2010). In these volumes the Covello & Merkhofer (1993) methodology was modified by including risk evaluation as a component of disease risk management, and the latter was incorporated into disease risk analysis. The basic structure of qualitative disease risk analysis advocated by the OIE is set out in Table 10.1. Murray *et al.* (2004) were largely concerned with the risks of infectious agent hazard transfer from source to destination environments and set out a series of questions through which to determine whether a pathogenic agent was a hazard. If (i) the animal being imported was considered a potential vehicle for the infectious agent, (ii) the same agent (including strains of that agent) was absent in the importing country (the destination) and (iii) the agent was capable of causing harm, then it was classified as a hazard.

The evolution of methods in disease risk analysis for the translocation of wild animals lags behind that for domestic animals probably because (i) our understanding of the epidemiology of infectious disease in wild animals is comparatively poor, (ii) there has been a lack of investment in disease risk analyses by wildlife professionals and (iii) moving wild animals into ecosystems is inherently more complex than moving domestic animals into relatively controlled environments under human care.

Approaches to qualitative disease risk analysis for translocations of wild animals

Davidson & Nettles (1992) provided a platform for disease risk analysis of free-living wild animal translocations through their evaluation of the disease risks of translocating wildlife for hunting purposes. These authors recognized that wild animals essentially represent a biological package constituting the host and their parasites (viruses, bacteria, fungi, protozoans, helminths and arthropods) and they set out two principal problems that could arise from translocation: (i) the introduction of an exotic parasite to a new geographic area and (ii) the artificial intensification of a parasite native to a destination site due to changes in population dynamics at the site with resultant disease. Davidson & Nettles (1992) used a qualitative two-tiered scoring system, while restricting hazards to infectious agents (parasites), including, first, a guess for the probability that a parasite would become established at a release site, based on the epidemiological characteristics of the parasite, and, second, the

Table 10.1   The summarized components of risk analysis as proposed by the World Organization for Animal Health (OIE) (adapted from Bruckner *et al.*, 2010, and set out to apply to an animal movement).

	Risk analysis		
Hazard identification	Risk assessment	Risk management	Risk communication
Identifying the hazards likely to be associated with the animal under consideration	Entry assessment – to estimate the likelihood of the animal introducing the hazard	Risk evaluation – to determine if the risk estimate is greater than the acceptable risk level	Developing a risk communication strategy and implementing it at all stages of the risk analysis
	Exposure assessment – to estimate the likelihood of susceptible animals being exposed to the hazard	Option Evaluation	
	Consequence assessment – to estimate the likely magnitude of potential biological, environmental and economic consequences associated with the entry, establishment or spread of the hazard and the likelihood of their occurrence	Implementation	
	Risk estimation – summarizing the entry, exposure and consequence assessments	Monitoring and review	

likelihood that the parasite would be pathogenic. The product of the pathogenicity and establishment guesses was used to describe the overall assessment, while recognizing the uncertainty in the evaluation. Davidson & Nettles (1992) appreciated the need for experimental evaluation of their risk analysis system but as far as we are aware no evaluation was published. Corn & Nettles (2001) published an assessment of the risks of translocating deer (*Cervus elaphus*) across the United States of America from West to East for hunting purposes, essentially using Davidson & Nettles' (1992) methodology.

Leighton (2002) proposed a method of qualitative disease risk analysis for the translocation of wild animals for multiple purposes (e.g. conservation, rehabilitation, hunting) based on the method published by the OIE (1999) and largely restricting the analysis to source hazards. Leighton (2002) advocated that a DRA should be conducted by a different team of professionals to the person or committee that would assess the overall costs and benefits of the translocation, to avoid bias in decision making due to conflicts of interests, such as personal beliefs or political or public pressures. Leighton (2002) advocated a six-step process in disease risk analysis, as follows: (i) the detailed description of the translocation plan; (ii) identification of the health hazards; (iii) the probability that the health hazard will occur (by expert judgement) and the magnitude of any negative consequences if it did; (iv) an overall statement of risk and the uncertainty in making this decision; (v) assessment of associated hazards and risks; and (vi) an outline of methods to reduce risk. Leighton (2002) included infectious agents and diseases in the source and destination environments as hazards, but chose to place particular emphasis on infectious agents present in the source ecosystem but absent from the destination ecosystem.

Neimanis & Leighton (2004) carried out a comprehensive qualitative disease risk analysis for the translocation of Eastern wild turkeys (*Meleagris gallopavo*) from Ontario to Nova Scotia, Canada, using a method similar to that proposed by Leighton (2002) in which the source and destination hazards were analysed. The assessment was based primarily on a literature review, but also from unpublished reports, opinions and 'anecdotal information solicited from wildlife health personnel'. Initially 'a comprehensive list of disease-causing agents that wild turkeys are susceptible to, and locations where infection and exposure' had been recorded within the United States of America was drafted (122 parasites were included). Thirty-nine 'disease-causing agents' were eliminated from further consideration if one or more of the following was true: '(a) never documented in wild turkeys, (b) not documented in source or destination ecosystems, i.e., occurs well out of source and destination ranges, and (c) not a significant pathogen for the poultry industry or for native galliform birds already in the source ecosystem'. Another 72 'disease-causing agents' were eliminated from the list because they posed 'no substantial risk' and the reasons for their elimination was specified for each agent in the report, e.g. because there had been no reports of clinical disease attributed to the agent. Further analysis of the remaining 11 disease-causing agents was based on a qualitative assessment of the probability that the 'disease-causing agent' would be introduced to the destination environment and susceptible species would be exposed, and the potential magnitude of any negative consequences to the 'health of susceptible species, the ecosystem and the human economy' should this introduction occur (based on expert judgement). The product of the probability of introduction and the magnitude of negative consequences was designated as negligible (one infectious agent), low (six), medium (none) or high (none). Four infectious agents were designated 'unknown and not assessable at this time' risk. Likewise an assessment was made of the probability that translocated turkeys would encounter health hazards at the destination and the magnitude of any negative consequences. Sources of uncertainty related to both evaluations were listed in order to convey the limitations of the analysis. Environmental contaminants were mentioned as a potential hazard in the report but no other non-infectious diseases were considered. In the final part of the report a list of methods was set out to reduce the risk of those disease-causing agents designated as greater than negligible risk.

Armstrong et al. (2003) and Miller (2007) developed methodology for undertaking qualitative DRA in endangered species programmes. This included (i) a Disease List and Project Diagram, (ii) tools for ranking or prioritizing hazards, (iii) tools for collecting information on hazards and (iv) a method for qualitative disease risk analysis based on the Covello & Merkhofer (1993) standard. Potential sources of information for hazard identification noted by Armstrong et al. (2003) were population health viability analyses (PVA), disease surveys, literature searches and unpublished reports, such as the Species Survival Plan Veterinary Advisor Protocols. Prioritizing hazards can be achieved through a method set out by Miller (2007) called 'paired ranking', in which a list of infectious agent hazards is repeatedly compared and ranked according to measures of importance to assessing disease risk. Such measures include the pathogenicity of the disease or the prevalence of the agent in the source population. When a disease is ranked higher than another under an individual parameter it is scored and the total score for each disease is used as a relative weighting of its importance to the reintroduction programme. Another method described by Miller (2007) consisted of scoring hazards on a set of criteria, e.g. the 'likelihood of susceptibility', 'likelihood of exposure' and 'severity for population', and the addition of the scores provided an overall score of the 'estimated significance to programme'.

Approaches to quantitative disease risk analysis for reintroduction of wild animals

Ballou (1993) was perhaps the first author to address the need for quantitative approaches to disease risk analysis and described how PVA could be used to assess the risks of specific diseases to reintroduction programmes. These models have the advantage that they can incorporate multiple other factors affecting the risk to a population, such as its genetic makeup, and analyse how changes in these factors, such as management action, affect the overall risk.

Armstrong et al. (2003) and Miller (2007) provided a set of quantitative tools for disease risk analysis specifically for reintroductions for conservation purposes. In their simplest form, visual models can be created to depict the complexity of a system, e.g. factors affecting the epidemiology of a specific infectious agent in a reintroduced population, and Miller (2007) described two tools to create such models (similar to the conceptual model stage discussed in Armstrong & Reynolds, this volume, Chapter 6). This can then be converted into a quantitative model to assess the probability that a reintroduced animal will be infected with a specific infectious agent. Miller (2007) reported use of quantitative modelling by Murayama et al. (2006) to estimate the risk of a Tsushima leopard cat (*Felis bengalensis euptileura*) becoming infected with feline immunodeficiency virus through being bitten by a feral cat, in the context of a proposed reintroduction programme. More intricate modelling of a disease in a population can be provided by OUTBREAK, an epidemiological modelling package that can be linked to the PVA package VORTEX, thus combining quantitative epidemiology with modelling of population dynamics (Miller, 2007). Although VORTEX has been used frequently to model reintroduced populations (Armstrong & Reynolds, this volume, Chapter 6), there are no published examples of the use of OUTBREAK to quantify the disease risks of a reintroduction programme, as far as we are aware.

Fernández et al. (2006) used a spatially explicit individual-based population model to carry out a PVA for a proposed reintroduction of wild boar (*Sus scrofa*) to Denmark and, by comparing the predicted distribution of wild boar to the distribution of domestic pig farms, analysed the risk of disease transmission from wild boar to domestic pigs based solely on the distance between the boars and pigs. Diseases of risk were considered as a single entity and there was no consideration of the differing transmission routes, or other epidemiological parameters, of parasites.

Fundamental difficulties in assessing the disease risks of wild animal translocations

A crucial difficulty with assessing the disease risks of free-living wild animal translocations is that the components of the biological package are not known for any free-living species in entirety because (i) the number of parasites harboured by each species is large and (ii) studies on parasite ecology and disease in wild animals are in their infancy. These two factors probably increase the risk that an alien parasite will be transferred from the source to destination environments. Our poor understanding of free-living wild animal parasites contrasts with the situation in domestic animals, where infectious agents have been studied for close to 150 years and there is relatively good knowledge of the parasites of the 10 or so host species involved. Gathering information on the parasites of the thousands of mammal, bird, reptile, amphibian, invertebrate and plant species that we might wish to translocate in the future is a much harder task for reintroduction biologists, but is a crucial project given the known disease risks of the introduction of alien pathogens (see Ewen *et al.*, this volume, Chapter 9). The human intestine has in the order of 800 species of bacteria (Bäckhed *et al.*, 2005) and therefore, given the many other viral, fungal, protozoan and possibly helminth species present, parasite numbers may possibly be in the order of 1000 or more. It is likely that parasite species numbers in other animals are similar, but the proportion of these parasites that are novel to the wild animals in the destination environment and that are pathogenic when they encounter species in the destination environment remains largely unknown. In the qualitative disease risk analyses that have been carried out for specific translocations, and have been reported, a large number of parasites have been assessed. For example, Neimanis & Leighton (2004) considered 122 parasites and did detailed qualitative DRA for 83 of these parasites. Quantitative approaches to the analysis of this number of infectious agents are unrealistic given our current understanding and it is therefore perhaps not surprising that no comprehensive quantitative disease risk analyses for a reintroduction has yet been published. Even the quantitative analysis of the six infectious agents designated as a risk by Neimanis & Leighton (2004) represents a formidable challenge.

Changes to ecosystems as a consequence of translocations can take many years to develop. For example, we know that the epidemic in Africa induced by the introduction of rinderpest virus lasted many years and, following the initial epidemic, persistent infection influenced population sizes of ungulates with knock-on effects on carnivores (Dobson & Hudson, 1985). Similarly, the effects of squirrelpox virus on the UK ecosystem are still being felt today, over 130 years since the probable date of the virus introduction (Sainsbury et al., 2008). Therefore in disease monitoring of translocations, as a component of disease risk management, we face dilemmas in choosing appropriate methods and timescales. Squirrelpox virus was unknown at the time of grey squirrel translocation to the UK; how do we attempt to monitor such unknown parasites in current reintroduction programmes, and over what timescale?

In disease risk analysis methods developed for domestic animals, apparently the principal concern has been to assess the risks of introducing novel infectious agents with the animals being translocated (e.g. Murray et al., 2004). In reintroductions undertaken for conservation purposes, host–parasite encounters at several stages of the reintroduction pathway are equally of concern because of their potential to cause disease in the highly valuable endangered populations being transported. These stages include: (i) during transit when the translocated host may contact novel parasites; (ii) at the destination where both parasites and non-infectious agents may cause disease at the population level, and parasites at the destination may be novel to the newcomers; (iii) during transit due to changes in the pathogenicity of commensal parasites carried between the source and destination sites influenced by stressors; and (iv) both pathogenic and commensal parasites introduced by the translocated hosts and contracted by hosts at the destination site. The influence of stressors on commensal parasite–host interactions during reintroduction is covered in Ewen et al. (this volume, Chapter 9).

To be considered a hazard under Murray et al.'s (2004) system of disease risk analysis, an infectious agent must be novel in the importing country and be known to cause harm. Parasites novel to a specific country were considered more likely to be associated with disease outbreaks because domestic animals in that country may not have developed an immune response to the parasite. In contrast, interactions between free-living wild animals and their parasites are less likely to respect country borders because these borders do not necessarily represent a barrier to either animal movement and/or the spread of parasites among populations. In an ecosystem, geographical (such as large rivers or mountain ranges), ecological (such as niche separation) and evolutionary (such as the absence of taxonomic groups) barriers are more likely to prevent the spread of animals and their parasites, and therefore if a translocation crosses such a barrier the animals transferred are more likely to introduce a novel parasite to the new environment.

Current approaches to qualitative disease risk analysis for reintroduction programmes

Following the publication of the detailed guidelines on disease risk analysis by the World Organisation for Animal Health (Murray *et al.*, 2004), Sainsbury *et al.* (2009), Sainsbury & Vaughan-Higgins (submitted) and Masters & Sainsbury (2011) used the latter as a basis from which to devise a method for disease risk analysis for translocations for conservation purposes by (i) using a holistic approach to hazard inclusion, (ii) defining hazards according to the interactions between the translocated host's immunity and the parasites with which the host interacts, (iii) defining hazards according to geographical, ecological and evolutionary barriers crossed, (iv) using post-release monitoring to detect disease and therefore gain the opportunity to detect novel or unknown parasites not classified as hazards at the time of the reintroduction and (v) considering both non-infectious and infectious hazards. In translocations for conservation purposes involving valuable animals, Sainsbury *et al.* (2009), Sainsbury & Vaughan-Higgins (submitted) and Masters & Sainsbury (2011) recognized a succession of 'hazard points' during the translocation pathway at which the reintroduction could be affected by hazards, not limited to infectious agents (Table 10.2). On the assumption that translocations that cross a geographical, ecological and evolutionary boundary are more likely to give rise to source, transport and destination hazards, all parasites novel to the destination environment, or novel to the reintroduced species in that environment, were considered capable of causing harm on account of their novelty and the likely absence of an effective immune response in hosts resident at the destination site. In contrast, Murray *et al.* (2004) required infectious agents to be known causes of harm to be defined as a hazard. Hazards were

Table 10.2 Definition of hazards according to the point on the translocation pathway at which the host–parasite interaction occurred (after Sainsbury *et al.*, 2009; Sainsbury & Vaughan-Higgins, submitted; Masters & Sainsbury, 2011).

Hazard	Definition
Source hazard	The translocated animal was considered a potential vehicle for the parasite and the parasite (or a strain of this parasite) was novel at the destination site
Transport hazard	Either (i) an infectious agent (or a strain of this infectious agent), present on route from the source to destination, which is exotic to the destination, for which the translocated animal might be a vehicle for introduction to the destination, or (ii) an infectious agent present on route from source to destination, which can infect the translocated animal and cause disease if precipitated by transit stressors
Carrier hazard	A commensal infectious agent to which the source population has co-adapted and co-evolved, but when the host is subjected to stressors, such as those associated with reintroduction, or factors that affect parasite dynamics, such as alterations in host density, cause disease in animals in transit or in the destination environment. Such hazards are effectively '*carried*' from the source to the destination population
Host immunodeficiency hazard	A commensal infectious agent to which the host population has co-adapted and co-evolved but, when the host population is subjected to stressors, such as those associated with reintroduction, or factors that affect parasite dynamics, such as alterations in host density, cause disease in reintroduced animals in the destination environment. The hosts are infected by such hazards at the *destination* environment
Population hazard	An infectious or non-infectious agent present at the destination for which there was evidence of potential effects at the population level, including effects on small populations
Destination hazard	An infectious agent absent at the source but present at the destination

evaluated by disease risk assessment according to the approach described by Murray *et al.* (2004), as shown by the example in Table 10.3.

Previously undetected infectious agents, which are likely to be introduced to the destination with the translocated animal, may remain undetected in the destination environment unless they cause disease in the translocated animals or similar species. Therefore post-release health monitoring of the translocated animals, and similar taxonomic groups, in the destination environment was advocated by Sainsbury & Vaughan-Higgins (submitted) as an approach to detect harmful alien parasites introduced with the reintroduced species.

Recently the New Zealand Department of Conservation (DoC) published its own DRA methodology for conservation translocations, which they have called the disease risk assessment tool (DRAT) (McInnes *et al.*, 2011). This follows a similar structure to earlier versions of qualitative DRA and considers both source and destination hazards, but with some changes. Most importantly, the new DRAT process follows a linear decision tree which (i) helps fast-track translocations that are deemed low risk and (ii) encourages surveillance monitoring and data sharing. DRAT achieves this by placing value judgements at the start of the process. It has been recognized that certain sites and/or species are very special in New Zealand and therefore increased care is needed to reduce disease risks. More formal DRAs are mandatory in these cases. There are, however, a large number of conservation translocations that are deemed of low risk (either the species is not listed as 'nationally critical' or the source and release locations are not 'DoC priority sites') and the DRAT can allow translocations to proceed with little or no disease management. Often low-risk sites are connected or at least present within the same mosaic of rural and agricultural lands of the main North and South Islands. This could be similar to stating that geographical, ecological and evolutionary boundaries are not being crossed and hence there is a low or unlikely disease risk (see above). The translocation disease management process is comprised of both the DRAT and an accompanying disease management workbook (DMW) that provides the required information about priority sites and species, current disease fact sheets and other important information. It is also clearly noted that the available information on wildlife disease is limited and that this restricts how accurate the DRAT and DMW can be. Importantly, the DRAT and DMW is meant to provide a translocation practitioner with a greater sense of understanding, involvement and justification of the risk assessment process.

Table 10.3   Disease risk analysis for the source hazard, West Nile virus, for the reintroduction of white-tailed sea eagles (*Haliaeetus albicilla*) to the UK from Eastern Europe (modified from Sainsbury *et al.*, 2009).

Source hazard	Justification for hazard status	Risk assessment				Risk management	
		Release assessment	Exposure assessment	Consequence assessment	Risk estimation	Risk evaluation	Risk options
West Nile virus (WNV)	Although antibodies to West Nile virus have been detected in birds in the UK (Buckley, *et al.*, 2003), their presence has not recently been verified, the virus has not been detected and no cases of disease due to this virus have been detected. From a precautionary perspective it would be preferable to assume West Nile virus is not present in the UK. West Nile virus has been detected in birds from Eastern Europe, e.g. in Slovakia (McLean and Ubico, 2007). No	Serological surveys in Eastern Europe suggest that there is a low likelihood that WTSE, like other birds, will be infected with WNV through contact with ornithophilic mosquitoes, and the latter are present in Eastern Europe (McLean and Ubico, 2007). Fatal infection in raptors (including red-tailed hawks (*Buteo jamaicensis*) and great horned owls (*Bubo virginianus*) has been reported (Saito *et al.*, 2007) but other bird Orders, including Passeriformes, are more susceptible to the infection and the disease (McLean and Ubico, 2007).	Falconiformes are known to develop a sufficient viraemia for infection to be transmitted to mosquitoes (Defra, 2009) and viraemia has a duration of approximately one week and so the arrival of a viraemic WTSE is possible. Since other bird species, particularly passerines, are highly susceptible to West Nile virus infection there is a high likelihood that these species will be exposed from ornithophilic mosquitoes (which are present in the UK) in contact with WTSE. There is a high probability that highly susceptible bird	There is a high probability that disseminated infection would occur if the virus is introduced because many passerine birds will be in the vicinity of WTSE at the release site. West Nile virus has given rise to epidemic disease in Passeriformes in the USA, where birds were naive to infection (McLean and Ubico, 2007) and, assuming the epidemiological parameters are similar in the UK, epidemic disease would be predicted. However, antibodies to WNV in UK bird populations have been detected without signs of epidemic disease.	The likelihood of release through importation in a WTSE is low but the likelihood of exposure of susceptible species to infection are high. Evidence suggests that the likelihood of a significant epidemic disease is low. Therefore the overall risk level is considered to be low	Preventive measures should be employed to reduce the disease risks	WTSE will only be translocated to England after a health examination shows each bird is fit and healthy, reducing the probability that a bird infected with WNV is transported. WTSE will be serologically tested for WNV immediately following translocation to gather information about the epidemiology of this disease in Eastern European WTSE

(Continued)

Table 10.3 (*Cont'd*)

Source hazard	Justification for hazard status	Risk assessment				Risk management	
		Release assessment	Exposure assessment	Consequence assessment	Risk estimation	Risk evaluation	Risk options
		cases of WNV disease have been reported in birds in Eastern Europe, which suggests that disease is rare. However, viraemia may occur without disease. Therefore there is a low likelihood of infection in a translocated WTSE	species will be infected. There is a high probability of dissemination of WNV through susceptible bird species because, at the time of importation in the summer, ornithophilic mosquitoes will be common. Humans are susceptible to infection and there is a low probability that they may be exposed through vector-borne transmission (Zeller and Schuffenecker, 2004)	Such evidence suggests that differing epidemiological parameters (possibly cross-protection from other flaviviruses; Gubler, 2007, cited by Defra, 2009) in the UK (and incidentally also in Continental Europe) have reduced the likelihood of disease outbreaks. An epidemic would have a major economic, environmental and biological impact, as witnessed by the effect of the WNV outbreak in North America over the last 10 years (McLean and Ubico, 2007), but the evidence suggests that there is a low risk of this happening in the UK			

A suggested approach for future qualitative and quantitative DRA

Adopting disease risk assessment methods from human and domestic animal systems has been useful for developing methods for assessing the probability of translocation failure due to disease and/or accidental co-introduction of infectious agents to release site ecosystems. However, there are major challenges for the effective use of these DRA tools. We advocate two additions to the DRA process: (i) better methods for dealing with uncertainty and (ii) risk analysis that is integrated into structured decision making.

There are many uncertainties because the DRA (assessing the probability of co-introduction of a disease-causing agent and the magnitude of negative consequences) is based on variable information, most often of poor quality and frequently relying on expert judgement (see above). We view a key area for advancing these tools as taking better account of this uncertainty. This applies equally to all DRA methods. Where information on parasites is limited, the qualitative DRA tool is obviously the method of choice. This situation probably includes the vast majority of translocations. Traditional approaches to quantitative DRA methodology have been limited in applicability given the poor knowledge of system states and the need for detailed information on any given pathogen and its host.

Regardless of whether DRA tools are quantitative or qualitative, and regardless of the degree of associated uncertainty, it is often not obvious how to use the DRA in making a reintroduction decision. The DRA provides information that is clearly relevant to a decision, yet there is seldom an explicit mechanism for translating it into one. Structured decision making provides such a mechanism (Figure 10.2). Thus, we advocate formal treatment of the disease risks of reintroduction programmes as decision problems using structured decision-making approaches (as a component of all decision problems in a given reintroduction programme rather than in isolation to them). We also advocate adaptive management for problems that are recurrent (multiple decisions over time) and characterized by substantial uncertainty that can be reduced through iteration, evaluation and learning. Structured decision making and adaptive management are both discussed in depth in Nichols & Armstrong (this volume, Chapter 7) and McCarthy *et al.* (this volume, Chapter 8).

Traditional DRA has contributed to management decisions in reintroduction programmes, but it has done so in a manner that is not transparent and

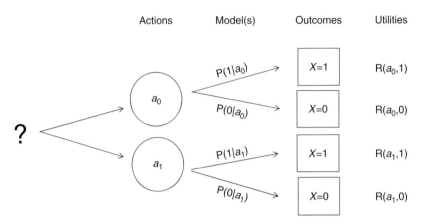

Figure 10.2 **Structured decision making involves anticipating outcomes, assigning utilities and estimating the probability of different outcomes occurring. This last component is often called 'risk analysis'. This figure shows the simplest possible decision scenario, where there are only two possible actions and two possible outcomes under each action. For example, the actions could be to screen (a_1) or not screen (a_0) for a parasite during translocation, and the outcomes could be that the parasite is transmitted to the destination site ($X = 1$) or not transmitted ($X = 0$). Utilities (R) would probably be assigned for each of the four possibilities based on the cost of the screening, the likely impact on the translocated organisms and the likely impact of the parasite on the ecosystem at the destination site. One decision criterion would be to select the action with the highest average utility, given by $\bar{R}(a_j) = \sum_x p(x|a_j)R(a_j, x)$ (adapted from Williams et al., 2002).**

is unlikely to be optimal. Authoritative judgement calls, scoring systems and linear decision trees offer limited utility for optimization or for future learning and adjustment. Current DRA applications typically rely on judgement of uncertainty (e.g. ranking overall risk as low, medium or high), and once completed there is little follow-through or appraisal of effectiveness or outcome. While it is recommended that DRAs are repeated for each translocation to account for any new information, there is no inherent system that encourages recording this new information (both within and across projects), integrating it with prior information and using it to make future decisions. Any new information is therefore frequently ad hoc in nature and faces many criticisms that also commonly face surveillance monitoring, not the least of which is lack of efficiency (see Nichols & Williams, 2006). We believe that the next important step in using DRA approaches is to incorporate the information that they contain directly into a decision process associated with the entire reintroduction programme.

A decision-theoretic approach to reintroduction decisions requires several components: explicit objectives, a set of management alternatives, models that project the consequences of these alternatives, a monitoring programme and some algorithm for translating these components into an actual decision (recommended action to take). Current DRAs do produce models projecting the consequences of management alternatives. Explicit objectives are typically expressed in terms of 'utilities' or quantitative expressions that allow us to place a value on any specific management outcome. In the case of a reintroduction programme, these utilities can include the value of the new population and the source population, the value (cost) of negative ecosystem impacts (including reintroduction of native parasites) and monetary costs of different management actions. An example of a set of alternative decisions in translocation might be: (i) translocate without regard to any parasite sampling or treatment, (ii) do not translocate, (iii) translocate conditional on parasite sampling of source population and/or release ecosystem or (iv) translocate individuals conditional on quarantine, parasite sampling and/or treatment.

The models to be used in a decision problem should project the consequences of the various management actions for the key elements of the objectives (the variables on which utilities are based). For example, models might include: (i) effects of carried parasites on vital rates of hosts at both release site and source location, (ii) effects of quarantine and treatment on carried parasites of released individuals, (iii) effects of quarantine and treatments directly on released individuals (e.g. stress), (iv) effects of release-site parasites on translocated species, (v) probability that the carried parasite is already at the release site or is native to the release site, (vi) probability that a new parasite carried by released individuals becomes established in the ecosystem, (vii) projected ecosystem impacts if a new parasite is established and (viii) probability of successful screening for each screening method. Some of these model components are already the focus of DRAs (see above). However, uncertainty in any of these relationships or probabilities can be incorporated into the decision process in the form of multiple discrete models or as parametric uncertainty in a key parameter of a general model. This is what is currently missing from DRA. In either case, confrontation of model-based predictions with monitoring data permits learning and reduction of this uncertainty (commonly referred to as *structural uncertainty*; see Nichols

& Armstrong, this volume, Chapter 7). In addition, previous data can greatly reduce the uncertainty at the start of a new programme if analysed carefully (see McCarthy *et al.*, this volume, Chapter 8).

The monitoring programme associated with a reintroduction decision process should be designed to provide data for four primary uses. The first is for the purpose of making state-dependent decisions. The appropriate action to implement will frequently be a function of current conditions and the system state. The second use is simply to assess the progress of the management programme. For example, we would like to know whether the reintroduction programme is succeeding in producing a viable new population. The third use of monitoring is for learning, specifically through comparison with model-based predictions, leading to reduction of both structural uncertainty and uncertainty in parameters estimates. Once again, the emphasis will be on developing a monitoring programme that meets the specific needs of the decision process of which it is a part (Nichols & Armstrong, this volume, Chapter 7). The final component of a decision process is an algorithm that uses information from all of the previously described components to come to an optimal decision based on the information available. In many current decision processes, this algorithm is the mental thought process used by the decision maker. Although there is nothing wrong with this approach, it may lead to suboptimal decisions and, perhaps more importantly, it is not transparent. Perhaps most importantly, SDM makes it possible to identify reasons for differences in opinion among stakeholders, and this can be achieved even in the absence of the many analytical challenges of optimization algorithms. Optimization algorithms use the objective function, knowledge of system dynamics (model(s)) and the estimated current state of the system (from monitoring) to select the best choice from the set of possible management actions. The iterative cycle of adaptive management is depicted in McCarthy *et al.* (see this volume, Figure 8.1 in Chapter 8). At each decision point, an algorithm is used to select the best management action based on the set of decision components. The action is implemented and drives the system to a new state. The new system state is identified via monitoring and compared with the predictions of each model. Degrees of faith in the models are modified, increasing for those that predicted well and decreasing for those that predicted poorly.

The above approach has not yet been applied to DRA and has also not yet been applied to reintroduction in general, although ideas for doing so are detailed in Chapters 7 and 8. It remains a challenge to incorporate this approach into DRA, but we view this as a rewarding investment. In qualitative

DRAs this likely means taking the method already used and making two changes. First, we would like to confront uncertainty, acknowledge it throughout the assessment and permit it to evolve (hopefully decrease) over time. This contrasts with the current situation in which we might make a decision based on judgement early in the process and then continue to use a 'point estimate' of risk based on previous data. In effect we should guard against statements such as 'pathogen a is of low threat because it has not been detected at the source location'. We would translate such statements into models that include details of our uncertainty in this claim (related to detection probability against previous sampling effort, test sensitivity, etc.). Second, uncertainty reduction becomes the focus of targeted monitoring to inform and update these models and/or our degrees of faith in them. Currently there is little impetus to follow through the effects of any decisions made from the DRA approach.

Quantitative methods of DRA are more closely aligned to structured decision making and should normally already acknowledge uncertainty. Key distinctions between quantitative DRA and adaptive management are: (1) it is not clear how uncertainty translates into a management decision in current DRA and (2) there is no formal approach to learning and reducing uncertainty in existing DRAs. Some DRAs may have included adaptive management as we define it here, but there are no clear examples of this in the literature to our knowledge.

A benefit of quantitative methods is that models are based on more detailed data with which to make more reliable predictions; the cost is in obtaining this detailed data. For example, understanding disease dynamics within a population requires knowledge of state variables including the abundance of susceptible (N_S), infected (N_I) and recovered (N_R) animals along with total abundance ($N = N_S + N_I + N_R$). In reality only a sample of the population will be surveyed at any time point to determine disease state (S, I or R). There will be different estimation problems including not detecting all animals in any state ($n_V \neq N_V$), and that state-specific detection probabilities are not necessarily equal ($p_S \neq p_I \neq p_R$). In such a situation, appropriate monitoring might include capture–recapture with an assessment of the animal state at each capture. Resulting data could be used with multistate modelling to estimate state-specific survival and transition probabilities in the presence of unequal detection probabilities and to account for disease state misclassification. Resulting estimates could be used to compare against predictions made about how the disease dynamics will change in both source and release locations and to estimate any wider effects of disease in the release ecosystem and our ability to detect whether screening, quarantine and treatment can effectively remove co-transfer of a pathogen if this is the objective. Again appropriate monitoring and feedback can inform and update models under adaptive management.

Conclusions

Qualitative and quantitative approaches to disease risk analysis for reintroduction programmes have been proposed but only a small number of qualitative risk analyses, primarily based on a system developed to assess risks to human health, have been published in this field. There remain some serious difficulties in carrying out effective disease risk analyses because: (i) the number of hazards to consider are numerous, (ii) hazards can hit at several points along the reintroduction pathway, (iii) some hazards are unknown at the outset of reintroduction and are difficult to detect, (iv) it may be years before a disease outbreak as a consequence of reintroduction is 'visible' and (v) the decisions made by disease risk analyses are not easily deciphered. Methods have been proposed and are in use in attempts to address these difficulties, including detailed post-release monitoring programmes and systems to prioritize and classify hazards to structure the disease risk analysis more clearly. Problems remain, particularly that decisions are based on poor quality and quantity of data, issues of uncertainty are not transparent and results are not fed back into the system to improve decision making. We have set out a framework for structured decision making and adaptive management to go some way to addressing these difficulties. In summary, we suggest that decision theory and adaptive management will allow sound decisions in the face of uncertainty.

Acknowledgements

Ideas in this chapter developed from discussions with Tony Mitchell-Jones, Ian Carter, Jim Foster, Chris Lloyd, Iain McGill, Yedra Feltrer, Fieke Molenaar, Katie Colville, Rebecca Vaughan-Higgins, Nic Masters and Kate McInnes. All authors participated and discussed ideas within the highly rewarding 2011 CBSG Disease Risk Analysis Tool Development Workshop, with a key figure being Richard Jakob-Hoff. The authors especially thank Jim Nichols for providing comments and editing earlier versions of this chapter.

References

Armstrong, D.P. & Reynolds, M.H. (2011) Modelling reintroduced populations: the state of the art and future directions. In *Reintroduction Biology: integrating science and management*, eds J.G. Ewen, D.P. Armstrong, K.A. Parker & P.J. Seddon, Chapter 6. Wiley-Blackwell, Oxford, UK.

Armstrong, D., Jakob-Hoff, R. & Seal, U.S. (eds) (2003) *Animal Movements and Disease Risk: A Workbook*. Conservation Breeding Specialist Group (SSC/IUCN), Apple Valley, Minnesota, USA.

Bäckhed, F., Ley, R.E., Sonnenburg, J.L. *et al.* (2005) Host–bacterial mutualism in the human intestine. *Science*, 307, 1915–1920.

Ballou, J.D. (1993) Assessing the risks of infectious disease in captive breeding and reintroduction programs. *Journal of Zoo and Wildlife Medicine*, 24, 327–335.

Bruckner, G., MacDiarmid, S., Murray, N. *et al.* (2010) *Handbook on Import Risk Analysis for Animals and Animal Products*. Office International des Epizooties, Paris, France.

Buckley, A., Dawson, A., Moss, S.R. *et al.* (2003) Serological evidence of West Nile virus, Usutu virus and Sindbis virus infection of birds in the UK. *Journal of General Virology*, 84, 2807–2817.

Corn, J.L. & Nettles, V.F. (2001) Health protocol for translocation of free-ranging elk. *Journal of Wildlife Diseases*, 37, 413–426.

Covello, V.T. & Merkhofer, M.W. (1993) *Risk Assessment Methods: approaches for assessing health and environmental risks*. Plenum Press, New York, USA.

Davidson, W.R. & Nettles, V.F. (1992) Relocation of wildlife: identifying and evaluating disease risks. *Transactions of the North American Wildlife and Natural Resources Conference*, 57, 466–473.

Defra (2009) *West Nile Virus: potential risk factors and the likelihood for the introduction of the disease into the United Kingdom*, authors K. Patel, M. Lopez, H. Roberts & M. Sabirovic, version 2, revised 24 February 2009, p. 34. International Animal Health, London, UK.

Dobson, A.P. & Hudson, P.J. (1985) Microparasites: observed patterns in wild animal populations. In *Ecology of Infectious Diseases in Natural Populations*, eds B.T. Grenfell & A.P. Dobson. Cambridge University Press, Cambridge, UK.

Ewen, J.G., Acevedo-Whitehouse, K., Alley, M. *et al.* (2011) Empirical consideration of parasites and health in reintroduction. In *Reintroduction Biology: integrating science and management*, eds J.G. Ewen, D.P. Armstrong, K.A. Parker & P.J. Seddon, Chapter 9. Wiley-Blackwell, Oxford, UK.

Fernández, N., Kramer-Schadt, S. & Thulke, H. (2006) Viability and risk assessment in species restoration: planning reintroductions for the wild boar, a potential disease reservoir. *Ecology and Society*, 11, 6.

Griffith, B., Scott, M.J., Carpenter, J.W. *et al.* (1993) Animal translocations and potential disease transmission. *Journal of Zoo and Wildlife Medicine*, 24, 231–236.

Gubler, D.J. (2007) The continuing spread of West Nile virus in the Western hemisphere. *Emerging Infections*, 45, 1039–1046.

Karesh, W.B., Cook, R.A., Bennett, E.L. *et al.* (2005) Wildlife trade and global disease emergence. *Emerging Infectious Diseases*, 11, 1000–1002.

Leighton, F. A. 2002. Health risk assessment of the translocation of wild animals. *Revised Scientific Technology, Office of International Epizootics (OIE)*, 21, 187–195.

Masters, N. & Sainsbury, A.W. (2011) *Disease Risk Analysis for the Wild to Wild Translocation of the Smooth Snake (Coronella austriaca) within the United Kingdom*. Natural England, Peterborough, UK.

McCarthy, M.A., Armstrong, D.P. & Runge, M.C. (2011) Adaptive management of reintroduction. In *Reintroduction Biology: integrating science and management*, eds J.G. Ewen, D.P. Armstrong, K.A. Parker & P.J. Seddon, Chapter 8. Wiley-Blackwell, Oxford, UK.

McInnes, K. (2011) *Translocation Disease Risk Management Process*. New Zealand Department of Conservation, Wellington, New Zealand.

McLean, R.G. & Ubico, S.R. (2007) Arboviruses in birds. In *Infectious Diseases of Wild Birds*, eds N.J. Thomas, D.B. Hunter & C.T. Atkinson, pp. 17–62. Blackwell Publishing, Iowa, USA.

Miller, P.S. (2007) Tools and techniques for disease risk assessment in threatened wildlife conservation programmes. *International Zoo Yearbook*, 41, 38–51.

Murayama, A., Traylor-Holzer, K., Reed, D. *et al.* (eds) (2006) Tsushima leopard cat conservation planning workshop report. IUCN/SSC Conservation Breeding Specialist Group, Apple Valley, Minnesota.

Murray, N., Macdiarmid, S.C., Wooldridge, M. *et al.* (2004) *Handbook on Import Risk Analysis for Animals and Animal Products*. Office of International Epizootics (OIE).

Neimanis, A.S. & Leighton, F.A. (2004) *Health Risk Assessment for the Introduction of Eastern Wild Turkeys (Meleagris gallopavo silvestris) into Nova Scotia*. Canadian Cooperative Wildlife Health Centre, University of Saskatchewan, Canada.

Nichols, J.D. & Williams, B.K. (2006) Monitoring for conservation. *Trends in Ecology and Evolution*, 21, 668–673.

Nichols, J.D. & Armstrong, D.P. (2011) Monitoring for reintroductions. In *Reintroduction Biology: integrating science and management*, eds J.G. Ewen, D.P. Armstrong, K.A. Parker & P.J. Seddon, Chapter 7. Wiley-Blackwell, Oxford, UK.

Office International des Epizooties (OIE) (1999) *International Animal Health Code*. Office International des Epizooties, Paris, France.

Sainsbury, A.W. & Vaughan-Higgins, R.J. (submitted) Analyzing disease risks associated with translocations for conservation purposes. *Conservation Biology*.

Sainsbury, A.W., Deaville, R., Lawson, B. *et al.* (2008) Poxviral disease in red squirrels *Sciurus vulgaris* in the UK: spatial and temporal trends of an emerging threat. *EcoHealth*, 5, 305–316.

Sainsbury, A.W., Vaughan-Higgins, R. & Curtiss, P.K. (2009) *Disease Risk Analysis for the Reintroduction of the White Tailed Sea Eagle (Haliaeetus albicilla) to England.* Natural England, Peterborough, UK.

Saito, K.E., Sileo, L., Green, D.E. *et al.* (2007) Raptor mortality due to West Nile virus in the United States, 2002. *Journal of Wildlife Diseases*, 43, 206–213.

Williams, B.K., Nichols, J.K. & Conroy, M.J. (2002) *Analysis and Management of Animal Populations.* Academic Press, New York, USA.

Zeller, H.G. & Schuffenecker, I. (2004) West Nile virus: an overview of its spread in Europe and the Mediterranean basin in contrast to its spread in the Americas. *European Journal of Clinical Microbiological Infectious Diseases*, 23, 147–156.

$$11$$

The Genetics of Reintroductions: Inbreeding and Genetic Drift

Lukas F. Keller[1], Iris Biebach[1], Steven R. Ewing[2] and Paquita E.A. Hoeck[1,3]

[1] Institute of Evolutionary Biology and Environmental Studies, University of Zürich, Zürich, Switzerland
[2] Royal Society for the Protection of Birds, Bedfordshire, United Kingdom
[3] Institute for Conservation Research, San Diego, California, United States of America

'Several authors therefore began to question whether inbreeding was of any importance in the wild. This scepticism about the importance of inbreeding depression from researchers outside the animal and plant breeding communities was, however, not new. Using the term "interbreeding" to describe inbreeding, Darwin had already remarked ". . . that any evil directly follows from any degree of close interbreeding has been denied by many persons; but rarely by any practical breeder . . ." (Darwin, 1868, vol. II, p. 116).'

Page 362

Introduction

Reintroduction is a form of conservation translocation that aims to establish self-sustaining populations by releasing wild or captive-bred individuals into a species' former range, following its extirpation or extinction in the wild (IUCN,

Reintroduction Biology: Integrating Science and Management. First Edition.
Edited by John G. Ewen, Doug P. Armstrong, Kevin A. Parker and Philip J. Seddon.
© 2012 Blackwell Publishing Ltd. Published 2012 by Blackwell Publishing Ltd.

1998; Seddon *et al.*, this volume, Chapter 1). To be successful in the long term, reintroduced populations need to be able to sustain themselves. This, in turn, requires that they are capable of adaptation to future environmental change. To be able to reach this goal, reintroduction biologists need to carefully consider ecological, environmental, disease, demographic, and genetic aspects that will increase the likelihood of success.

This chapter focuses on two key genetic aspects of reintroduction biology, inbreeding (and its fitness consequences, known as inbreeding depression) and loss of genetic variation. These two genetic processes are particularly important in reintroductions because all reintroduced populations experience bottlenecks, i.e. periods of small population size, sometimes several in a row. The first bottleneck often occurs before conservation measures are taken, when species are endangered and hence, by definition, small. In some cases, a captive breeding population is then founded with a number of wild-caught individuals – the second bottleneck. Finally, animals are released back into their former range – the third bottleneck. Thus, all reintroduced populations have experienced periods when their population size was small. In small populations, inbreeding and loss of genetic variation are particularly pronounced (e.g. Hedrick & Miller, 1992) and hence reintroduced populations are particularly vulnerable to these two processes.

In this chapter, we focus on the establishment and persistence phase, i.e. on the population released into its former range (Seddon *et al.*, this volume, Chapter 1). We do not consider captive breeding programmes, which are an important component of many reintroductions to provide sufficient numbers of individuals for release. In captivity, genetic processes such as adaptation to captivity occur (Frankham, 2008; Williams & Hoffman, 2009) in addition to inbreeding and loss of genetic variation. Genetic issues in captivity are of central importance to reintroduction biology (e.g. Robert, 2009). Space constraints have led us to neglect them here, not a belief that these genetic processes are less important than inbreeding and genetic drift. Finally, genetics can be used to monitor reintroduced populations and to assess the progress towards creating a self-sustaining population (e.g. Johnson *et al.*, 2010b). This important aspect has recently been covered elsewhere (Schwartz *et al.*, 2007; Laikre *et al.*, 2009).

This chapter presents the genetic concepts required to understand and manage levels of inbreeding and loss of genetic variation in reintroduced populations. The following two chapters (Groombridge *et al.*, Chapter 12, and Jamieson & Lacy, Chapter 13) discuss the practical applications of these genetic concepts to reintroductions. We start with a brief historical overview of the role of inbreeding and inbreeding depression in conservation biology. We then discuss different meanings of the term 'inbreeding' and how inbreeding is related to loss of genetic variation through genetic drift. The next sections address the estimation of inbreeding depression and the causes of variations in inbreeding depression. We also ask whether inbreeding effects measured at the level of the individual translate into effects at the population level. Finally, we discuss loss of genetic variation and the corresponding loss of evolutionary potential.

Inbreeding in conservation biology – a short history

Inbreeding depression, the reduction in vigour, reproductive success, survival, growth, disease resistance, etc., of individuals produced by matings between relatives is one of the oldest observations in evolutionary genetics and has been studied for over 200 years. Pioneering experiments in plants on inbreeding and inbreeding depression by Kölreuter (1766), Knight (1799) and others were followed by Charles Darwin's book-length exploration of the consequences of inbreeding and cross-breeding in plants (Darwin, 1876). However, mechanistic explanations of the causes of inbreeding depression remained elusive until the rediscovery of Mendel's laws of inheritance in 1900. This rediscovery allowed the formulation of rigorous theories of the consequences of inbreeding based on the mechanisms of heredity, and led to a series of in-depth experiments on inbreeding with both animals and plants in the first half of the 20th century (see Wright, 1977, pp. 6–137 for a more detailed account).

These extensive experiments showed that, not in every case but on average, inbreeding reduces fitness (Wright, 1977). Consequently, when the field of conservation biology emerged, the overwhelming evidence of inbreeding depression in the field of animal and plant breeding led to concerns that inbreeding may have similar effects in the small populations typical of endangered species (e.g. Soulé, 1980). While evidence from zoo populations seemed to support this view (e.g. Ralls et al., 1979), comparable data from natural populations were scarce. Several early studies of inbreeding effects in the wild failed to find evidence of inbreeding depression (summarized in Thornhill, 1993) and no natural population had at that stage been observed to decline or go extinct due to inbreeding (Caro & Laurenson, 1994). Several authors therefore began to question whether inbreeding was of any importance in the wild. This scepticism about the importance of inbreeding depression from researchers outside the animal and plant breeding communities was, however, not new. Using the term 'interbreeding' to describe inbreeding, Darwin had already remarked '... that any evil directly follows from any degree of close interbreeding has been denied by many persons; but rarely by any practical breeder...' (Darwin, 1868, vol. II, p. 116).

The critical questions asked in the early to mid-1990s were very beneficial: they led to substantial efforts to measure the magnitude of inbreeding depression directly in free-living animals and plants. This research substantiated Darwin's dictum: inbreeding depression occurs commonly in nature and can be severe (see reviews in Crnokrak & Roff, 1999; Hedrick & Kalinowski, 2000; Keller & Waller, 2002). It occurs in many different traits, including such complex traits as senescence (Keller *et al.*, 2008), and it can lead to increased extinction risks (e.g. Saccheri *et al.*, 1998; Gaggiotti, 2003). As is the case in agricultural and laboratory systems, and as expected by theory (Moorad & Wade, 2005), there is a great deal of variation in the observed magnitude of inbreeding depression in the wild (see below for more details). Inbreeding depression is often, but not always, more pronounced in the wild than in the laboratory (Jiménez *et al.*, 1994; Joron & Brakefield, 2003; Kristensen *et al.*, 2008), but otherwise inbreeding depression in nature follows patterns that are similar to those discovered in agricultural and laboratory systems. Therefore, inbreeding depression cannot be ignored in conservation and particularly in reintroduction programmes where managers may have some control over the build-up of inbreeding (Jamieson & Lacy, this volume, Chapter 13).

What is inbreeding?

Confusingly, the term 'inbreeding' has been used to describe various related, yet not identical, phenomena leading to considerable misunderstandings in the literature (Jacquard, 1975). In all cases, the term refers to situations where relatives mate. The various definitions differ in the reference population that is used when calculating relatedness and thus inbreeding. It is beyond the scope of this book to explain the various definitions in detail. The interested reader is referred to Jacquard (1975) and Keller & Waller (2002). Some of the confusion stems from the fact that researchers who use molecular markers to estimate inbreeding generally use a different definition of 'inbreeding' than researchers who use pedigree information. Box 11.1 outlines basic concepts of and differences among two of the five meanings of 'inbreeding' (Jacquard,

1975): inbreeding under random mating and inbreeding as non-random mating. Box 11.2 outlines ways in which we can obtain quantitative measures of the occurrence of inbreeding. In the following, we will use 'inbreeding' whenever we refer to overall inbreeding, F_{IT} (see Box 11.1). When we refer to specific meanings of the term 'inbreeding', we will use the specific terminology outlined in Boxes 11.1 and 11.2 (also see Box 11.3).

Box 11.1 Inbreeding

Since inbreeding refers to mating among relatives, measures of inbreeding quantify the relatedness among parents, either for individual pairs or for the average of a population (see also Box 11.2). The relatedness of the parents in turn depends on the relatedness of their parents, and so on *ad infinitum*. Since it is impossible to trace all ancestors all the way back, inbreeding can never be measured on an absolute scale but only relative to the information available. Thus, inbreeding is always measured relative to something, and it is crucial to be aware of this relativity to avoid confusing different meanings of 'inbreeding' (Jacquard, 1975). The base population relative to which one calculates inbreeding is often called the reference population. This might be the founders of a population whose relatedness to each other is unknown and therefore often assumed to be zero.

The degree of inbreeding is commonly measured by the 'inbreeding coefficient'. Although originally defined as the correlation among uniting gametes (Wright, 1922b), the inbreeding coefficient is now commonly defined as the probability, F, that two gametes which unite to produce a zygote carry identical copies of the same ancestral allele. The two alleles are identical by descent because they derive from the same allele in a common ancestor (Malécot, 1948; Keller & Waller, 2002). Thus, F measures the probability of identity by descent.

Wright's inbreeding coefficient (Wright, 1922b) is defined at the individual level and can be estimated using different approaches (see Box 11.2). However, quite often the data necessary to calculate individual inbreeding coefficients are not available, yet we are interested in the average inbreeding level of a population, a common situation in reintroduced populations (Jamieson, 2011). When considering average inbreeding at the population level, the total inbreeding can be broken down into two components: inbreeding due to random mating and inbreeding due to non-random mating. Wright's well-known 'F-statistics' (F_{IT}, F_{ST} and F_{IS}; see Wright, 1969, pp. 294–295) allows this decomposition and thus serves to give biologists insights into the different processes creating inbreeding (see Crow, 1980, for a detailed derivation and Jacquard, 1975, for more details).

Inbreeding under random mating (F_{ST}) represents the amount of inbreeding that occurs in finite populations under random mating relative to the inbreeding present in the founders of the population. Thus, the founders of a population serve as the reference population in this kind of inbreeding. Since all real populations are finite, this type of inbreeding always occurs, the more so the smaller a population. In fact, F_{ST} increases every generation proportionally to twice the reciprocal of the effective size (N_e) of the population ($\Delta F_{ST} = \frac{1}{(2N_e)}$; see Crow & Kimura, 1970, p. 101). Starting at $F_{ST\_t=0} = 0$, the average expected F_{ST} in a population at any given time can be calculated as $F_{ST\_t=g} = 1 - (1 - \frac{1}{2N_e})^g$ (Crow & Kimura, 1970, p. 331), where g is the number of generations that have elapsed since the foundation of the population and N_e is the effective population size (see Box 11.3). F_{ST} only considers inbreeding occurring due to random mating and is thus also a measure of the amount of genetic drift occurring in a population (see text and Box 11.3). Most readers will be more familiar with F_{ST} as a measure of allele frequency divergence among populations or, in other words, population subdivision. Reassuringly, the two definitions of F_{ST} as a measure of inbreeding and as a measure of population subdivision are the same (Crow & Kimura, 1970, p. 107).

Inbreeding as non-random mating (F_{IS}) measures the degree of relatedness among mates relative to the relatedness expected if mating occurred at random in a population of the same size. In other words, F_{IS} is a measure of inbreeding that uses a randomly mating population of the same size as the reference population. A population's F_{IS} is greater than zero if parents are on average more closely related than two mates chosen at random from the population. Vice versa, F_{IS} is negative if relatives show a tendency to avoid each other as mates. Provided no other mechanisms are at work (such as population subdivision or immigration), F_{IS} corresponds to the deviation from Hardy–Weinberg expectations (Wright, 1969).

In real populations, total inbreeding often consists of both the random and the non-random components. Imagine, for example, a small population where distant relatives mate simply by chance, yet close relatives (such as parents and their offspring or siblings) avoid mating with each other if possible. This is a situation commonly observed in small, isolated human populations. In such situations, inbreeding as non-random mating (F_{IS}) is negative, yet inbreeding under random mating (F_{ST}) is positive. The two components of inbreeding can be combined as $F_{IT} = F_{ST} + F_{IS} - F_{ST}F_{IS}$. Thus, F_{IT} combines the inbreeding that occurs due to small population size and that which occurs due to non-random mating. When both F_{ST} and F_{IS} are non-zero, it is a matter of their relative strength whether total inbreeding (F_{IT}) is positive, negative or zero. An illustration of these principles can be found in examples from livestock breeding. For example, in standard-bred horses, F_{ST} was 0.043 due to the small population size, but mating among close relatives was somewhat avoided, leading to an F_{IS} of -0.003. Overall, F_{IT} was therefore 0.04, lower than F_{ST} (Wright, 1977, p. 551).

In conservation and reintroduction biology, we are most often interested in F_{IT}. However, whether F_{IT} is mostly due to F_{ST} or F_{IS} is important for several biological processes, e.g. for the occurrence of purging (see main text). Thus, a decomposition of F_{IT} into its components F_{ST} and F_{IS} yields important insights.

Box 11.2 **Measuring inbreeding**

Some of the confusion surrounding the use of the term 'inbreeding' stems from the fact that different techniques lend themselves more readily to the estimation of the different forms of inbreeding presented in Box 11.1 and that some do so at the population but not at the individual level. Recall that F_{IS} is the deviation from Hardy–Weinberg expectations (assuming the absence of other factors causing deviations, a crucial caveat). This deviation can easily be calculated from molecular genetic information, using many freely available software packages. Therefore, it is inbreeding as non-random mating (F_{IS}) that is estimated most often using *molecular marker data* (e.g. Leimu & Mutikainen, 2005; Charman *et al.*, 2010; Ficetola *et al.*, 2011). Sometimes, studies employing this approach state something like 'F_{IS} was zero, hence we found no evidence for inbreeding'. From the information given in Box 11.1 it follows that $F_{IS} = 0$ only means that there was no avoidance or preference of relatives as mates and that this does not imply that total inbreeding (F_{IT}) was also zero. Hence, quantifying inbreeding only by F_{IS} is of limited use in conservation and reintroduction biology.

More interestingly for conservation applications, molecular markers can also be used to estimate F_{ST} (Vitalis *et al.*, 2001; Biebach & Keller, 2010). The estimation procedures are more involved and are easier if $F_{IS} = 0$, but information on F_{ST} can provide important insights into average inbreeding levels in populations and how they are affected by management actions. For example, Biebach & Keller (2010) used this approach to show how the number of animals released and the initial growth of the reintroduced populations determined current levels of average inbreeding in 41 reintroduced ibex populations.

An important caveat is that molecular markers are very suitable to estimate inbreeding at the population level (i.e. the average inbreeding of a population; see, for example, Biebach & Keller, 2010), but they do not lend themselves equally well to estimates of individual inbreeding coefficients. The reason is that marker characteristics such as heterozygosity are poorly correlated with individual inbreeding coefficients (also see Groombridge *et al.*, this volume, Chapter 12) unless there is a lot of inbreeding, a lot of variation in inbreeding among individuals (Szulkin *et al.*, 2010) or information from a large number of loci (Weir *et al.*, 2006).

Pedigrees, on the other hand, are very well suited to estimating inbreeding at the individual and population level, with many software packages available to do the job. Pedigree analysis yields estimates of total inbreeding (F_{IT}, if averaged over the population) that are relative to the most distant known ancestors of the individuals in the pedigree (the founders) who, for lack of better knowledge, are assumed to be non-inbred and unrelated to each other. In other words, pedigree inbreeding is relative to the available pedigree information. If additional pedigree information is found, for example, because formerly unknown stud-book records are discovered, estimated pedigree inbreeding will change because the available pedigree information changes. Note, however, that the levels of inbreeding calculated after the inclusion of more pedigree records are not more exact or closer to reality than the original ones. They simply reflect different pedigree information (Jacquard, 1975, p. 342).

One of the drawbacks of pedigree-based estimates of inbreeding is that they are sensitive to missing pedigree information (Haig & Ballou, 2002). However, algorithms exist to accommodate incomplete pedigree data in the calculation of inbreeding coefficients (van Raden, 1992). Ewing et al. (2008) give an example of the application of this method to the calculation of inbreeding coefficients in a reintroduced population of Mauritius kestrels (*Falco punctatus*) and show how inbreeding would be underestimated by 10–20% if the issue of incomplete pedigree data were ignored.

Sometimes, it may be known that a population has been totally isolated and of a certain population size before the time when the first pedigree records are available (e.g. Jamieson et al., 2003). In that case, this additional background inbreeding F_{bg} (which is of the random mating type, F_{ST}) can be calculated using the equation given for F_{ST} in Box 11.1, and added to the pedigree inbreeding: $F_{overall} = F_{ped} + F_{bg} - F_{ped}F_{bg}$ (Crow, 1980).

Not surprisingly, given the different aspects of inbreeding they measure and the different reference populations used, inbreeding calculated from pedigrees (which measures F_{IT}) can be very different from inbreeding as non-random mating (F_{IS}), particularly in small populations where matings among close relatives will occur even under random mating. This scenario, small population size and random mating, will be a common situation in reintroduction biology, where population size will be low and the number of potential mates thus often limited. This will lead to high estimates of average inbreeding from pedigrees (F_{IT}), yet low estimates of F_{IS}. Indeed, a number of studies of reintroduced populations have found such a pattern (e.g. Biebach & Keller, 2009; Jamieson et al., 2009).

Box 11.3 **Effective population size**

Genetic drift and inbreeding (F_{ST}) both occur at a rate proportional to $\frac{1}{(2N)}$ (see Box 11.1). However, this only holds true if the population represented by N follows a number of rather restrictive assumptions: no mutations, no migration, no selection, equal number of females and males, Poisson distribution of offspring number, constant population size, etc. (Allendorf & Luikart, 2007, p. 147–170). No natural population is likely to meet all these assumptions of an idealized population. How, then, can we predict the rate of genetic drift or the increase in inbreeding (F_{ST}) in a real population? The answer is the *effective population size*, a concept central to population genetics (Charlesworth, 2009). The basic idea behind the concept of the effective population size is that we retain the simple equations relating the increase in inbreeding or the amount of genetic drift to the size of an idealized population, but we replace the idealized population size (N) with the effective population size (N_e). The effective population size is then defined as the size of an idealized population that would give rise to the same variance in allele frequencies and the same rate of inbreeding as the population under study (Allendorf & Luikart, 2007, p. 148). In the words of Sewall Wright, the inventor of the concept, the effective population size is '. . . whatever must be substituted for N in the basic formula . . .' to describe the actual change in heterozygosity or inbreeding (Wright, 1969, p. 211).

There are different ways of defining the effective population size, depending on the genetic process one wants to describe (Crow & Kimura, 1970; Lande & Barrowclough, 1987; Caballero, 1994; Lacy, 1995). The *variance effective population size* (N_{ev}) describes the expected genetic drift in allele frequencies, while the *inbreeding effective population size* (N_{ei}) describes the rate of inbreeding per generation.

In a population of constant size, the two measures of the effective population size will be the same. However, in a growing or declining population, the two effective population sizes will not be the same (Crow & Kimura, 1970, pp. 361–364; Caballero, 1994). This stems from the fact that the amount of genetic drift is a function of the size of the progeny generation, while the rate of inbreeding is a function of the number of individuals in the grandparental generation (Crow & Kimura, 1970,

p. 361). Growing populations are the goal of any reintroduction project, and in growing populations the progeny generations are generally larger than their corresponding grandparental generations. Thus, N_{ev} is expected to be larger than N_{ei}. In other words, in growing populations the rate of inbreeding (F_{ST}) is expected to be larger (because N_{ei} is smaller) than the rate of genetic drift. Indeed, in a study of a reintroduced population of the Mauritius kestrel, Ewing *et al.* (2008) found that N_{ev} was nearly twice as large as N_{ei}, suggesting that the rate of inbreeding was nearly twice the rate of loss of genetic variation due to drift. Thus, although we can ignore this distinction in many fields of biology, in reintroduction biology we need to keep in mind that $N_{ev} \neq N_{ei}$.

As was the case with inbreeding (Box 11.2), this distinction is important because different techniques lend themselves more readily to the estimation of the different effective population sizes. While pedigrees allow estimation of either N_{ei} or N_{ev} (Templeton, 2006; see Ewing *et al.*, 2008, for an example), molecular markers can be used to estimate N_{ev} in various ways, either from a single sample or from temporally spaced samples (see Wang, 2005, for a review). Obtaining estimates of N_{ei} from molecular data, on the other hand, is more difficult. Thus, attention needs to be paid to the definition of N_e when comparing estimates of effective population sizes with observed changes in inbreeding or allele frequency variation (e.g. Biebach & Keller, 2010).

Genetic drift

In addition to inbreeding depression, loss of genetic variation is a major concern in reintroduced populations (Allendorf & Luikart, 2007, pp. 117–142). Random genetic drift is the main process by which managed populations lose genetic variation (Lacy, 1987), with adaptation to captivity adding further to this loss (Frankham, 2008). Genetic drift refers to random changes in allele frequencies from generation to generation, which occur in finite populations due to the random sampling of parental alleles during Mendelian inheritance (Frankham *et al.*, 2002, pp.178–186). This random sampling of parental alleles is akin to a coin toss: each time an offspring is produced, a parental allele has a probability of $1/2$ to be passed on to this offspring. This creates random variation in how often a particular allele is passed on to the next generation, and this in turn leads to random variation in allele frequencies from generation to generation. This random variation is greater, the smaller the population (Figure 11.1). Thus, small populations experience more genetic drift than large populations. In the long run, these random changes from one generation to the next will lead to some alleles being lost completely, and hence to a loss of overall genetic variation. Since the changes in allele frequencies are random, different populations will lose different alleles. As a consequence, genetic drift not only leads to the loss of genetic variation within populations but also increases the amount of genetic divergence among populations (Gaggiotti & Couvet, 2004). Thus, reintroduction projects that establish several reintroduced populations from a single source population will shift genetic variation from within a (source) population to variation among (descendant) populations (O'Ryan *et al.*, 1998; Biebach & Keller, 2009).

While genetic drift and inbreeding are closely related concepts that can both be measured by inbreeding coefficients (Box 11.1), they are not identical. Extreme situations can illustrate this. For example, in an infinitely large population in which relatives prefer each other as mates (i.e. with non-random mating), it is possible to have inbreeding but no genetic drift (Kristensen & Sørensen, 2005). However, the opposite (genetic drift without inbreeding) is impossible because inbreeding is a direct consequence of the finite population size that causes genetic drift. Only under random mating do genetic drift and inbreeding go hand in hand (see Box 11.1), because under random mating both the increase in inbreeding and the loss in heterozygosity due to genetic drift are directly proportional to the reciprocal of twice the effective population size ($\frac{1}{2N_e}$; Allendorf & Luikart, 2007, p. 124; see Box 11.3 for an explanation of effective population size). A second difference is that inbreeding without inbreeding depression reduces heterozygosity but does not change allele frequencies (Hedrick, 2000, p. 181). Genetic drift, in contrast, changes allele frequencies, albeit randomly. Thus, genetic drift, and not inbreeding, is the cause of loss of genetic variation in small populations (Templeton, 2006, pp. 108–109). If these distinctions between inbreeding and genetic drift seem confusing, you are in good company: Sewall Wright and R.A. Fisher, two of the founders of mathematical population genetics, never agreed on the meaning of inbreeding and random genetic drift (Crow, 2010).

Despite historical precedence for misunderstandings and disagreements, distinguishing between genetic drift and inbreeding is essential to an understanding of inbreeding effects in small populations (see also Box 11.1). For

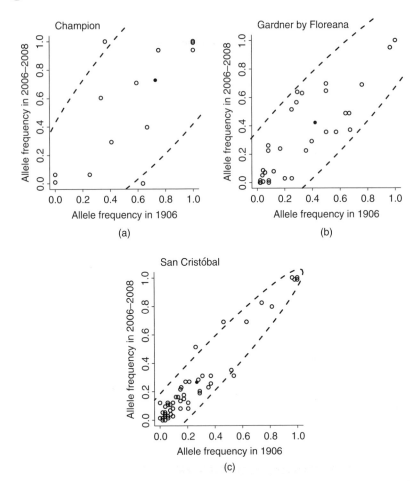

Figure 11.1 Small populations experience more genetic drift than large populations. Each panel compares the allele frequencies at 16 microsatellite loci (open circles) in mockingbird populations in Galápagos (*Mimus trifasciatus*, Figure 11.2, and *Mimus melanotis*) in 1906 with the allele frequencies in the same populations approximately one hundred years later. As expected from theory and as witnessed by the width of the 95% normal probability contour ellipses, the smallest population (Champion) shows the greatest differences in allele frequencies, followed by the medium-sized population on Gardner by Floreana and the large population on San Cristóbal. Note that the average allele frequency across all loci (black dot) did not change between 1906 and 2006–2008 in either case, as expected from a random process. (Data from Hoeck *et al.*, 2010.)

Figure 11.2   **Floreana mockingbird, Galápagos (photo: Paquita Hoeck).**

example, as we will see below, these differences matter for purging: the selective removal of alleles causing inbreeding depression is more likely when inbreeding occurs through non-random mating with little genetic drift than when inbreeding goes hand in hand with genetic drift. A final important caution about equating inbreeding (F_{ST} in Box 11.1) and genetic drift: the prediction that both are related to the effective population size (see Box 11.3) in the same way only applies to populations in equilibrium. If a population is not in equilibrium, e.g. because a population is growing, genetic drift is less pronounced than the increase in inbreeding. The opposite holds true in declining populations (Box 11.3). Immigration can also lead to deviations from equilibrium and hence genetic drift and inbreeding can become disassociated with immigration (e.g. Keller *et al.*, 2001).

Bottlenecked populations are expected to experience substantial genetic drift and inbreeding (Nei *et al.*, 1975). Because all reintroduced populations will have experienced one or several bottlenecks, reintroduced populations tend to show evidence of both inbreeding and genetic drift. For example, genetic drift led to the loss of genetic variation within and increased differentiation among reintroduced Alpine ibex (*Capra ibex ibex*) populations in Switzerland (Biebach & Keller, 2009). Over a hundred years after the beginning of this successful reintroduction project, the genetic structure of Alpine ibex was determined primarily by conservation actions of the past rather than the biology of ibex. Furthermore, Alpine ibex populations had accumulated average levels of inbreeding close to those expected from one generation of half-sib mating (Biebach & Keller, 2010). As expected from theory, variations in founder group size and in early population growth rates explained the substantial variation in inbreeding levels among ibex populations (Biebach & Keller, 2010).

Causes and estimates of inbreeding depression

Why does inbreeding reduce vigour and fitness of inbred individuals? The answer lies in the fact that relatives are more likely to share the same alleles at a certain locus than non-relatives. Thus, parents that inbreed will produce offspring that are more homozygous (Crow, 1980). This increased homozygosity means that deleterious recessive alleles are more often expressed. Deleterious recessive alleles are common in most populations and their increased expression is the main, albeit not only, cause of inbreeding depression (Charlesworth & Willis, 2009).

Inbreeding depression is measured by comparing trait values of inbred individuals to that of outbred individuals. This is commonly achieved by regressing (log-transformed) trait values on individual inbreeding coefficients (e.g. Keller & Waller, 2002; Charlesworth & Willis, 2009). The slope of this regression is a measure of the inbreeding load (also known as the number of lethal equivalents; see Keller & Waller (2002) for references and explanations), i.e. of the rate at which trait values decline with increasing inbreeding: high inbreeding loads indicate strong inbreeding depression, while low inbreeding loads indicate low inbreeding depression. Estimates of inbreeding loads can be used to compare the strength of inbreeding depression across life history stages, traits, environments or populations (e.g. Keller, 1998; Keller *et al.*, 2002; Jamieson *et al.*, 2007; Jamieson & Lacy, this volume, Chapter 13). Often the focus is on fitness traits, as exemplified by the terminology ('inbreeding load' suggests a fitness effect). However, the concept of inbreeding load extends to all traits.

Obtaining reliable estimates of the inbreeding load is difficult, as Darwin already noted in his work on domesticated animals and plants: 'The evil results from close [inbreeding] are difficult to detect, for they accumulate slowly...' (Darwin, 1868, vol. II, p. 114). For one, detecting statistically significant levels of inbreeding depression requires large sample sizes, possibly several hundred individuals, and a reasonable sample of highly inbred individuals (Kalinowski & Hedrick, 1999; Keller & Waller, 2002). Importantly for applications in reintroduction biology, the sample size requirements are larger the smaller the number of founder individuals (Kalinowski & Hedrick, 1999). Consequently, many datasets from reintroduced species will lack the statistical power necessary to confirm statistically even substantial inbreeding loads. This does not mean that it is impossible to obtain statistically significant estimates of inbreeding depression in reintroduced populations: some studies have successfully done so (e.g. Jamieson *et al.*, 2003, 2007; Swinnerton *et al.*, 2004; Fredrickson *et al.*, 2007; Brekke *et al.*, 2010). A rather sobering example comes from some of our work on free-living song sparrows (*Melospiza melodia*) on Mandarte Island in Canada. An analysis of 21 years of data revealed that lifetime reproductive success of males declined with inbreeding (Keller, 1998), but the effect was not statistically significant. Although the magnitude of the effect was large (a 10% increase in inbreeding decreased male lifetime reproductive success by 25%), statistical significance was only revealed when we included 28 years of data in a later reanalysis (Keller *et al.*, 2006).

This example reiterates the fact that statistical and biological significance are not the same. If one finds no statistically significant evidence for inbreeding depression, this may indicate a lack of statistical power rather than an absence of inbreeding depression. In other words, a non-significant effect does not imply that the effect is weak or absent. Conversely, a significant effect does not imply that the effect is strong. This statistical issue needs to be considered when planning a reintroduction or when studying levels of inbreeding depression in a reintroduced population, especially because many reintroduced populations are not ideally suited for the estimation of inbreeding depression. Given these issues of statistical power when quantifying inbreeding depression, and given the overwhelming evidence for the existence of inbreeding depression in general, it may be best to assume that reintroduced species will exhibit inbreeding depression whether or not it has actually been demonstrated in a particular case (Jamieson & Lacy, this volume, Chapter 13).

Variation in inbreeding depression

While inbreeding depression in fitness traits is nearly ubiquitous on average, it is highly variable among populations and species (e.g. Ralls *et al.*, 1988; Keller *et al.*, 2002; Laws & Jamieson, 2011), among traits (e.g. Lacy *et al.*, 1996; Keller *et al.*, 2006), life-history stages (e.g. Husband & Schemske, 1996), environmental conditions (e.g. Marr *et al.*, 2006; Szulkin & Sheldon, 2007) and even among founder lineages within the same population (e.g. Lacy *et al.*, 1996). As noted above, variation in inbreeding depression is as much a 'rule' as its presence on average. This is not surprising. Since inbreeding depression is ultimately caused by mutations that create deleterious recessive alleles, we would expect populations, traits and individuals to differ simply by chance in the number of deleterious recessive alleles they carry. Indeed, studies that decomposed total inbreeding depression into components due to individual founders have generally observed founder-specific variation in inbreeding depression (e.g. Waller, 1984; Lacy *et al.*, 1996; Willis, 1999; Fox, 2005; Gulisija *et al.*, 2006), suggesting that the deleterious recessive alleles are distributed unevenly among individuals. Such among-founder variations will also lead to among-population variations in the magnitude of inbreeding depression. Selection against deleterious recessive alleles (purging, see below) will cause additional among-population variations in the inbreeding load.

There is a large body of evidence that inbreeding depression varies with environmental conditions (e.g. Keller *et al.*, 2002; Marr *et al.*, 2006; Szulkin & Sheldon, 2007; Waller *et al.*, 2008; reviewed in Armbruster & Reed, 2005). Environmental stress may exacerbate inbreeding depression if some alleles are only deleterious under certain environmental conditions (e.g. Vermeulen *et al.*, 2008) if the genes that affect stress tolerance are also affected by inbreeding (Kristensen *et al.*, 2005), if selection against deleterious mutations depends on environmental conditions (Kristensen *et al.*, 2008) or if the degree of phenotypic plasticity depends on inbreeding (Lerner, 1954). There seems to be a general pattern across a wide range of taxa that increasing environmental stress increases inbreeding depression (Armbruster & Reed, 2005; Marr *et al.*, 2006). Despite this general trend, the relationship between environmental stress and inbreeding depression is very variable among populations or lineages (Armbruster & Reed, 2005) and among traits within a single population (Marr *et al.*, 2006). In fact, one quarter of the studies reviewed by Armbruster & Reed (2005) found a pattern opposite to the general expectation, i.e. a decrease in inbreeding depression with increasing environmental severity. A possible explanation for these findings is the phenotypic variation hypothesis proposed by Waller *et al.* (2008). Under this hypothesis, levels of phenotypic variation set an upper boundary to the amount of selection that can occur. Thus, environments that increase variation among inbred and outbred individuals increase inbreeding depression, while environments that decrease this variation reduce inbreeding depression. It will be fascinating to see how often the phenotypic variation hypothesis can explain associations between environmental conditions and the magnitude of inbreeding depression in the wild. Whatever the causes, environmental variation in the magnitude of inbreeding depression remains an important topic for conservation biologists: if inbreeding depression is particularly pronounced when environmental conditions are harsh, inbreeding depression may have a major impact on population persistence times (Bijlsma *et al.*, 2000).

Finally, inbreeding depression varies substantially among traits. While much of that variation is not well understood and probably due to stochastic variation in the load of deleterious mutations, there is a general pattern that fitness-related traits show much higher inbreeding depression than other traits, such as morphological (DeRose & Roff, 1999) or cognitive characters (Nepoux *et al.*, 2010). This seems to be a consequence of the proportionally higher levels of non-additive genetic variance in fitness traits (Roff & Emerson, 2006). Another frequently observed trend is that inbreeding depression in fecundity is stronger than in survival (e.g. Lacy *et al.*, 1996, and references therein). This pattern would be expected if deaths caused by inbreeding occur most often when many genes are first expressed, i.e. in the prezygotic or embryonic stages. However, it is unclear whether this pattern also holds in the wild. At least one study of a wild bird species found no such pattern (Keller, 1998).

In summary, inbreeding depression is ubiquitous on average, as is variation in its magnitude among individuals, populations, environments and traits. Some of this variation shows consistent patterns but much of it appears to be a consequence of chance variation in the load of deleterious recessive alleles.

Purging and fixation

Inbreeding depression is a form of selection, with the magnitude of homozygosity being the trait under selection: more inbred, and hence more homozygous, individuals have lower fitness than less inbred individuals (for a discussion of how to equate inbreeding to estimates of heterozygosity from molecular marker data, see Groombridge *et al.*, this volume, Chapter 12). This lower fitness of inbred individuals is expected to lead to a reduction in the frequency of the deleterious recessive alleles that cause inbreeding depression (Allendorf & Luikart, 2007, pp. 323–329). As a consequence, the inbreeding load is expected to be smaller in populations that have experienced periods of inbreeding in their recent evolutionary history. This process has become known as 'purging' (Frankham *et al.*, 2002, p. 295).

Inspired by successful purging in agricultural and laboratory populations (Wright, 1922a; Comstock & Winters, 1944), early conservation geneticists asked whether intentional inbreeding could be used to purge the inbreeding load in an endangered species to such a degree that further inbreeding would no longer be a major concern (Templeton & Read, 1984). Empirical and theoretical research has since answered this question in the negative. Although there is no doubt that purging can occur (Byers & Waller, 1999; Crnokrak & Barrett, 2002), as a strategy to reduce the inbreeding load purging is unlikely to be effective and predictable enough. Consequently, purging is not advisable as a management tool (e.g. Frankham *et al.*, 2001; Miller & Hedrick, 2001; Leberg & Firmin, 2008). In fact, it may make matters worse.

Why is purging unlikely to be effective enough to make it a management tool in conservation? First, population genetic theory predicts that drift can counteract selection. If drift is sufficiently strong, random changes in allele frequency may be larger than those caused by selection, thus rendering selection inefficient. The efficiency of selection is therefore a balance between the harmfulness of an allele and the degree of genetic drift (Wright, 1931; Hedrick, 2000, pp. 328–329). In other words, selection on a detrimental allele is not only a function of its harmfulness but also of the effective population size, which quantifies the amount of drift (see Box 11.3). The bigger a population, the less harmful an allele needs to be in order to be removed efficiently by selection. This leads to purging being more effective in populations with large effective population sizes than in those with low effective population sizes. This pattern has been confirmed experimentally (see references in Leberg & Firmin, 2008; but see Mikkelsen *et al.*, 2010) and is sometimes referred to as purging under 'slow' and 'fast' inbreeding because a large effective population size leads to 'slow' inbreeding (F_{ST}, see Box 11.1) and a low effective population size to 'fast' inbreeding. Reintroduced populations and other populations of conservation concern with their typically small effective population sizes (e.g. Ewing *et al.*, 2008) are those in which purging is least likely to be effective (Leberg & Firmin, 2008).

Only very detrimental alleles will be under substantial selection in such populations. Since it has been estimated that a substantial proportion of the inbreeding load is due to mildly deleterious alleles (Charlesworth & Willis, 2009), this implies that much of the inbreeding load will not be removed by purging.

In line with theory, empirical studies in zoo populations have found little evidence for purging (e.g. Ballou, 1997; Boakes *et al.*, 2007), as have a meta-analysis of plant studies (Byers & Waller, 1999) and some experimental studies (e.g. Lacy & Ballou, 1998; Leberg & Firmin, 2008; Mikkelsen *et al.*, 2010). These findings are somewhat in contrast to the successful purging that has been shown to occur in a number of other experimental populations (e.g. Reed & Bryant, 2001; Crnokrak & Barrett, 2002; Swindell & Bouzat, 2006). The apparent contradiction can in part be explained by the fact that theory predicts purging to be more effective when inbreeding occurs due to non-random mating (F_{IS}) rather than under random mating (F_{ST}, see Box 11.1; Wang, 2000; Glémin, 2003, 2007). The reasons for this pattern are complex (Glémin, 2003) and have to do with the fact that with random mating variation in homozygosity is stochastic while with non-random mating an excess of homozygotes is produced systematically. In accordance with this prediction from theory, those purging studies reviewed in Byers and Waller (1999) that gave the clearest evidence of purging were those in which inbreeding had been created by non-random mating. In experiments and in agricultural applications, non-random mating is common (Wang, 2000), but except for the occasional avoidance of very close inbreeding (e.g. Ewing *et al.*, 2008) most populations of conservation concern appear to breed more or less randomly ($F_{IS} \approx 0$; e.g. see Biebach & Keller, 2009; Jamieson *et al.*, 2009).

Second, like inbreeding depression, purging may depend on environmental conditions (Bijlsma *et al.*, 1999, 2000; Reed & Bryant, 2001; Armbruster & Reed, 2005). Since some alleles are only deleterious under certain environmental conditions (e.g. Vermeulen *et al.*, 2008), they will only be purged by selection when those particular conditions arise (Bijlsma *et al.*, 1999). If these environmental conditions do not arise, the alleles are temporarily 'hidden' from selection and, thus, remain in the population. Alleles conferring susceptibility to novel or rare diseases may be among the most important ones that fall into this category (Ross-Gillespie *et al.*, 2007). When the diseases are not present, alleles conferring susceptibility to them may accumulate over time without negatively affecting the individuals that carry them (e.g. van

Oosterhout et al., 2007). However, when a disease is ultimately encountered, the alleles will be expressed and will reduce the fitness of their carriers (also see Ewen et al., this volume, Chapter 9). Thus, a population may appear to be purged when the disease is not present, yet experience significant inbreeding depression when the pathogen is encountered. This effect has been documented in captive populations of naked mole-rats, where substantial inbreeding depression occurred during a novel coronavirus outbreak despite apparent previous purging (Ross-Gillespie et al., 2007).

Both of these factors combine to make purging through intentional inbreeding not a viable management tool. Furthermore, the results suggest that a history of inbreeding alone (e.g. because a species is an island endemic) does not necessarily imply reduced inbreeding depression. Experimental results in white-footed mice (Lacy et al., 1996; Lacy & Ballou, 1998), naked mole-rats (Ross-Gillespie et al., 2007) and a number of other species have shown this clearly.

However, why would purging through intentional inbreeding make matters worse? The answer is fixation, i.e. the loss of all but one allele at a locus (Wang et al., 1999). Fixation leads to complete homozygosity in a population at a particular locus. Intentional inbreeding with its associated reduction of the effective population size increases the likelihood that alleles will become fixed through genetic drift (e.g. Wang et al., 1999; Hedrick, 2000, p. 258). Importantly, because selection is less effective in small populations, even detrimental alleles can become fixed (e.g. Hedrick, 1994), thus reducing the average fitness of a population. Fixation of detrimental alleles is of great conservation concern because it reduces fitness and thus, potentially, population viability (Leberg & Firmin, 2008), and no management strategy other than genetic rescue through the introduction of genetic variation from the outside can revert such fixation (see Groombridge et al., this volume, Chapter 12). Perhaps the best-known example of near-fixation of deleterious traits followed by a management strategy that employed genetic rescue is the case of the Florida panther (*Puma concolor coryi*; see Johnson et al., 2010b). As exemplified by the Florida panther, fixation can be a major threat to the viability of small populations but probably only at very low population sizes (Lande, 1998; Whitlock et al., 2003). Avoiding fixation is an additional motivation for achieving a sizeable population within a few generations in reintroduction programmes.

Furthermore, fixation is another reason why estimates of inbreeding depression within a population may give misleading estimates of the negative effects of inbreeding (Bataillon & Kirkpatrick, 2000; Keller & Waller, 2002). When a population becomes fixed for a detrimental allele, inbreeding will not produce more homozygous offspring than outbreeding, since all individuals in the population are already homozygous for this allele. Thus, inbreeding depression will be apparently absent, even though the population will experience a reduced mean fitness due to the fixation of this allele (e.g. Fredrickson *et al.*, 2007). Consequently, low inbreeding depression can either indicate a truly low inbreeding load or it can indicate that detrimental alleles have become fixed. The two scenarios have fundamentally different management implications (management action is required in the latter but not the first case), yet they cannot be distinguished unless one has data on the fitness of crosses between populations (Glémin *et al.*, 2003). Thus, in addition to the statistical difficulties involved in confirming significant inbreeding depression, fixation provides another reason why reintroduction strategies should minimize inbreeding, even in the absence of apparent inbreeding depression in the population under study. In fact, in the context of fixation, the finding of significant inbreeding depression is a better sign than a lack of inbreeding depression: it is evidence that a population has not become fixed, and hence still retains genetic variation.

Inbreeding effects on population dynamics

Our discussion of inbreeding depression has so far been limited to traits of individuals, such as individual fitness components like survival and fecundity. However, inbreeding depression in individual fitness components is only of importance to conservation efforts if reductions in individual fitness translate into reduced population growth rates or reduced carrying capacities. As conservation biologists we are primarily interested in preserving populations, not individuals, and hence we are above all interested in effects at the population level (Keller *et al.*, 2007).

Inbreeding can potentially reduce population growth rates and increase extinction risk. This has been demonstrated in experimental fish, plant and insect populations (Leberg, 1993; Newman & Pilson, 1997; Bijlsma *et al.*, 2000) and in a wild butterfly meta-population (Saccheri *et al.*, 1998). Increases in population growth rates following experimentally restored immigration into inbred populations (e.g. Vilà *et al.*, 2003; Hogg *et al.*, 2006) also provide evidence for inbreeding effects at the population level, although reversal of fixation would lead to a similar pattern.

However, many small and inbred populations apparently do not experience reduced population growth rates (e.g. Broders *et al.*, 1999; Jamieson *et al.*, 2007), at least not in the short term (see Figure 13.1 in Jamieson & Lacy, this volume, Chapter 13, for an example where a model predicts negative population growth rates in the long term). Why would inbreeding depression in individual fitness not always translate into a reduced population growth rate? The strongest argument is that of soft selection (Wallace, 1975; Saccheri & Hanski, 2006; Keller *et al.*, 2007), i.e. that the probability of survival of an individual depends on the presence or absence of other individuals. For example, a certain proportion of juveniles might die (or emigrate) in a territorial species simply because all territories are occupied by stronger competitors. Had there been fewer strong competitors, all juveniles might have acquired territories. In other words, selection was 'soft' because it depended on the presence of other individuals. If we translate this example into the context of inbreeding depression, soft selection implies that the least inbred individuals get the breeding territories. In a large population these might be completely outbred individuals. In a very small population, however, the least inbred individuals might be appreciably inbred, albeit less so than anyone else. In the absence of any more outbred competitors, these inbred individuals may produce enough offspring so that inbreeding depression in individual fitness has negligible effects on population size (Wallace, 1975). Thus, if soft selection predominates in natural populations, inbreeding may reduce population growth rates less than it reduces individual fitness (Keller *et al.*, 2007).

Soft selection is expected to be prominent in populations that exhibit high levels of density and frequency dependence (for more details see Keller *et al.*, 2007). The degree to which inbreeding depression will affect the growth of reintroduced populations is thus likely to depend on the degree of density dependence, which is probably low in many reintroduced populations (e.g. Grøtan *et al.*, 2008; Ewen *et al.*, 2011; Armstrong *et al.*, 2005), at least at the beginning. Later, as the population reaches carrying capacity this may change and soft selection may start to dominate. Thus, the potential impact of inbreeding on the population growth rate of reintroduced species may change over time and also with environmental conditions (see Keller *et al.*, 2007, for details).

Much of this is, however, still conjecture because we lack sufficiently detailed data. Identifying and quantifying the conditions under which population growth rates are depressed by inbreeding remains one of the major challenges for conservation genetics today.

Loss of genetic variation

We have seen above that inbreeding and genetic drift are related but not identical processes. Genetic drift leads to the loss of genetic variation, which is the raw material of evolutionary change. If a population is facing a new environment, such as introduced predators, disease or competitors, or a changing abiotic environment, such as increasing temperatures, evolutionary adaptation can only occur if the species harbours sufficient genetic variation for selection to operate on (Hedrick, 2001). Hence, if populations lose genetic variation, they may lose their evolutionary potential, i.e. their ability to adapt quickly to changes in their environment (Willi *et al.*, 2006; Allendorf & Luikart, 2007, p. 355). For example, Frankham *et al.* (1999) have shown that Drosophila lines which had lost a substantial portion of their heterozygosity were less able to adapt to changing environmental conditions (in terms of increasing salt stress in their environment) than control lines which had not lost genetic variation.

Since all populations will experience environmental change in their future, reintroduction biologists want to ensure that reintroduced populations retain high evolutionary potential (Groombridge *et al.*, this volume, Chapter 12, and Jamieson & Lacy, this volume, Chapter 13). Environment permitting, this will require effective population sizes in the low thousands in the long term (Willi *et al.*, 2006; Jamieson & Lacy, Chapter 13), considerably more than what would be necessary if avoiding inbreeding was the only concern (Lande, 1995; Whitlock, 2000; Willi *et al.*, 2006). Thus, in the long term avoiding the loss of evolutionary potential is probably the greater challenge for reintroduction biologists than avoiding too much inbreeding (see also Groombridge *et al.*, Chapter 12).

Avoiding the loss of evolutionary potential will often require a way of monitoring the loss of genetic variation in reintroduced populations. This poses several problems. One concerns the kind of genetic variation that should be monitored: should we measure neutral or adaptive variation (Hedrick, 2001)? As reintroduction biologists we are primarily interested in preserving adaptive variation and variation that may become adaptive in the future. Ideally, therefore, we would be able to measure adaptive variation but this is often difficult. Quantitative genetic methods are well suited to measure adaptive genetic variation (Lynch, 1996; Frankham *et al.*, 2002, pp. 97–125; Allendorf & Luikart, 2007, pp. 257–284), but they require a lot of data, probably often more than what is available in many reintroduced populations.

Adaptive genetic variation can also be monitored using molecular techniques, but this is conceptually and technically difficult. New molecular methods are changing this (see Groombridge *et al.*, Chapter 12), but at the time of writing, adaptive variation could rarely be assessed at more than a handful of loci in reintroduced species. Thus, very often all we can do is to measure neutral genetic variation and in the following we focus on this type of genetic variation. Care needs to be taken, however, when neutral variation is used as an indicator of adaptive variation because the relationship between adaptive and neutral variation is complex (Hedrick, 2001).

Which measures should we use to quantify genetic variation? As outlined in Box 13.1 of Jamieson & Lacy (this volume, Chapter 13), different measures of genetic variation are used in conservation genetics. When using molecular markers, two metrics are often employed: heterozygosity (H_e) and number of alleles (A). The two measures differ in that H_e predicts short-term selection response, while A predicts long-term evolutionary potential (James, 1970; Allendorf & Luikart, 2007, p. 111). Furthermore, loss of H_e depends only on the effective population size (H_e is lost at a rate of $\frac{1}{2N_e}$ regardless of the absolute level of heterozygosity), while the loss of A depends additionally on the number and frequencies of the alleles in the population (Kimura, 1955; Allendorf & Luikart, 2007, pp.123–129). This makes A more sensitive to genetic drift than H_e if rare alleles are present in a population. Conversely, genetic drift has less of an effect on A than on H_e if rare alleles have already been lost, e.g. due to a previous bottleneck. In the context of reintroduced populations that often experience serial bottlenecks, this means that genetic drift is expected to have different effects on H_e and A in the first bottlenecks than in subsequent ones. Indeed, as expected from theory, H_e declined by a similar amount with each bottleneck that occurred in the reintroduction of Alpine ibex, while the loss of A was higher in the first than in subsequent bottlenecks (Biebach & Keller, 2009). Taken together, these considerations make it clear that heterozygosity and number of alleles are complementary measures and that genetic variation should be quantified using both (Allendorf & Luikart, 2007, p. 111).

Summary

Reintroduced populations experience both the effects of inbreeding and genetic drift and reintroduction planning therefore needs to account for both. The main challenge that reintroduction programmes face is to achieve high enough effective population sizes so as to minimize inbreeding and the loss of genetic variation necessary for future adaptive evolution. In the long run, the latter will be likely to require effective population sizes in the low thousands.

This chapter provided a summary of the theory underlying some of the thinking on inbreeding and genetic drift in conservation genetics. This body of theory combined with empirical verifications provides a basis for management decisions, which are discussed in more detail in Groombridge *et al.* (this volume, Chapter 12) and Jamieson & Lacy (this volume, Chapter 13).

Acknowledgements

The authors thank Philipp Becker, John Ewen, Jim Groombridge, Ian Jamieson, Bob Lacy, Franziska Lörcher, Pirmin Nietlisbach, Erik Postma, Don Waller and Peter Wandeler for helpful discussions or comments on earlier drafts of the manuscript, Barbara Oberholzer for help with the references and the Swiss National Science Foundation, the Natural Environmental Research Council (UK), the Carnegie Trust (Scotland), the Basler Stiftung für Biologische Forschung (Switzerland), the Swiss Federal Office for the Environment and the Forschungskredit of the University of Zurich for supporting their work on various species that have been, or will be, reintroduced.

References

Allendorf, F.W. & Luikart, G. (2007) *Conservation and the Genetics of Populations.* Blackwell, Malden, Massachusetts, USA.

Armbruster, P. & Reed, D.H. (2005) Inbreeding depression in benign and stressful environments. *Heredity*, 95, 235–242.

Armstrong, D.P., Davidson, R.S., Perrott, J.K. *et al.* (2005) Density-dependent population growth in a reintroduced population of North Island saddlebacks. *Journal of Animal Ecology*, 74, 160–170.

Ballou, J.D. (1997) Ancestral inbreeding only minimally affects inbreeding depression in mammalian populations. *Journal of Heredity*, 88, 169–178.

Bataillon, T. & Kirkpatrick, M. (2000) Inbreeding depression due to mildly deleterious mutations in finite populations: size does matter. *Genetical Research*, 75, 75–81.

Biebach, I. & Keller, L.F. (2009) A strong genetic footprint of the re-introduction history of Alpine ibex (*Capra ibex ibex*). *Molecular Ecology*, 18, 5046–5058.

Biebach, I. & Keller, L.F. (2010) Inbreeding in reintroduced populations: the effects of early reintroduction history and contemporary processes. *Conservation Genetics*, 11, 527–538.

Bijlsma, R., Bundgaard, J. & van Putten, W.F. (1999) Environmental dependence of inbreeding depression and purging in *Drosophila melanogaster*. *Journal of Evolutionary Biology*, 12, 1125–1137.

Bijlsma, R., Bundgaard, J. & Boerema, A.C. (2000) Does inbreeding affect the extinction risk of small populations? Predictions from *Drosophila*. *Journal of Evolutionary Biology*, 13, 502–514.

Boakes, E.H., Wang, J. & Amos, W. (2007) An investigation of inbreeding depression and purging in captive pedigreed populations. *Heredity*, 98, 172–182.

Brekke, P., Bennett, P.M., Wang, J.L. *et al.* (2010) Sensitive males: inbreeding depression in an endangered bird. *Proceedings of the Royal Society of London B*, 277, 3677–3684.

Broders, H.G., Mahoney, S.P., Montevecchi, W.A. *et al.* (1999) Population genetic structure and the effect of founder events on the genetic variability of moose, *Alces alces*, in Canada. *Molecular Ecology*, 8, 1309–1315.

Byers, D.L. & Waller, D.M. (1999) Do plant populations purge their genetic load? Effects of population size and mating history on inbreeding depression. *Annual Review of Ecology and Systematics*, 30, 479–513.

Caballero, A. (1994) Developments in the prediction of effective population size. *Heredity*, 73, 657–679.

Caro, T.M. & Laurenson, M.K. (1994) Ecological and genetic factors in conservation: a cautionary tale. *Science*, 263, 485–486.

Charlesworth, B. (2009) Effective population size and patterns of molecular evolution and variation. *Nature Reviews Genetics*, 10, 195–205.

Charlesworth, D. & Willis, J.H. (2009) Fundamental concepts in genetics – the genetics of inbreeding depression. *Nature Reviews Genetics*, 10, 783–796.

Charman, T.G., Sears, J., Green, R.E. *et al.* (2010) Conservation genetics, foraging distance and nest density of the scarce great yellow bumblebee (*Bombus distinguendus*). *Molecular Ecology*, 19, 2661–2674.

Comstock, R.E. & Winters, L.M. (1944) A comparison of the effects of inbreeding and selection on performance in swine. *Journal of Animal Science*, 3, 380–389.

Crnokrak, P. & Barrett, S.C.H. (2002) Perspective: purging the genetic load: a review of the experimental evidence. *Evolution*, 56, 2347–2358.

Crnokrak, P. & Roff, D.A. (1999) Inbreeding depression in the wild. *Heredity*, 83, 260–270.

Crow, J.F. (1980) The estimation of inbreeding from isonymy. *Human Biology*, 52, 1–14.

Crow, J.F. (2010) Wright and Fisher on inbreeding and random drift. *Genetics*, 184, 609–611.

Crow, J.F. & Kimura, M. (1970) *An Introduction to Population Genetics Theory*. Burgess Publishing Company, Minneapolis, Minnesota, USA.

Darwin, C. (1868) *The Variation of Animals and Plants under Domestication*. J. Murray and Co., London, UK.

Darwin, C. (1876) *The Effects of Cross and Self Fertilization in the Vegetable Kingdom*. J. Murray and Co, London, UK.

DeRose, M.A. & Roff, D.A. (1999) A comparison of inbreeding depression in life-history and morphological traits in animals. *Evolution*, 53, 1288–1292.

Ewen, J.G., Acevedo-Whitehouse, K., Alley, M. *et al.* (2011) Empirical consideration of parasites and health in reintroduction. In *Reintroduction Biology: integrating science and management*, eds J.G. Ewen, D.P. Armstrong, K.A. Parker & P.J. Seddon, Chapter 9. Wiley-Blackwell, Oxford, UK.

Ewing, S.R., Nager, R.G., Nicoll, M.A.C. *et al.* (2008) Inbreeding and loss of genetic variation in a reintroduced population of Mauritius kestrel. *Conservation Biology*, 22, 395–404.

Ewen, J.G., Thorogood, R. & Armstrong, D.P. (2011) Demographic consequences of adult sex ratio in a reintroduced hihi population. *Journal of Animal Ecology*, 80, 448–455.

Ficetola, G.F., Garner, T.W.J., Wang, J.L. *et al.* (2011) Rapid selection against inbreeding in a wild population of a rare frog. *Evolutionary Applications*, 4, 30–38.

Fox, C.W. (2005) Problems in measuring among-family variation in inbreeding depression. *American Journal of Botany*, 92, 1929–1932.

Frankham, R. (2008) Genetic adaptation to captivity in species conservation programs. *Molecular Ecology*, 17, 325–333.

Frankham, R., Ballou, J.D. & Briscoe, D.A. (2002) *Introduction to Conservation Genetics*. Cambridge University Press, Cambridge, UK.

Frankham, R., Gilligan, D.M., Morris, D. *et al.* (2001) Inbreeding and extinction: effects of purging. *Conservation Genetics*, 2, 279–285.

Frankham, R., Lees, K., Montgomery, M.E. *et al.* (1999) Do population size bottlenecks reduce evolutionary potential? *Animal Conservation*, 2, 255–260.

Fredrickson, R.J., Siminski, P., Woolf, M. *et al.* (2007) Genetic rescue and inbreeding depression in Mexican wolves. *Proceedings of the Royal Society of London B*, 274, 2365–2371.

Gaggiotti, O. E. (2003) Genetic threats to population persistence. *Annales Zoologici Fennici*, 40, 155–168.

Gaggiotti, O.E. & Couvet, D. (2004) Genetic structure in heterogeneous environments. In *Evolutionary Conservation Biology*, eds R. Ferrière, U. Dieckmann & D. Couvet, pp. 229–243. Cambridge University Press, Cambridge, UK.

Glémin, S. (2003) How are deleterious mutations purged? Drift versus nonrandom mating. *Evolution*, 57, 2678–2687.

Glémin, S. (2007) Mating systems and the efficacy of selection at the molecular level. *Genetics*, 177, 905–916.

Glémin, S., Ronfort, J. & Bataillon, T. (2003) Patterns of inbreeding depression and architecture of the load in subdivided populations. *Genetics*, 165, 2193–2212.

Groombridge, J.J., Raisin, C., Bristol, R. *et al.* (2011) Genetic consequences of reintroductions and insights from population history. In *Reintroduction Biology: integrating science and management*, eds J.G. Ewen, D.P. Armstrong, K.A. Parker & P.J. Seddon, Chapter 12. Wiley-Blackwell, Oxford, UK.

Grøtan, V., Sæther, B.E., Filli, F. *et al.* (2008) Effects of climate on population fluctuations of ibex. *Global Change Biology*, 14, 218–228.

Gulisija, D., Gianola, D., Weigel, K.A. *et al.* (2006) Between-founder heterogeneity in inbreeding depression for production in Jersey cows. *Livestock Science*, 104, 244–253.

Haig, S.M. & Ballou, J.D. (2002) Pedigree analyses of wild populations. In *Population Viability Analysis*, eds S.R. Beissinger & D.R. McCullough, pp. 388–405. The University of Chicago Press, Chigago, Illinois, USA.

Hedrick, P.W. (1994) Purging inbreeding depression and the probability of extinction – full-sib mating. *Heredity*, 73, 363–372.

Hedrick, P.W. (2000) *Genetics of Populations*. Jones and Bartlett, Boston, Massachusetts, USA.

Hedrick, P.W. (2001) Conservation genetics: where are we now? *Trends in Ecology & Evolution*, 16, 629–636.

Hedrick, P.W. & Kalinowski, S.T. (2000) Inbreeding depression in conservation biology. *Annual Review of Ecology and Systematics*, 31, 139–162.

Hedrick, P.W. & Miller, P.S. (1992) Conservation genetics – techniques and fundamentals. *Ecological Applications*, 2, 30–46.

Hoeck, P.E.A., Bollmer, J.L., Parker, P.G. *et al.* (2010) Differentiation with drift: a spatio-temporal genetic analysis of Galapagos mockingbird populations (*Mimus* spp.). *Philosophical Transactions of the Royal Society of London B*, 365, 1127–1138.

Hogg, J.T., Forbes, S.H., Steele, B.M. *et al.* (2006) Genetic rescue of an insular population of large mammals. *Proceedings of the Royal Society of London B*, 273, 1491–1499.

Husband, B.C. & Schemske, D.W. (1996) Evolution of the magnitude and timing of inbreeding depression in plants. *Evolution*, 50, 54–70.

IUCN (1998) *IUCN Guidelines for Reintroductions*. Prepared by the IUCN/SSC Reintroductions Specialist Group, IUCN, Gland, Switzerland, and Cambridge, UK.

Jacquard, A. (1975) Inbreeding: one word, several meanings. *Theoretical Population Biology*, 7, 338–363.

James, J.W. (1970) The founder effect and response to artificial selection. *Genetical Research*, 16: 241–250.

Jamieson, I.G. (2011) Founder effects, inbreeding, and loss of genetic diversity in four avian reintroduction programs. *Conservation Biology*, 25, 115–123.

Jamieson, I.G. & Lacy, R.C. (2011) Managing genetic issues in reintroduction biology. In *Reintroduction Biology: integrating science and management*, eds J.G. Ewen, D.P. Armstrong, K.A. Parker & P.J. Seddon, Chapter 13. Wiley-Blackwell, Oxford, UK.

Jamieson, I.G., Roy, M.S. & Lettink, M. (2003) Sex-specific consequences of recent inbreeding in an ancestrally inbred population of New Zealand takahe. *Conservation Biology*, 17, 708–716.

Jamieson, I.G., Taylor, S.S., Tracy, L.N. *et al.* (2009) Why some species of birds do not avoid inbreeding: insights from New Zealand robins and saddlebacks. *Behavioral Ecology*, 20, 575–584.

Jamieson, I.G., Tracy, L.N., Fletcher, D. *et al.* (2007) Moderate inbreeding depression in a reintroduced population of North Island robins. *Animal Conservation*, 10, 95–102.

Jiménez, J.A., Hughes, K.A., Alaks, G. *et al.* (1994) An experimental-study of inbreeding depression in a natural habitat. *Science*, 266, 271–273.

Johnson, W.E., Onorato, D.P., Roelke, M.E. *et al.* (2010a) Genetic restoration of the Florida panther. *Science*, 329, 1641–1645.

Johnson, J.A., Talbot, S.L., Sage, G.K. *et al.* (2010b) The use of genetics for the management of a recovering population: temporal assessment of migratory peregrine falcons in North America. *PlosOne*, 5 (11), e14042, doi:10.1371/journal.pone.0014042.

Joron, M. & Brakefield, P.M. (2003) Captivity masks inbreeding effects on male mating success in butterflies. *Nature*, 424, 191–194.

Kalinowski, S.T. & Hedrick, P.W. (1999) Detecting inbreeding depression is difficult in captive endangered species. *Animal Conservation*, 2, 131–136.

Keller, L.F. (1998) Inbreeding and its fitness effects in an insular population of song sparrows (*Melospiza melodia*). *Evolution*, 52, 240–250.

Keller, L.F. & Waller, D.M. (2002) Inbreeding effects in wild populations. *Trends in Ecology and Evolution*, 17, 230–241.

Keller, L.F., Biebach, I. & Hoeck, P.E.A. (2007) The need for a better understanding of inbreeding effects on population growth. *Animal Conservation*, 10, 286–287.

Keller, L.F., Grant, P.R., Grant, B.R. *et al.* (2002) Environmental conditions affect the magnitude of inbreeding depression in survival of Darwin's finches. *Evolution*, 56, 1229–1239.

Keller, L.F., Jeffery, K.J., Arcese, P. *et al.* (2001) Immigration and the ephemerality of a natural population bottleneck: evidence from molecular markers. *Proceedings of the Royal Society of London B*, 268, 1387–1394.

Keller, L.F., Marr, A.B. & Reid, J.M. (2006) The genetic consequences of small population size: inbreeding and loss of genetic variation. In *Conservation and*

Biology of Small Populations, eds J.N.M. Smith, L.F. Keller, A.B. Marr & P. Arcese, pp. 113–137. Oxford University Press, New York, USA.

Keller, L.F., Reid, J.M. & Arcese, P. (2008) Testing evolutionary models of senescence in a natural population: age and inbreeding effects on fitness components in song sparrows. *Proceedings of the Royal Society of London B*, 275, 597–604.

Kimura, M. (1955) Random genetic drift in multi-allelic locus. *Evolution*, 9, 419–435.

Knight, T. (1799) An account of some experiments on the fecundation of vegetables. *Philisophical Transactions of the Royal Society of London*, 89, 195–204.

Kölreuter, J.G. (1766) *Vorläufige Nachricht von einigen das Geschlecht der Pflanzen betreffenden Versuchen und Beobachtungen*. Leipzig, Germany.

Kristensen, T.N. & Sørensen, A.C. (2005) Inbreeding – lessons from animal breeding, evolutionary biology and conservation genetics. *Animal Science*, 80, 121–133.

Kristensen, T.N., Barker, J.S.F., Pedersen, K.S. *et al.* (2008) Extreme temperatures increase the deleterious consequences of inbreeding under laboratory and semi-natural conditions. *Proceedings of the Royal Society of London B*, 275, 2055–2061.

Kristensen, T.N., Sørensen, A.C., Sørensen, D. *et al.* (2005) A test of quantitative genetic theory using *Drosophila* – effects of inbreeding and rate of inbreeding on heritabilities and variance components. *Journal of Evolutionary Biology*, 18, 763–770.

Lacy, R.C. (1987) Loss of genetic diversity from managed populations: interacting effects of drift, mutation, immigration, selection, and population subdivision. *Conservation Biology*, 1, 143–158.

Lacy, R.C. (1995) Clarification of genetic terms and their use in the management of captive populations. *Zoo Biology*, 14, 565–577.

Lacy, R.C. & Ballou, J.D. (1998) Effectiveness of selection in reducing the genetic load in populations of *Peromyscus polionotus* during generations of inbreeding. *Evolution*, 52, 900–909.

Lacy, R.C., Alaks, G. & Walsh, A. (1996) Hierarchical analysis of inbreeding depression in *Peromyscus polionotus*. *Evolution*, 50, 2187–2200.

Laikre, L., Nilsson, T., Primmer, C.R. *et al.* (2009) Importance of genetics in the interpretation of favourable conservation status. *Conservation Biology*, 23, 1378–1381.

Lande, R. (1995) Mutation and conservation. *Conservation Biology*, 9, 782–791.

Lande, R. (1998) Risk of population extinction from fixation of deleterious and reverse mutations. *Genetica*, 103, 21–27.

Lande, R. & Barrowclough, G.F. (1987) Effective population size, genetic variation, and their use in population management. In *Viable Populations for Conservation*, ed. M.E. Soulé, pp. 87–123. Cambridge University Press, Cambridge, UK.

Laws, R.J. & Jamieson, I.G. (2011) Is lack of evidence of inbreeding depression in a threatened New Zealand robin indicative of reduced genetic load? *Animal Conservation*, 14, 47–55.

Leberg, P.L. (1993) Strategies for population reintroduction: effects of genetic variability on population growth and size. *Conservation Biology*, 7, 194–199.

Leberg, P.L. & Firmin, B.D. (2008) Role of inbreeding depression and purging in captive breeding and restoration programmes. *Molecular Ecology*, 17, 334–343.

Leimu, R. & Mutikainen, P. (2005) Population history, mating system, and fitness variation in a perennial herb with a fragmented distribution. *Conservation Biology*, 19, 349–356.

Lerner, I.M. (1954) *Genetic Homeostasis*. Oliver and Boyd, Edinburgh, UK.

Lynch, M. (1996) A quantitative-genetic perspective on conservation issues. In *Conservation Genetics: case histories from nature*, eds J.C. Avise & J.L. Hamrick, pp. 471–501. Chapman & Hall, New York, USA.

Malécot, G. (1948) *Les Mathématiques de l'Hérédité*. Masson & Cie, Paris, France.

Marr, A.B., Arcese, P., Hochachka, W.M. *et al.* (2006) Interactive effects of environmental stress and inbreeding on reproductive traits in a wild bird population. *Journal of Animal Ecology*, 75, 1406–1415.

Mikkelsen, K., Loeschcke, V. & Kristensen, T.N. (2010) Trait specific consequences of fast and slow inbreeding: lessons from captive populations of *Drosophila melanogaster*. *Conservation Genetics*, 11, 479–488.

Miller, P.S. & Hedrick, P.W. (2001) Purging of inbreeding depression and fitness decline in bottlenecked populations of *Drosophila melanogaster*. *Journal of Evolutionary Biology*, 14, 595–601.

Moorad, J.A. & Wade, M.J. (2005) A genetic interpretation of the variation in inbreeding depression. *Genetics*, 170, 1373–1384.

Nei, M., Maruyama, T. & Chakraborty, R. (1975) The bottleneck effect and genetic variability in populations. *Evolution*, 29, 1–10.

Nepoux, V., Haag, C.R. & Kawecki, T.J. (2010) Effects of inbreeding on aversive learning in *Drosophila*. *Journal of Evolutionary Biology*, 23, 2333–2345.

Newman, D. & Pilson, D. (1997) Increased probability of extinction due to decreased genetic effective population size: experimental populations of *Clarkia pulchella*. *Evolution*, 51, 354–362.

O'Ryan, C., Harley, E.H., Bruford, M.W. *et al.* (1998) Microsatellite analysis of genetic diversity in fragmented South African buffalo populations. *Animal Conservation*, 1, 85–94.

Ralls, K., Ballou, J.D. & Templeton, A. (1988) Estimates of lethal equivalents and the cost of inbreeding in mammals. *Conservation Biology*, 2, 185–193.

Ralls, K., Brugger, K. & Ballou, J. (1979) Inbreeding and juvenile mortality in small populations of ungulates. *Science*, 206, 1101–1103.

Reed, D.H. & Bryant, E.H. (2001) Fitness, genetic load and purging in experimental populations of the housefly. *Conservation Genetics*, 2, 57–62.

Robert, A. (2009) Captive breeding genetics and reintroduction success. *Biological Conservation*, 142, 2915–2922.

Roff, D.A. & Emerson, K. (2006) Epistasis and dominance: evidence for differential effects in life-history versus morphological traits. *Evolution*, 60, 1981–1990.

Ross-Gillespie, A., O'Riain, M.J. & Keller, L.F. (2007) Viral epizootic reveals inbreeding depression in a habitually inbreeding mammal. *Evolution*, 61, 2268–2273.

Saccheri, I. & Hanski, I. (2006) Natural selection and population dynamics. *Trends in Ecology and Evolution*, 21, 341–347.

Saccheri, I., Kuussaari, M., Kankare, M. *et al.* (1998) Inbreeding and extinction in a butterfly metapopulation. *Nature*, 392, 491–494.

Schwartz, M.K., Luikart, G. & Waples, R.S. (2007) Genetic monitoring as a promising tool for conservation and management. *Trends in Ecology and Evolution*, 22, 25–33.

Soulé, M.E. (1980) Thresholds for survival: maintaining fitness and evolutionary potential. In *Conservation Biology: an evolutionary–ecological perspective*, eds M.E. Soulé & B.A. Wilcox, pp. 151–169. Sinauer Associates, Inc., Sunderland, Massachusetts, USA.

Swindell, W.R. & Bouzat, J.L. (2006) Reduced inbreeding depression due to historical inbreeding in *Drosophila melanogaster*: evidence for purging. *Journal of Evolutionary Biology*, 19, 1257–1264.

Swinnerton, K.J., Groombridge, J.J., Jones, C.G. *et al.* (2004) Inbreeding depression and founder diversity among captive and free-living populations of the endangered pink pigeon *Columba mayeri*. *Animal Conservation*, 7, 353–364.

Szulkin, M. & Sheldon, B.C. (2007) The environmental dependence of inbreeding depression in a wild bird population. *PlosOne*, 2 (10), e1027, doi: 10.1371/journal.pone.0001027.

Szulkin, M., Bierne, N. & David, P. (2010) Heterozygosity-fitness correlations: a time for reappraisal. *Evolution*, 64, 1202–1217.

Templeton, A. R. (2006) *Population Genetics and Microevolutionary Theory*. Wiley-Liss, Hoboken, New Jersey, USA.

Templeton, A.R. & Read, B. (1984) Factors eliminating inbreeding depression in a captive herd of Speke's gazelle (*Gazella spekei*). *Zoo Biology*, 3, 177–199.

Thornhill, N.W. (1993) *The Natural History of Inbreeding and Outbreeding: theoretical and empirical perspectives*. University of Chicago Press, Chicago, Illinois, USA.

van Oosterhout, C., Smith, A.M., Hanfling, B. *et al.* (2007) The guppy as a conservation model: implications of parasitism and inbreeding for reintroduction success. *Conservation Biology*, 21, 1573–1583.

van Raden, P.M. (1992) Accounting for inbreeding and crossbreeding in genetic evaluation of large populations. *Journal of Dairy Science*, 75, 3136–3144.

Vermeulen, C.J., Bijlsma, R. & Loeschcke, V. (2008) QTL mapping of inbreeding-related cold sensitivity and conditional lethality in *Drosophila melanogaster*. *Journal of Evolutionary Biology*, 21, 1236–1244.

Vilà, C., Sundqvist, A., Flagstad, ø. *et al.* (2003) Rescue of a severely bottlenecked wolf (*Canis lupus*) population by a single immigrant. *Proceedings of the Royal Society of London Series B*, 270, 91–97.

Vitalis, R., Dawson, K. & Boursot, P. (2001) Interpretation of variation across marker loci as evidence of selection. *Genetics*, 158, 1811–1823.

Wallace, B. (1975) Hard and soft selection revisited. *Evolution*, 29, 465–473.

Waller, D. M. (1984) Differences in fitness between seedlings derived from cleistogamous and chasmogamous flowers in *Impatiens capensis*. *Evolution*, 38, 427–440.

Waller, D.M., Dole, J. & Bersch, A.J. (2008) Effects of stress and phenotypic variation on inbreeding depression in *Brassica rapa*. *Evolution*, 62, 917–931.

Wang, J.L. (2000) Effects of population structures and selection strategies on the purging of inbreeding depression due to deleterious mutations. *Genetical Research*, 76, 75–86.

Wang, J.L. (2005) Estimation of effective population sizes from data on genetic markers. *Philosophical Transactions of the Royal Society of London B*, 360, 1395–1409.

Wang, J.L., Hill, W.G., Charlesworth, D. *et al.* (1999) Dynamics of inbreeding depression due to deleterious mutations in small populations: mutation parameters and inbreeding rate. *Genetical Research*, 74, 165–178.

Weir, B.S., Anderson, A.D. & Hepler, A.B. (2006) Genetic relatedness analysis: modern data and new challenges. *Nature Reviews Genetics*, 7, 771–780.

Whitlock, M.C. (2000) Fixation of new alleles and the extinction of small populations: drift load, beneficial alleles, and sexual selection. *Evolution*, 54, 1855–1861.

Whitlock, M.C., Griswold, C.K. & Peters, A.D. (2003) Compensating for the meltdown: the critical effective size of a population with deleterious and compensatory mutations. *Annales Zoologici Fennici*, 40, 169–183.

Willi, Y., van Buskirk, J. & Hoffmann, A.A. (2006) Limits to the adaptive potential of small populations. *Annual Review of Ecology Evolution and Systematics*, 37, 433–458.

Williams, S.E. & Hoffman, E.A. (2009) Minimizing genetic adaptation in captive breeding programs: a review. *Biological Conservation*, 142, 2388–2400.

Willis, J.H. (1999) The role of genes of large effect on inbreeding depression in *Mimulus guttatus*. *Evolution*, 53, 1678–1691.

Wright, S. (1922a) The effects of inbreeding and crossbreeding on guinea pigs. III. Crosses between highly inbred families. United States Department of Agriculture, Bulletin 1121, Washington, DC, USA.

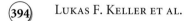

Wright, S. (1922b) Coefficients of inbreeding and relationship. *American Naturalist*, 56, 330–338.

Wright, S. (1931) Evolution in Mendelian populations. *Genetics*, 16, 0097–0159.

Wright, S. (1969) *Evolution and the Genetics of Populations*, vol. 2, *The Theory of Gene Frequencies*. University of Chicago Press, Chicago, Illinois, USA.

Wright, S. (1977) *Evolution and the Genetics of Populations*, vol. 3, *Experimental Results and Evolutionary Deductions*. University of Chicago Press, Chicago, Illinois, USA.

Genetic Consequences of Reintroductions and Insights from Population History

Jim J. Groombridge[1], Claire Raisin[1], Rachel Bristol[1,2] and David S. Richardson[2,3]

[1]Durrell Institute of Conservation and Ecology, University of Kent, United Kingdom
[2]Nature Seychelles, Victoria, Mahé, Republic of Seychelles
[3]School of Biological Sciences, Norwich Research Park, University of East Anglia, United Kingdom

'*Despite the genetic concerns associated with small populations, such as loss of genetic diversity and inbreeding (Frankham et al., 2002; Keller & Waller, 2002; Frankham, 2005), few cases exist where a known level of severe genetic impoverishment has led to translocation being ruled out as a viable option.*'

Page 396

Introduction

Conservation translocations (including 'reintroductions' and 'introductions', see Seddon *et al.*, this volume, Chapter 1) are an important tool for recovering endangered species, and are likely to remain so as long as intensive conservation management methods remain at the heart of species conservation. A multitude of considerations are required before initiating a reintroduction programme (see other contributions in this book). Keller *et al.* (this volume, Chapter 11)

Reintroduction Biology: Integrating Science and Management. First Edition.
Edited by John G. Ewen, Doug P. Armstrong, Kevin A. Parker and Philip J. Seddon.
© 2012 Blackwell Publishing Ltd. Published 2012 by Blackwell Publishing Ltd.

focus on population genetic theory and how it relates to translocation biology. Here we compliment that theory by concentrating on the evidence, showing that genetic factors play a significant role in extinction risk in threatened species and that it is, therefore, important that genetic management is a part of any translocation strategy.

The amount and type of genetic diversity that is available for translocation will be governed by the genetic constitution of the source population, which in turn is heavily influenced by a species' population history. Therefore, while historical population profiles of endangered species and future translocation strategies to conserve them may at first seem disparate topics, they are intrinsically linked. This chapter attempts to illustrate the importance of this relationship. It will also discuss how other factors need to be considered when planning the genetic management of a translocated population, including: (i) the sensitivity of the methods used to measure genetic parameters and (ii) which individuals to select for translocation. Choosing individuals based on genetic parameters (e.g. a low inbreeding coefficient and high, or unique, genetic variability) is not straightforward. For example, should you choose 'high-value' individuals for release and accept their loss from the source population? Furthermore, other phenotypic qualities that may help translocated populations to establish also need to be considered, e.g. reproductive performance.

It is clear then that genetic considerations for translocation programmes are rarely straightforward. However, our understanding of the genetics of small population biology can inform us regarding two fundamental decisions in translocation programmes: i.e. *how many* individuals should be translocated and the number of subsequent translocations that might be needed to secure long-term viability. Furthermore, the benefits of 'genetic rescue' – the introgression of novel genetic material into genetically impoverished populations to alleviate detrimental effects of inbreeding – have been elegantly demonstrated in small natural populations (Madsen *et al.*, 1999; Tallmon *et al.*, 2004). Tools like this now illustrate how genetic management can have a substantial impact upon the success of translocations.

Despite the genetic concerns associated with small populations, such as loss of genetic diversity and inbreeding (Frankham *et al.*, 2002; Keller & Waller, 2002; Frankham, 2005), few cases exist where a known level of severe genetic impoverishment has led to translocation being ruled out as a viable option. For example, when conservation managers were considering the fate of the po'ouli (*Melamprosops phaeosoma*), a critically endangered Hawaiian bird whose population comprised only three birds in 2000, captive breeding for eventual reintroduction was one of a series of options (Groombridge *et al.*, 2004; VanderWerf *et al.*, 2006). Unfortunately, efforts to save the po'ouli failed before numbers reached a point where reintroduction became feasible (the species is now listed by the IUCN as 'possibly extinct'). Despite their often low chance of success (Beck *et al.*, 1994; Wolf *et al.*, 1996; Fischer & Lindenmeyer, 2000), translocations have continued to be used as a major tool in endangered species management. Indeed, severely bottlenecked species provide some of the best-known examples of species recovery via reintroduction programmes. Such celebrated examples include the Seychelles warbler (*Acrocephalus sechellensis*) (Komdeur, 1994; Richardson *et al.*, 2006), Mauritius kestrel (*Falco punctatus*) (Jones *et al.*, 1995; Nicoll *et al.*, 2004), New Zealand's Chatham Island black robin (*Petroica traverse*) (Ardern & Lambert, 1997) and the Guam rail (*Rallus owstoni*) (Jenkins, 1979; Haig *et al.*, 1990). These, and numerous other examples, illustrate how endangered species and their translocated populations can recover dramatically even when these species might be considered to be beyond hope genetically.

Importantly, it is now widely accepted that loss of genetic diversity and increased levels of inbreeding in a population can lead to problems associated with inbreeding depression, a reduced ability to adapt and, consequently, an increased extinction risk (Saccheri *et al.*, 1998; Frankham, 2005). Application of population genetic theory, as set out in Chapter 11 by Keller *et al.*, suggests these problems will also be present in translocated populations. Therefore, ensuring that such populations receive what genetic diversity remains in the source population is important, because these problems may contribute to the generally low rates of success observed across translocations (Beck *et al.*, 1994; Fischer & Lindenmayer, 2000). Beyond this, reductions in genetic variation can have significant consequences for the longer-term ability of populations to adapt ('evolutionary potential') (Frankham *et al.*, 1999).

Molecular studies of island species have proven particularly informative for assessing the genetic effects of translocations. A good example comes from the South Island saddleback (*Philesturnus carunculatus carunculatus*), a New Zealand passerine. This species was recovered from only 36 individuals – sourced from a dwindling population with historically low levels of genetic variation – to a population of over 1200 birds (Hooson & Jamieson, 2003). In this case, loss of genetic variation as a consequence of the founder effects of sequential translocations was shown to be minimal (Taylor & Jamieson, 2008). However, instances such as this may merely illustrate that a limited amount of neutral genetic diversity in the original source population means there is little variation left to lose. Another important caveat is that genetic data from neutral markers do not necessarily reflect levels of adaptive variation, which is more likely to be important for evolutionary potential and the long-term viability of translocated populations (see Keller *et al.*, this volume, Chapter 11, for discussion). Ironically, examples such as these described above may lead to the interpretation that there is little need to worry about the genetic consequence of translocations where historic genetic diversity is thought to have been very low. In this context, knowledge of historical levels of genetic diversity can provide additional insight, as discussed in detail later in this chapter.

Despite the advances in our understanding of reintroduction genetics, it can be difficult to draw general conclusions from specific case studies. Numerous studies have demonstrated clear negative consequences related to the loss of genetic diversity and elevated rates of inbreeding in restored or reintroduced populations. While the contrasting evidence from these studies may be due to underlying differences in the statistical power of the marker sets used to detect genetic effects (a topic discussed by Keller *et al.* in Chapter 11), they illustrate the need to summarize general trends in the genetic consequences observed across translocated populations and species.

This chapter therefore aims to:

- Review the empirical literature for the genetic patterns described in Chapter 11 by Keller *et al.*, such as inbreeding depression, and how this has influenced reintroduction success.
- Describe how such effects can be measured and what genetic markers are available to measure them.
- Examine how knowledge of a species' history can help understand genetic patterns observed in reintroductions.
- Consider observed genetic patterns alongside the issue of 'how many individuals to translocate' (discussed by Jamieson & Lacy, this volume, Chapter 13).
- Discuss lessons learnt and how they might be applied in reintroduction biology.

This chapter does not detail the genetic problems that often arise when selecting individuals for reintroduction from captive stocks where genetic adaptation to captivity may be a major concern. This topic has been dealt with in detail elsewhere (King, 1939; Lewis & Thomas, 2001; Frankham, 2008; Robert, 2009).

The genetic consequences of translocation

The key parameters that determine the genetic diversity captured when individuals are selected for translocation from a source population are: the number of individuals, the proportion of those that contribute genetically to the next generation (founders) and the number and frequency of alleles within their collective genomes. These parameters define the founding genetic diversity of the new population, which is then shaped by the effects of genetic drift, natural selection, gene-flow, migration and the subsequent accumulation of new mutations. Of course, the impact of these processes is influenced by the fact that translocated populations are relatively small, which predisposes them to losing genetic diversity more rapidly than would be expected in a widespread non-threatened population. Consequently, one might expect a translocated population to show a loss of genetic diversity in comparison to the captive/wild source, purely as a result of genetic drift. This effect will be tempered only by increases in size as the population becomes established, or by gene flow as the population expands and connects with neighbouring populations (Frankham *et al.*, 2002).

Within this context the choice of individuals is important in determining both the short- and longer-term genetic consequences of a translocation. Preferred strategies vary from selecting individuals that are genetically 'surplus' to the source population to choosing the most productive breeders (Frankham *et al.*, 2002). The genetic trade-offs inherent in such decision making have long been debated (Earnhardt, 1999), and it remains true today that no single strategy is universally optimal. The initial translocations of Guam rail consisted of genetic lineages over-represented in captivity (Haig *et al.*, 1990), whereas translocations of Bali mynah (*Leucopsar rothschildi*) focused on captive individuals in their reproductive prime (Collins *et al.*, 1998). The choice of strategy will depend on the specific programme goal, e.g. redressing genetic over-representation in captivity or promoting rapid population growth in the wild – the latter being a strategy that is likely to slow loss of genetic diversity in translocated populations (Allendorf & Luikart, 2007).

Loss, gain or no change in genetic diversity

A brief review of the emerging field of reintroduction biology indicates an early awareness of the need for translocation programmes to ensure adequate genetic diversity in reintroduced populations (Haig *et al.*, 1990; Hughes, 1991; Leberg, 1993; Nader *et al.*, 1999; Harmon & Van Den Bussche, 2000). However, surprisingly few studies have thoroughly monitored translocated populations specifically for changes in key genetic parameters (i.e. heterozygosity, allelic richness, rate and level of inbreeding) relative to their source population (Seddon *et al.*, 2007). Table 12.1 summarizes some of those studies. Studies that have assessed genetic variability in source populations in advance of translocations ('pre-translocation studies') have also been included here, to illustrate how genetic information has been used to guide translocation plans and to provide a contrast to 'post-translocation' genetic studies. Table 12.1 is not a complete inventory of all genetic studies on translocated populations (studies have been selected so as to capture a range of observed outcomes), but it does allow some overall observations to be made.

From our review (Table 12.1) there appears to be no clear overall pattern in the trajectory of population genetic diversity following translocation, though perhaps there is a downward trend compared to the source population. Although at least half of the studies reported reduced levels of genetic diversity and/or increased levels of inbreeding in populations relative to their source populations, a number of studies detected little or no effect. The reasons put forward for this include: (i) low levels of diversity in the ancestral population prior to translocation (Taylor & Jamieson, 2008; Komdeur *et al.*, 1998), (ii) immigration from neighbouring free-living populations (Le Gouar *et al.*, 2008), (iii) rapid population expansion (VonHoldt *et al.*, 2008; Wisely *et al.*, 2003, 2008) and (iv) a large number of founders (Miller *et al.*, 2009). Some of those studies on New Zealand bird species, such as the New Zealand robins and saddlebacks (Jamieson *et al.*, 2008; Taylor & Jamieson, 2008; Jamieson, 2011), are discussed in more detail in Jamieson & Lacy (this volume, Chapter 13). In another example, high levels of diversity at microsatellite loci were observed in two large translocated populations of the wild turkey, *Meleagris gallopavo silvestri*, in Indiana, with one showing allelic richness equivalent to the source population (Harmon & Van Den Bussche, 2000). These results demonstrate that a loss of genetic diversity is avoidable in translocation. However, as previously stated, it may be that the source populations in these studies had little diversity to begin with due to prior effects of population decline, and therefore little variation to lose. The prior effects of historical bottlenecks are discussed later in this chapter.

There is also little evidence for reduced population persistence resulting from genetic diversity loss. For example, a study of North Island robins

Table 12.1 **Studies involving the genetic monitoring of reintroduced populations.**

Species	Marker used	Conclusions	Study reference
Pre-reintroduction assessment of genetic diversity			
Guam rail (*Rallus owstoni*)	Allozymes	Choice of genetic management can lead to either loss or gain in genetic diversity	Haig *et al.* (1990)
Crested ibis (*Nipponia nippon*)	MHC	Low MHC diversity. Recommended MHC-diverse individuals to found reintroduction	Zhang *et al.* (2006)
Scarlet macaw (*Ara macao*)	DNA fingerprinting	High levels of genetic diversity amongst founders of captive programme	Nader *et al.* (1999)
Seychelles warbler (*Acrocephalus sechellensis*)	Microsatellites MHC class 1	Low levels of neutral diversity but considerable higher MHC diversity. Higher numbers of individuals need to be translocated to ensure MHC variation is captured	Richardson & Westerdahl (2003); Hansson & Richardson (2005)
Seychelles white-eye (*Zosterops modestus*)	DNA fingerprinting	Low diversity within each of 2 remnant populations, but considerable diversity between. Recommended mixed stock introductions to recombine diversity.	Rocamora & Richardson (2003)
Post-reintroduction assessment of genetic diversity			
Seychelles warbler	DNA fingerprinting	Low levels of neutral diversity but no significant loss during reintroduction	Komdeur *et al.* (1998)
Mauritius kestrel (*Falco punctatus*)	Pedigree records	High levels of inbreeding and loss of diversity in reintroduced population	Ewing *et al.* (2008)

(*Continued*)

Table 12.1 (Cont'd)

Species	Marker used	Conclusions	Study reference
Lesser kestrel (*Falco naumanni*)	Microsatellites	Wild v. reintro = No diff in obs-H, F_{IS}. Reintro showed <H and> F_{IS} than captive source	Alcaide *et al.* (2010)
South Island saddleback (*Philesturnus carunculatus carunculatus*)	Microsatellites	No significant loss of alleles from source to sequentially translocated populations	Taylor & Jamieson (2008)
South Island Saddleback (*Philesturnus carunculatus carunculatus*) & North island robin (*Petroica longipes*)	Microsatellites	Dramatic historical genetic loss but little recent loss despite population crash (saddleback). Genetic diversity low in ancestral populations (robin), minimal loss for modern populations.	Taylor *et al.* (2007)
North Island Robin (*Petroica longipes*)	Pedigree data	Moderate effects of inbreeding depression, low level effects	Jamieson *et al.* (2007)
South Island Saddleback (*Philesturnus carunculatus carunculatus*)	Pedigree records	Populations founded from fewer individuals experienced more inbreeding and loss of diversity (unequal sex ratios and skewed breeding success also contributed). Rate of inbreeding and loss	Jamieson (2011)
North Island Robin (*Petroica longipes*)	Pedigree records		Jamieson (2011)
Takahe (*Porphyrio hochstetteri*)	Pedigree records	of genetic diversity varied between species during reintroduction phase, but	Jamieson (2011)
Stewart Island Robin (*Petroica australis rakiura*)	Pedigree records	were reflected in final population size.	Jamieson (2011)
Hihi	Microsatellites	Loss of diversity and divergence and inbreeding depression	Brekke *et al.* (2011) Brekke *et al.* (2010)

South Island Robin	DNA fingerprinting	Reduced genetic diversity	Ardern & Lambert (1997)
Black Robin	DNA fingerprinting	Low genetic diversity	Ardern & Lambert (1997)
South Island robin (*Petroica australis australis*)	MHC, minisatellite DNA	Moderate diversity at MHC Class II B. Low genetic diversity relative to source population	Miller & Lambert (2004)
Chatham Island black robin (*Petroica traversi*)	MHC, minisatellite DNA	Monomorphic at MHC Class II B. Low genetic diversity.	Miller & Lambert (2004)
North island saddleback (*Philesturnus carunculatus rufusater*)	DNA fingerprinting, microsatellites and isozymes	Pattern of reduced allelic diversity with severity of bottleneck. Loss of genetic variability through reintroduction	Lambert *et al.* (2005)
Wild turkey (*Meleagris gallopavo silvestris*)	DNA fingerprinting	Level of genetic diversity in reintroduced populations similar to other outbred populations	Harmon & Van Den Bussche (2000)
Wild turkey (*Meleagris gallopavo silvestris*)	Microsatellites, mtDNA sequence	High genetic diversity in reintro populations (augmented by geneflow from nearby subpops)	Latch & Rhodes (2005)
Bearded vulture (*Gypaetus barbatus*)	DNA fingerprinting	Low genetic diversity in the free-living population	Negro & Torres (1999)
Bearded vulture (*Gypaetus barbatus*)	Microsatellites	High genetic diversity in the captive population compared to the reintroduced population	Gautschi *et al.* (2003)

(Continued)

Table 12.1 (*Cont'd*)

Species	Marker used	Conclusions	Study reference
Greater prairie chicken (*Tympanuchus cupido pinnatus*)	Microsatellites	Neutral genetic variation restored to historic levels by translocations, which removed detrimental variation linked to inbreeding depression and increased fitness	Bouzat *et al.* (2010)
Griffon vulture (*Gyps fulvus*)	Microsatellites	Minimal loss of genetic diversity due to high immigration rate with neighbouring populations	Le Gouar *et al.* (2008)
Yellowstone grey wolves (*Canis lupis*)	Microsatellites and pedigree records	High levels of genetic diversity and low levels of inbreeding	VonHoldt *et al.* (2008)
Black-footed ferret (*Mustela nigripes*)	Microsatellites and pedigree records	Minimal loss of genetic diversity where reintroduced populations grew rapidly	Wisely *et al.* (2003, 2008)
Tuatara (*Sphenodon*)	Microsatellites	Loss of genetic diversity minimized by high founder number, while male sex bias exacerbated rate of loss	Miller *et al.* (2009)
Arabian oryx	Microsatellites	Simultaneous effects of inbreeding depression and out-breeding depression on juvenile survival	Marshall & Spalton (2000)
Eurasian beavers	MHC, DNA fingerprinting	Low genetic variability in successfully reintroduced and rapidly expanded Swedish population	Ellegren *et al.* (1993)
Mexican wolves	pedigree	Inbreeding depression and genetic rescue	Fredrickson *et al.* (2007)

| Alpine ibex | Microsatellites | Loss of heterozygosity and standardized number of alleles across serial bottlenecks/reintroductions. Most ancestral genetic diversity now present in reintroduced population. Genetic 'footprint' of reintroduction history remains after 100 years of reintroduction activity. In addition to founder number, early population growth rate was important determinant of inbreeding rate. | Biebach & Keller (2009) Biebach & Keller (2010) |

by Jamieson *et al.* (2007) revealed how increasing levels of inbreeding had a measurable fitness consequence but a negligible impact on population persistence over a reasonable time frame. While results like this suggest that effects of inbreeding might not necessarily always impact on population persistence, it is important to be aware that the impact of inbreeding and lack of adaptive potential may become apparent only over the long term (Keller & Waller, 2002) or under specific adverse conditions (e.g. Richardson *et al.*, 2004). An important message from the robin study (one that may hold true for many instances where inbreeding effects are observed to be small) is that frequencies of close inbreeding are usually so low in most populations (but see Richardson *et al.*, 2004) that their demographic consequences are minimal in the short to medium term, but may eventually become a problem unless alleviated by immigration. A broader point is that most studies report genetic rather than demographic consequences related to these inbreeding signatures. Indeed, signatures of reduced genetic diversity are used more often to infer compromised populations rather than direct evidence derived from measures of individual or population level fitness.

The increasing frequency of genetic surveys on translocated populations (as shown in Table 12.1) is beginning to enable contrasts to be drawn across comparable species systems. One recent study by Jamieson (2011) does exactly this, using pedigrees from four translocated bird populations to ascertain how the different translocation parameters used influence rates of inbreeding and loss of genetic diversity. Jamieson (2011) shows, first, that populations founded from fewer individuals experienced more inbreeding and loss of diversity and, second, that these parameters varied extensively between species during the reintroduction phase, but once carrying capacity was reached differences in level of inbreeding were determined by the final population sizes. The comparative element of this study proved invaluable in detecting the influence of carrying capacity on determining long-term rates of inbreeding in reintroduced populations.

The findings of some studies demonstrate the kind of genetic consequences expected of translocations. For example, recent genetic and demographic work on New Zealand tuatara by Miller *et al.* (2009) demonstrated how reintroduced populations founded by larger groups retain a higher proportion of their heterozygosity and allelic diversity. Furthermore, the scale of bottleneck during translocation has been shown to be important in predicting failure of reintroduction programmes (Thevenon & Couvet, 2002). Elsewhere, surveys of genetic diversity in serially bottlenecked populations of dice snake (*Natrix tessellata*) revealed significant loss of genetic variation and increased frequency of scale abnormalities, implying that the effects of multiple bottlenecks incurred as a consequence of conservation management cannot be ignored (Gautschi *et al.*, 2002).

The increasing accessibility of molecular markers in conservation has not only made the genetic monitoring of reintroduced populations a great deal easier but has provided a way for the genetic consequences of a species' reintroduction programme to be evaluated. One recent example is the Mauritius parakeet (*Psittacula echo*), an endangered island endemic species successfully recovered from less than 20 individuals in 1980 to over 500 birds by 2010 (see Jones & Merton, this volume, Chapter 2). Recent genetic work has allowed a retrospective look at the impact of the translocation programme on genetic diversity within this restored population (see Box 12.1).

Box 12.1 **Genetic interpretation of reintroductions of the Mauritius parakeet**

Conservation managers need to be aware not only of existing genetic architecture in endangered species but also of the potential effects that intensive management interventions such as translocation might have on these patterns. The Mauritius parakeet (*Psittacula echo*) (Figure 12.1), an endangered island endemic species, illustrates the effect that intensive management can have on a populations' genetic architecture. The Mauritius parakeet declined to less than 20 birds during the 1980s (Duffy, 1993; Lovegrove *et al.*, 1995) before being successfully restored to ~ 500 individuals (Richards & Tollington, 2010) following a 30-year captive-breeding and reintroduction programme. During the most intensive period of management between 2000 and 2005, offspring were often removed from the wild nests of unexperienced parents and either reared in captivity for later release elsewhere or cross-fostered to more experienced pairs. In 2005 intensive management had to be drastically scaled back to limit the spread of an outbreak of Psittacine beak and feather disease among the recovering population, although sampling and monitoring continued. This management history enabled an assessment of population genetic structure, before and after the period of intensive management, to quantify its effect on the genetic architecture of the restored population. A suite of 22 microsatellite markers were used to genotype 500 individuals sampled across the species' range between 1995 and 2008. *STRUCTURE* analysis (Pritchard *et al.*, 2000) of the population before management began suggested the existence of several genetically distinct subpopulations, whereas after intensive management the population had become largely homogenized (Figure 12.2). The genetic structure detected prior to management was

Figure 12.1 **Mauritius parakeet (photo: Gregory Guida, Durrell Wildlife Conservation Trust).**

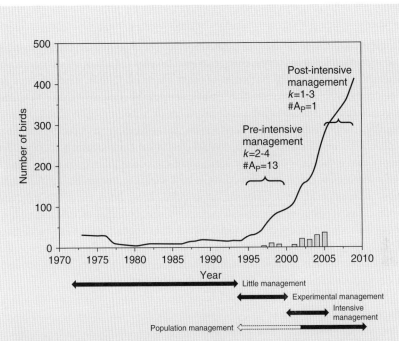

Figure 12.2 Census population size of the Mauritius parakeet from 1973 to 2009 and number of birds reintroduced since monitoring began (shaded bars) during the different phases of the programme. K = number of genetically distinct subpopulations/clusters determined by BAPS/STRUCTURE analyses, A_P = total number of private alleles for each of the two sampled periods (pre-intensive management 1995–2000 and post-intensive management 2005–2008) (Raisin *et al.*, 2011).

likely the result of fragmentation of the dwindling population, linked to habitat loss and the effects of genetic drift between population fragments. This case study illustrates the benefits, in terms of slowing the loss of genetic diversity from genetic drift, of genetic diversity being redistributed during management. Continued monitoring of the restored Mauritius parakeet population will reveal the rate at which a genetic structure might re-emerge in the restored populations once conservation actions are no longer required.

How should we measure these patterns?

Problems associated with small population size, such as loss of genetic diversity and increased levels of inbreeding, can lead to a loss of adaptive variation and declines in evolutionary potential (Frankham *et al.*, 2002). These issues are directly relevant to conservation translocations, particularly those that may be restricted to translocating small numbers of founding individuals. Consequently, measuring key parameters such as inbreeding in these populations is crucial to understand their potential effects on translocation success. The theory that underpins our understanding of the consequences of inbreeding is well established, but how inbreeding is best measured remains an important debate. In general, two different approaches have been used for inferring levels of inbreeding in a population. Longitudinal studies of marked natural populations have enabled individual inbreeding coefficients (f) to be derived directly from pedigree data for some species. For many others, however, this information is unavailable, and consequently genetic markers have been used to try to measure and assess the effects of inbreeding.

Marker-based methods

Studies using molecular markers to investigate inbreeding normally rely on the fact that inbreeding reduces heterozygosity. Unfortunately, marker heterozygosity overall is now believed to be a relatively weak instrument for detecting levels of inbreeding (Pemberton, 2004). Studies have indicated that multilocus heterozygosity estimates based on the relatively small numbers of markers used in most such studies (normally <30) may not accurately measure genome-wide heterozygosity (Balloux *et al.*, 2004; Slate *et al.*, 2004). In most populations larger numbers of loci (hundreds) may have to be screened to get an accurate measure of inbreeding unless variance in inbreeding is high. However, as variance in inbreeding is expected to be high in the small bottlenecked populations relevant to translocations, it may be feasible to get accurate measures with an intermediate number of markers (>30 but <100s) (Slate *et al.*, 2004). An alternative approach is to use the sharing of neutrally inherited loci across individuals within a population to assess relatedness between adults. These data can then be used to infer inbreeding directly within one contemporary generation and assess its affects downstream (e.g. Richardson *et al.*, 2004). Over the years, a number of marker-based methods to determine relatedness have been developed (Lynch, 1988; Lynch & Ritland,

1999; Queller & Goodnight, 1989; Wang, 2007a), but no single method outperforms others in all situations (Wang, 2007b). Although using genetic markers alone to assess levels of inbreeding may be problematic, the potential for loss of heterozygosity to cause inbreeding depression can still be assessed through heterozygosity-fitness correlations (see Box 9.2 in Ewen *et al.*, this volume, Chapter 9, for details). The evidence now suggests that such heterozygosity-fitness correlations are generally caused by associative overdominance effects (i.e. homozygosity at a marker locus reflects homozygosity at a linked functional locus that affects fitness) rather than genome-wide effects (see Box 9.2 in Ewen *et al.*, this volume, Chapter 9). Ultimately, the accuracy of any of these measures depends on how informative markers are used (Slate *et al.*, 2004; Wang, 2007a). The depth of coverage normally required means that genetic screening is unlikely to provide an easy means to estimate genome-wide variation and monitor the effects of inbreeding in endangered/translocated populations. However, in the absence of detailed pedigree and life history information, marker-based methods, if used and interpreted correctly, can still provide important information.

Pedigree-based measures

The construction of population pedigrees, which involves intensive field monitoring of uniquely marked individuals to collect data on parentage of offspring across generations, is rightly considered the best method for estimating inbreeding (Slate *et al.*, 2004; Pemberton, 2008). This approach is sometimes easy to integrate into routine monitoring of translocated populations. However, the accuracy and completeness of pedigree required to obtain accurate estimates, which involves the investment of considerable resources, is rarely available for wild populations (Pemberton, 2008). Another fundamental problem with pedigree analysis is that the majority of methods assume the base population to be unrelated and non-inbred, two characteristics unlikely to be true in bottlenecked populations (Waller, 1993; Ruiz-Lopez *et al.*, 2009). This problem can be addressed if information is available to reconstruct the historical trajectory of decline in population size, allowing accumulation of inbreeding in the base population (founders) of the pedigree. In addition, the estimates of inbreeding for each individual in the population are all relative to the base population, so accurate assessment requires a number of generations. Gaps in the pedigree or incorrect data regarding parentage can also cause problems. For example, the assumption that individuals of unknown origin are unrelated can lead to incorrect estimation of these coefficients (Marshall *et al.*, 2002; Markert *et al.*, 2004) and incomplete paternity information has the effect of underestimating levels of inbreeding (Overall *et al.*, 2005). More encouragingly, however, pedigree datasets do not have to contain a large number of generations to enable accurate measures of inbreeding to be calculated. While inbreeding accumulates across generations, it is matings between relatives during the most recent generations that has the greatest effect on inbreeding coefficients. Indeed, simulations suggest that a minimum of three generations of pedigree data are needed to yield estimates of *f* that correlate very well with more extensive pedigrees (Balloux *et al.*, 2004). This is good news for translocation programmes that have invested in marking individuals and setting up long-term population monitoring programmes, but is a potentially daunting prospect for those that do not have a well-characterized source population.

Which method should be used?

Whereas one might expect pedigree-based and marker-based measures of inbreeding to reflect one another, the two rarely correlate. Correlations between pedigree and molecular inbreeding coefficients are often weak in natural populations (Markert *et al.*, 2004; Overall *et al.*, 2005; Slate *et al.*, 2004) and there are a number of potential reasons for this. First, if the population has undergone a bottleneck, the founders themselves are likely to have reduced heterozygosity; therefore measures of genetic diversity based on heterozygosity, or rare alleles for example, may not decrease much more through time with continued inbreeding. This is one reason why marker-based inbreeding studies often underestimate effects of inbreeding depression; once inbreeding has resulted in fixation of a single allele at a particular locus, further inbreeding is no longer measurable at that locus (discussed in more detail by Keller *et al.*, this volume, Chapter 11). Second, if the founders happen to be from different lineages then subsequent breeding between them will have the effect of creating admixture, thereby increasing heterozygosity and decreasing the detectable level of inbreeding – two changes that would not be reflected in pedigree information. Third, founders are generally assumed to be non-inbred and unrelated to each other, which is unlikely to be the case for most recently bottlenecked populations. However, under certain circumstances correlations between the two methods are likely to arise (Ruiz-Lopez *et al.*, 2009), e.g. in endangered species that possess high levels of, and high variance in, inbreeding (e.g. see Hedrick *et al.*, 2001). Importantly, the strength of this association is dependent on the characteristics of the founding population and the amount of genetic diversity that has been lost since the founding (or bottleneck) event (Ruiz-Lopez *et al.*, 2009). Furthermore, when realistic levels of inbreeding are applied to the base population, the strength of the relationship can be expected to improve (Pemberton, 2008).

Overall, it is generally accepted that pedigrees should be used to estimate these parameters where possible (Slate *et al.*, 2004, Pemberton, 2008). However, gaps and missing relationships in the pedigree should be augmented, and genetic parentage verified, using molecular data where possible in order to enhance completeness and accuracy (Pemberton, 2008). Although this is something to aim for, it is unlikely to be achievable for the majority of reintroduced populations. In these cases, marker-based approaches can offer a realistic alternative, provided that enough independent markers are used.

What is the evidence for inbreeding depression in translocated populations?

Notable parallels exist between genetic characteristics of a population, such as population size, rate of inbreeding and loss of genetic diversity, and factors considered to be important determinants of reintroduction success (Leberg & Firmin, 2008). For instance, reintroducing wild-caught rather than captive-reared stock and releasing more rather than fewer individuals have been demonstrated to increase the probability of reintroduction success (Fischer & Lindenmayer, 2000; Wolf *et al.*, 1996). Consequently, the strategies chosen by translocation programmes may have a bearing on the extent to which inbreeding depression is expressed in reintroduced populations and their subsequent evolutionary viability. However, as discussed earlier, the evidence for consistent patterns of change in levels of genetic diversity in translocated populations is scant, and what evidence there is for inbreeding depression is dispersed across individual case studies. Of course, those studies that do demonstrate an effect are most often those based on good-quality information extracted from long-term, intensive monitoring efforts of translocated populations. These types of study take a long time to achieve and only now are multispecies comparisons based on such information becoming possible, paving the way for observing unifying trends in genetic effects (Jamieson, 2011).

As discussed above, detecting inbreeding depression is not easy, and this may explain why evidence for its effects remains relatively elusive. For example, it requires samples of usually more than 100 individuals with a high variance in inbreeding (some closely inbred individuals and some not) to provide resolution. Furthermore, larger sample sizes are required when the number of founders is small (Kalinowski & Hedrick, 2006). Problems of low statistical power are likely to apply to the majority of reintroduced populations (see Keller *et al.*, this volume, Chapter 11). In addition, patterns of inbreeding depression are likely to vary substantially among traits and are likely to be obscured by variations in environmental conditions among reintroduced populations (Keller and Waller, 2002; reviewed in Armbruster & Reed, 2005). Environmental stress increases inbreeding depression, while fitness traits and individual fecundity show stronger effects than morphological characters and individual survival (DeRose & Roff, 1999). Consequently, a finding of no significant inbreeding depression in a reintroduced population may reflect a lack of power to detect it rather than it not being present.

The difficulty of relating genetic evidence to reintroduction outcome is illustrated by the example of the Mauritius kestrel (Jones *et al.*, 1995; Nicoll *et al.*, 2004). The population of kestrels in the eastern Bambous mountains of Mauritius has been monitored closely since the population was founded in 1986. An analysis of the pedigree available for the population indicates that it has experienced one of the highest rates of inbreeding documented for a wild vertebrate population (2.6% per generation; Ewing *et al.*, 2008; see Keller *et al.*, this volume, Chapter 11), caused by the population crash to a single pair in 1974 and perhaps exacerbated by the small number (24) of founders that initiated the eastern population. Despite the high levels of inbreeding, survival of individuals within this population appears to be within normal levels (Nicoll *et al.*, 2004). This apparent robustness implies that genetic issues have had a minimal negative impact on short-term viability, but it remains to be seen how this genetic impoverishment (particularly the loss of allelic diversity) may have compromised this species' ability to adapt in the future. Alternatively, and as may be the case with many other translocated populations of endangered species, any detrimental genetic effects may be masked by the beneficial effects of the intensive management of the population, which has included supplemental feeding and provisioning of nest sites. Populations that are managed artificially in this way might not persist so well when assistance is withdrawn. Therefore, one effect of conservation management is that it may mask inbreeding depression in translocated populations, which may be why evidence of inbreeding depression in these populations is not immediately apparent.

Genetic variation at neutral and functional loci

One thing that is clear from the assessment of genetic variation/inbreeding in reintroduction programmes (Table 12.1) is that microsatellites have become the genetic markers of choice for such studies. It is easy to understand why; they are relatively cheap and easy to use, and can also facilitate the investigation of a whole range of questions relevant to the conservation of endangered species, e.g. migration rates and mating systems (for a full review of the use of genetic markers in conservation see Frankham *et al.*, 2010). However, it is important to understand the limitations that are intrinsic to these markers and the potential problems that may arise because of the way they are used or interpreted (e.g. Pemberton, 2004; Ljungqvist *et al.*, 2010; Vali *et al.*, 2008).

One fundamental problem is that microsatellite markers, and most other markers used in conservation genetics, measure variation within non-coding sections of the genome. Consequently, while they can be employed to assess various parameters relevant to population genetics, including genome-wide levels of variation (but see the earlier discussion), they may not reflect variation at functional loci important to the fitness of the species in question (Reed & Frankham, 2001; van Tienderen *et al.*, 2002; Bekessy *et al.*, 2003). Patterns of variation at expressed genes, where various types of selection can act, may differ from those at neutral sites (Frankham *et al.*, 2010). Therefore, there could be consequences to making decisions regarding translocations, or assessing their outcomes, based on neutral markers. For example, balancing selection, which occurs at some crucial functional loci (Spurgin & Richardson, 2010; Castric & Vekemans, 2004), can, at least to some degree, oppose drift-dependent loss of genetic variation (Hughes & Yeager, 1998). Consequently, variation at these loci may significantly exceed those observed at the microsatellite markers. If the number of founders required to translocate variation from a population is determined by observations (or theoretical models) of neutral variation, the number needed to retain the higher levels of functional variation could be underestimated. This could result in the loss of critical functional variation in the new populations. Such problems have not gone unnoticed (Jamieson & Lacy, this volume, Chapter 13) and studies are now underway to construct models that estimate the number of founders required for the retention of allelic variation at loci under selection. Another problem may arise because neutral markers may not reflect adaptive genetic differences that have arisen between recently separated populations because of rapid subsequent selection (Cohen, 2002). This could lead to an underestimation of the genetic variation available for future conservation management.

So what is the alternative to neutral markers? We could focus on identifying and characterizing variation at specific critical loci (chosen to measure variability at ecologically important traits) within endangered species. However, there are many problems associated with this approach, not least the difficulty of identifying which loci are critical to the fitness of a given species and ensuring that this holds true across different contexts, e.g. captive verses wild populations or different environmental conditions (Frankham et al., 2010). Recent developments in genomic techniques have, in theory, made it possible to investigate functional variation across the genome of any given species (Ouborg et al., 2010), but the time, expertise and finance required to do so will prohibit this for most species. There are, however, some specific loci already identified, such as those of the Major Histocompatibility Complex (MHC), that are obvious candidates in terms of managing variation at individual loci.

Genes of the MHC are a crucial component of the vertebrate immune system, where they play a major role in identifying pathogens and triggering an adaptive immune response (Hughes & Yeager, 1998). The evidence for associations between MHC variants and pathogen resistance is overwhelming (e.g. Jeffery & Bangham, 2000; Sommer, 2005; Worley et al., 2010) and the extraordinary variation observed within the MHC of most species is thought to be maintained by pathogen-mediated balancing selection (Hedrick, 2002; Jeffery & Bangham, 2000; Spurgin & Richardson, 2010). Variation at the MHC also influences other important biological traits, including mate choice, kin recognition and autoimmune disease (Edwards & Hedrick, 1998; Penn et al., 2002). Given the impact that pathogens can have on endangered populations, the retention of diversity at MHC genes has long been recognized as important in the conservation of vertebrates, both in terms of contemporary fitness and the ability to respond to future challenges (Hedrick et al., 1999, 2000; Hughes, 1991). However, it is only relatively recently that the development of suitable molecular techniques has allowed the proliferation of studies surveying MHC variation in endangered species (Sommer, 2005). Box 12.2 illustrates an example of the Seychelles warbler and how important MHC variation may be in small populations of endangered species. Other examples where MHC genes are now being used to monitor endangered species genetically include the Mexican wolf (*Canis lupus baileyi*) (Hedrick et al., 2000), Chatham Island black robin (Miller & Lambert, 2004) and the crested ibis (*Nipponia nippon*) (Zhang et al., 2006).

Box 12.2 **Functional and non-functional genetic variation within the Seychelles warbler**

The Seychelles warbler (*Acrocephalus sechellensis*; SW) (Figure 12.3) is assumed to have originally occurred on most Seychelles islands, which constituted a single large island during the last ice age (Collar & Stuart, 1985). However, between 1940 and 1968 it went through a severe genetic bottleneck with less than 29 birds remaining on the island of Cousin (Lousteau-Lalanne, 1968). As a result of conservation efforts the population has since recovered and now remains stable at a carrying capacity of c. 320 adult birds (Brouwer *et al.*, 2007). The low neutral variation detected by DNA fingerprinting (Komdeur, 1998) and microsatellite loci (Richardson *et al.*, 2001; Hansson & Richardson, 2005) is consistent with the genetic bottleneck this species has been through. However, functional genetic variation was also investigated by focusing on sequence variation within MHC Class I genes. Molecules encoded for by alleles at these loci play a key role in eliciting an adaptive immune response to intracellular pathogens in vertebrates (Hughes & Yeager, 1998) and individual differences in MHC diversity influence pathogen susceptibility (Doherty & Zinkernagel, 1975). Therefore it is easy to see why variation at these functional loci could be critical to the genetic health of a population. Interestingly, MHC variation in the SW was higher than expected given the neutral variation observed and appears to have been maintained by balancing selection (Richardson & Westerdahl, 2003). Further studies have since shown that MHC-dependent fertilizations occur in the SW; females paired to low MHC diversity males gain extra-pair fertilizations with higher MHC diversity males and by doing so they increase the MHC diversity of offspring (Richardson *et al.*, 2005). Finally, strong independent positive associations between both MHC diversity and specific alleles and survival have been found in this population (see Figure 12.4) (Brouwer *et al.*, 2010). Consequently, the interacting effects of sexual selection and pathogen-mediated viability selection appear to be important in maintaining MHC variation in the SW.

Three new populations of SW have been established by translocating birds from Cousin Island; 29 individuals were moved to both Aride and Cousine Island in 1988 and 1990 respectively (Komdeur, 1994). DNA

Figure 12.3　**Seychelles warbler (photo: Danny Ellinger).**

fingerprinting indicated that the majority of neutral genetic variation in the source population had been maintained in these new populations (0.96 and 0.99 respectively; Komdeur *et al.*, 1998). However, initial studies suggest that some MHC alleles may not have been retained (D.S. Richardson, unpublished data). To ensure a more complete retention of neutral and (the more diverse) MHC variation, the number of individuals translocated was increased to 58 when establishing a population on Denis in 2004 (Richardson *et al.*, 2006). Studies are now underway to compare exactly how much neutral and MHC variation has been retained in the new populations, and to determine whether any loss has been due to founder effects, subsequent genetic drift or differential selection. Given the association between MHC diversity/specific alleles and survival in the SW it is clear that maximizing the MHC variation in all populations may be important, both to enhance current survival and to maximize their ability to survive the arrival of any new pathogens.

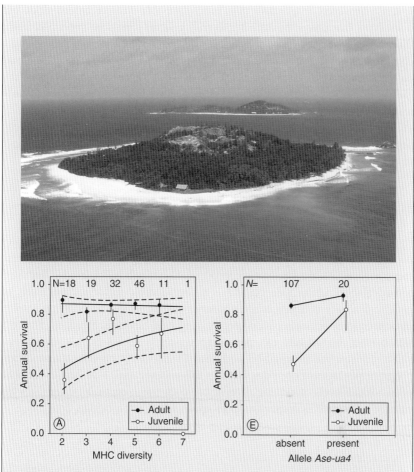

Figure 12.4   The effects of MHC parameters on annual survival probabilities in the Seychelles warbler population on Cousin Island (pictured in forefront) (photo: David S. Richardson). In the lower figures, (A) shows MHC diversity (number of alleles across duplicated MHC loci) and (E) the presence or absence of the specific MHC allele *Ase-ua4*. Numbers at the top indicate the number of offspring followed and ±95% confidence intervals are shown. The results show a clear effect of MHC diversity on juvenile survival but not adult survival, while the presence of *Ase-uea4* affects survival in both juveniles and adults. (Taken from Brouwer *et al.*, 2010).

It is, however, important that individual-locus management does not distract us from also maximizing overall genetic variation, especially if the threat to the species is from genome-wide inbreeding depression. Using a combination of genome-wide neutral markers and specific 'critical' loci may be the safest way to assess the threat posed by reduced genetic diversity in a given population. More studies are now needed to assess the effect of both historical bottlenecks and translocations on the relative levels of variation at critical functional loci, such as the MHC, and neutral loci. This should provide a better understanding of how to plan future translocation strategies to manage genetic variation overall.

Historical genetic signatures of endangered species

By their very nature, translocations of endangered species frequently involve the release of individuals from genetically depleted source populations, where some sort of historical reduction in population size has occurred. Therefore, historical population size can have a considerable influence on levels of genetic diversity within a contemporary population due to differing extents of population decline between species. Knowing how much genetic diversity was present historically can put currently observed levels in context. The value of historical information and how it might be applied to translocation programmes has not been discussed in this chapter until now because studies of historical levels of genetic diversity remain the exception rather than the rule for most species conservation programmes (Groombridge, 2007).

Their relevance to translocation programmes

Translocation efforts can benefit not only from knowing whether a species has maintained a stable population size over time or has endured a recent population crash, but also from an understanding of historical patterns of gene flow in the ancestral population. Many translocated populations originate from captivity and therefore experience two managed bottlenecks, one at founding of the captive population and another upon release of captive-bred individuals back into the wild. Both of these will influence the subsequent rate of inbreeding and loss of genetic diversity experienced by the establishing translocated population. However, prior to founding captive populations or undertaking any translocation, all endangered species will have experienced an initial bottleneck, the pattern and extent of which will have been determined by the population's decline to endangerment. This historical bottleneck will impose background levels of inbreeding and genetic diversity loss depending on its severity and duration. Characterizing the initial bottleneck can enable prediction of the relative magnitude of subsequent bottlenecks linked to conservation management.

Knowing the shape of the historical bottleneck can also help clarify whether a lack of evidence of inbreeding depression is due to low statistical power or a true lack of inbreeding depression (an issue described in Keller *et al.*, this volume, Chapter 11). For instance, low ancestral levels of diversity would indicate a long, protracted bottleneck, which can result in severe reduction in heterozygosity and accumulation of levels of inbreeding in the surviving population, with consequent expression of genetic load in those individuals that survive. In contrast, a shorter, sharper crash in genetic diversity from ancestrally high levels would suggest there has been less time for inbreeding and a loss of heterozygosity to accumulate. In these cases, an equivalent genetic load is less likely to be expressed as deleterious effects in those individuals that are used in conservation translocations. Some of these predictions have been demonstrated experimentally in *Drosophila* populations exposed to bottlenecks of varying intensity (England *et al.*, 2003). Furthermore, historical patterns of gene flow detected using ancestral populations can help to determine how translocated populations should be managed (or mixed) to restore recently lost connectivity between ancestral sites. How information on historical genetic diversity has been used to provide guidance on management decision (e.g. whether or not to apply genetic rescue) is illustrated in the following sections (see also Box 12.3).

Box 12.3 Knowledge of historical genetic diversity informs future reintroduction of Floreana mockingbirds

The critically endangered Floreana mockingbird of the Galapagos Islands was, during Charles Darwin's visit there in 1835, found across the island of Floreana and the two nearby tiny satellite islands. Today, this endemic species has been extirpated from the main island as a result of the impact of introduced mammals, and exists only as small populations on the satellite islands – around 20–50 individuals on Champion and 200–500 on Gardner-by-Floreana. Plans to restore the ecosystem on Floreana have paved the way for a programme to reintroduce this species to Floreana (Charles Darwin Foundation, 2008, 2009). However, the crucial issue for reintroduction managers is an evolutionary one: have the two remnant island populations been evolving independently since well before the period of human-influenced decline on Floreana? Measures of genetic differentiation using microsatellite markers (Hoeck et al., 2009a) showed clear divergence between the two contemporary populations (pairwise $F_{ST} = 0.533$), not unexpected given their small population sizes (Hoeck et al., 2009b). The effects of genetic drift over the past century were then demonstrated by comparing contemporary levels of genetic diversity with those detected in museum specimens collected between 1905 and 1906, revealing a 39% loss of heterozygosity on Champion (the smaller of the two islands) and a 6% loss on the larger island of Gardner-by-Floreana (Hoeck et al., 2009b). These data confirmed the erosion of genetic diversity within each of the two island populations since 1906, but the museum specimens also showed that these two populations were already significantly differentiated in 1906. The mockingbird had become extinct on Floreana 21 years earlier in 1885, so the question still remained – had this divergence begun since the onset of human impacts on the island or had the two satellite islands always harboured genetically differentiated populations well before this? The true answer to this question has profound consequences for the reintroduction of mockingbirds back on to Floreana. Reintroducing individuals from both islands to form a single flock on Floreana will risk causing problems of outbreeding depression if their genetic differentiation is a reflection of ancient separately evolving evolutionary pathways. However, the alternative, reintroducing individuals from only one of the satellite islands, risks overlooking valuable genetic diversity from the other and may expose the reintroduction programme to subsequent problems of inbreeding depression. An elegant coalescent analysis determined that the Gardner-by-Floreana population diverged from Floreana approximately 450 years ago and the Champion population diverged 270 years ago (Hoeck et al., 2009b), a scenario of three somewhat subdivided populations connected by recurrent gene flow. With the decline and eventual extinction of the Floreana population, gene flow between the satellite islands declined. Their former connection with Florenea was supported by inclusion in the analyses of microsatellite genotypes obtained for two mockingbird specimens collected on Floreana in 1835 by Darwin himself. This clarification, using genetic data from historical specimens, has helped biologists to understand contemporary genetic patterns clearly and to advocate a reintroduction on Floreana that includes individuals from both satellite islands, in order to maximize genetic diversity upon which selection can act, while conserving each of the island populations *in situ*. In practical terms, knowledge of the population's genetic history has helped to guide future reintroduction plans, which are likely to minimize negative genetic consequences and maximize the chances of a successful outcome.

Illustrative case studies

An example that illustrates how knowledge of a species' historical population genetics can help interpret translocation outcome is that of the Mauritius (*Falco punctatus*) and Seychelles kestrels (*Falco araea*). These two kestrel species are endemic to different Indian Ocean islands, but each have had well-documented population histories and conservation programmes that have included translocations.

Considered to be relatively widespread across the Seychelles by Victorian naturalists, the Seychelles kestrel was described in 1966 as 'critically endangered' (Vincent, 1966), with probably 'less than 30 birds' confined to the island of Mahé (Gaymer et al., 1969). This species was the focus of small-scale reintroduction efforts in the 1970-1980s, which were largely unsuccessful. However, the species did start to recover and by 1981 the population was estimated at approximately 370 pairs (Watson, 1981; Collar & Stuart, 1985). The population is thought to have remained relatively stable since (Kay et al., 2002). Microsatellite genotypes of 42 museum skins of Seychelles kestrels collected from 1862 to 1940 provided a comparison to contemporary genetic data. Estimates of temporal change in N_e suggest that this species has endured a recent and severe population crash, from an ancestral N_e of $\sim$ 320 individuals 150 years ago to an N_e of eight individuals (95% limits,

3.5–22.0) in 2004 Genetic Consequences of Reintroductions (Figure 12.5; see Groombridge *et al.*, 2009). The contemporary $N_e = 8$ estimate supports field reports of 'less than 30 birds' in the 1960s (Vincent, 1966; Gaymer *et al.*, 1969). A similar study examining temporal change in N_e for the Mauritius kestrel also used pre-bottleneck museum specimens and confirmed that ancestral N_e was far higher ($N_e = 941$) during the 100 or so years before this species'

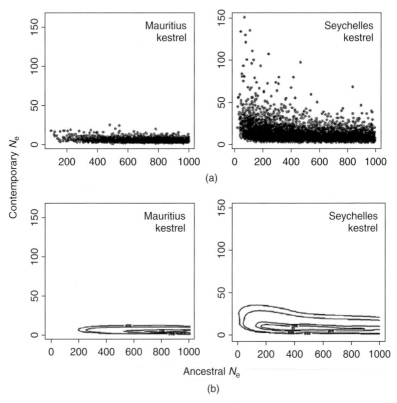

(a)

(b)

Figure 12.5 **Posterior distribution of the ancestral and contemporary effective population size** (N_e) **of the Seychelles kestrel and Mauritius kestrel, following the methods of Beaumont (2003). (a) The density of points is proportional to the probability density of population size at the two different times; hence an off-diagonal distribution indicates a change in** N_e **(note that the** *y* **axis extends only to 150). (b) 25–95% higher posterior density limits of the posterior distribution. The joint mode is plotted as a single open circle. (Taken from Groombridge *et al.*, 2009).**

population crashed to just four known birds in 1974. The crash subsequently resulted in a 57% reduction in heterozygosity and a 55% reduction in allelic diversity (Groombridge *et al.*, 2000) as well as a post-bottleneck N_e of four (Beaumont, 2003; Jones *et al.*, 1995; Groombridge *et al.*, 2001).

These two studies have enabled the outcomes of reintroduction efforts for both species to be reinterpreted. Translocation efforts for the Seychelles kestrel were largely unsuccessful. In 1977, thirteen birds were translocated from Mahé to Praslin, where the species had previously been extirpated. By 1980 the Praslin population had grown to 10 pairs (Watson, 1989), but it subsequently declined to just a handful of territories (Kay *et al.*, 2004). The species as a whole, however, has not shown any sign of problems connected to its recent genetic impoverishment and has recovered with the support of minimal conservation efforts. The largely unsuccessful translocations on Seychelles are in direct contrast to those of the Mauritius kestrel, where a strategy of captive breeding and large-scale releases (a total of 333 birds released by 1994) was successful in restoring that species to $\sim$ 800 birds by 2000 (Jones *et al.*, 1995; Groombridge *et al.*, 2000, 2001; Jones & Merton, this volume, Chapter 2). Both species experienced dramatic historical reductions in N_e, but each responded very differently to translocation efforts. Both species have historical bottleneck profiles that are comparable, suggesting similar levels of inbreeding in the translocated individuals from each source. One explanation is that the extent of inbreeding depression may have varied due to differences in the amount of opportunity for purging; whereas wild adults were used for translocation on Seychelles, the 15-year captive-breeding programme on Mauritius may have allowed deleterious alleles to be purged, lightening the genetic load of the reintroduced populations. However, empirical evidence for the beneficial effects of purging is weak (discussed in more detail in Keller *et al.*, this volume, Chapter 11) and its effects are unlikely to be felt over the short timescales considered here. Alternatively, timing and the age of individuals may have played a part. Translocation efforts on Seychelles involved adults and occurred when the remnant population was slowly rebounding, whereas those on Mauritius used mostly juveniles and began after many years of captive breeding (Jones *et al.*, 1995). The lengthy captive-breeding component of the translocations on Mauritius further sets them apart from those on Seychelles. Surprisingly, however, although problems associated with genetic adaptation to captivity would have been most likely on Mauritius, they have not become evident.

Timescales across which reductions in population size have occurred will of course vary between species according to their ecological history. For the majority perhaps, such as the kestrel species described here, populations will have begun to decline within the last few hundred years as a consequence of human activity (Groombridge *et al.*, 2000, 2009; Nichols *et al.*, 2001). However, for other species, declines may have begun much earlier. For example, in the Hawaiian goose (*Branta sandvicensis*), genetic data from historical material indicates a protracted bottleneck spanning many centuries of human habitation (Paxinos *et al.*, 2002). Extensive translocation programmes have been implemented for both the Mauritius kestrel and Hawaiian goose, so we might expect downstream evidence of inbreeding depression in their respective reintroduced populations to reflect their contrasting historical bottleneck profiles. While reintroduced Mauritius kestrels have so far shown very little effects of inbreeding depression, the signs of inbreeding depression expected in the reintroduced populations of Hawaiian goose have not materialized, despite the species' protracted bottleneck (Veillet, 2007).

Information on the genetic makeup of ancestral populations can also be used to inform future plans for translocations directly in instances where the extent of gene flow, or mixing, between subpopulations is unknown. An example is the use of historical genetic information to inform translocation plans for the Floreana mockingbird, *Mimus trifasciatus* (see Box 12.3). Decisions that act on historical information to mix recently separated populations in order to retain, or reinstate, genetic diversity are part of a continuum of creative strategies to manage populations genetically. At the extreme end of this are strategies involving intentional introgression of external genetic diversity into a population to mitigate effects of inbreeding depression and reinvigorate population growth ('genetic rescue'). This management tool has been developed within conservation biology for over a decade. Early experiments using *Drosophila* established that very few immigrants might be necessary to have a substantial positive effect, although the outcome was difficult to predict due to large variations in expected versus actual increase in genetic diversity (Ball *et al.*, 2000; Ingvarsson, 2001). Importantly, while the effects of immigrant genes can have positive (and often dramatic) effects on the trajectory of small populations, there is a risk that these benefits can be offset by the influence of genetically very divergent individuals and subsequent outbreeding depression (Tallmon *et al.*, 2004). Applying genetic rescue to reintroduction programmes requires some caution (see Jamieson & Lacy, this volume, Chapter 13, for more discussion of this topic).

Alongside the need to mitigate effects of inbreeding depression, the addition of variation that might be adaptive in future environments should help the long-term persistence of the translocated population (Hedrick, 2005). Despite the uncertainties surrounding the concept of genetic rescue, the evidence in favour of its use is compelling. One iconic example is that of the Florida panther, *Puma concolor coryi*, which was rescued through the introgression of novel genetic diversity from female panthers from the Texas population (Pimm *et al.*, 2006). Other case studies have documented substantial positive effects, including the recovery of an isolated population of Scandinavian grey wolves, *Canis lupis*, by a single immigrant (Vila *et al.*, 2003) and the restoration of a bighorn sheep, *Ovis canadensis*, population (Hogg *et al.*, 2006). More encouraging still is the genetic rescue of greater prairie chickens reported by Bouzat *et al.* (2010). In this case, translocations of novel genetic diversity via translocation efforts not only mitigated the effects of inbreeding depression and resulted in immediate increases in fitness but succeeded in restoring genetic diversity to ancestral levels.

Overall, there is some way to go before approaches such as genetic rescue become commonplace in translocation programmes, but the increasing number (and taxonomic diversity) of studies documenting positive genetic and demographic benefits suggests that there is important future scope for this and other kinds of creative thinking in reintroduction biology. What is clear is that only by comparing levels of genetic variation before, as well as after, such events will we be able better to gauge levels of introgresssion and effects of genetic rescue.

What lessons can we learn and how should we apply them?

The review of case studies in this chapter suggests that the evidence for deleterious genetic consequences in translocated populations is largely anecdotal, but influenced by a small (but growing) number of detailed studies that have utilized long-term datasets obtained from intensive field programmes. There is not yet any concrete evidence linking translocation success to genetic parameters but, as stated above, this may be for a number of reasons and may not reflect a lack of a direct connection.

The majority of genetic studies of translocated populations have usually been conducted *after* translocations have taken place (often because of logistical reasons or funding priorities), but the information they yield can be used to redress any imbalances observed. One example is the critically endangered Seychelles paradise flycatcher, *Terpsiphone corvina*, a species endemic to the Seychelles archipelago. The last population on La Digue Island was used as the source of 23 birds reintroduced to Denis Island in 2008 as a first step towards restoring this species. An initial survey using microsatellite markers has shown that 85% of allelic diversity was captured among the translocated individuals, compared to simulations that predicted a range of potential values between 55 and 90% (Tucker, 2010; Bristol *et al.*, in preparation). However, genetic surveys completed *before* translocations, such as those for the Seychelles white-eye, *Zosterops modestus* (Rocamora & Richardson, 2003), allow a more informed approach to ensure those groups of individuals released contain appropriate and/or representative levels of genetic diversity. Post-translocation studies do allow managers to decide whether additional translocation efforts are required. As the science of reintroduction biology develops, genetic studies prior to translocation should be encouraged and information integrated into the planning of species translocation programmes.

Species recovery initiatives with long-term population monitoring programmes have the potential to provide valuable genetic datasets punctuated by measurable reintroduction events. These case studies are likely to produce the resolution required to evaluate how important genetic diversity really is to translocation outcome. At the very least, medium- to long-term genetic monitoring of translocated populations enables an assessment of the effects of genetic drift and can allow for long-term genetic management planning, such as later supplementary 'top-up' translocations to be planned. Inferences drawn from our review of studies that have monitored genetic diversity in translocated populations (Table 12.1) suggests no clear pattern as yet with regards to loss of genetic diversity and levels of inbreeding, although the scale at which translocations are currently implemented (e.g. below optimal numbers of founders) and acknowledged limitations of our ability to detect inbreeding effects suggest that patterns are likely to emerge as this field of reintroduction science develops. Although many of the key parameters that determine the genetic makeup of translocated populations, e.g. individuals released and gene flow/migration among subpopulations, can be managed to minimize negative consequences, other parameters, such as reproductive skew and inbreeding avoidance, are less easily manipulated.

The predominant use of neutral markers to address genetic questions relevant to reintroductions is slowly changing to embrace functional loci. These markers promise a new perspective on how translocations are planned in terms of functional genetic variation, for instance in terms of defining optimal founder number. What is encouraging is how managers are increasingly engaging with geneticists to generate empirical genetic data from translocation programmes. Greater numbers of these types of studies promise a better grasp of the true genetic consequences of translocation. Alongside the use of genetic markers, pedigrees derived from detailed field monitoring of translocated populations are increasingly being used to address genetic management issues. However, while pedigrees are considered to outperform marker-based data, their usefulness relies on them being complete, a requirement that is not easily or frequently met. In these instances, genetic markers can be used to fill in, or confirm, missing pedigree data. Consequently, routine genetic sampling alongside field monitoring remains the most reliable option if the full extent of negative genetic consequences is to be detected – and therefore managed – in reintroduced populations.

Information on historical levels of genetic diversity can help to place current levels of genetic diversity observed in source and reintroduced populations in context. The theoretical issues discussed in Keller *et al.*, this volume, Chapter 11, advocate that managers should assume that the effects of inbreeding depression are present even when no effect is detected. That assumption can be refined by characterizing the historical bottleneck. Our development of reintroduction biology as a science is likely to benefit in the future from increased use of historical specimens and greater integration of creative management philosophies, such as genetic rescue.

Given the difficulties in detecting inbreeding depression in reintroduced populations, it is not surprising that many field managers might see little urgency in assuming its presence and managing its effects. Therefore, a pre-emptive policy of maximizing the number of individuals translocated makes good sense (as outlined by Jamieson & Lacy, this volume, Chapter 13). Using a case study of mohua, *Mohoua ochrocephala*, (see Box 13.3 in Jamieson & Lacy, Chapter 13), Jamieson & Lacy point out that while simulations based on observed allele frequencies suggest that successfully reintroducing 20–30 individuals is sufficient to capture >97.5% gene diversity, increasing this much beyond 60 individuals is likely to win little genetic benefit, but will of course enhance the likelihood of establishment and population growth. This guidance is likely to hold true for neutral diversity, but less so for diversity at functional loci such as MHC genes, where translocation of large numbers of individuals can help retain important rare alleles. Instilling these broad guidelines, now increasingly supported by empirical evidence, should provide a benchmark for future reintroduction efforts.

Summary

Genetic factors are likely to have consequences for long-term persistence of reintroduced populations, but scientifically robust methods to measure their effects have not yet become sufficiently widely used to enable broad yet accurate predictions. Currently, what guidance the science of reintroduction biology can offer to practitioners is limited to generalizations influenced by a small number of well-documented studies. As the science of reintroduction biology develops (Seddon *et al.*, 2007), limitations may emerge as to what factors can, and cannot, be manipulated to manage reintroduced populations sustainably for the long term. What is clear is that detecting negative genetic consequences in reintroduced populations is compromised not only by our ability to measure them but also by the comparatively small sample sizes (numbers of individuals) inherent in reintroductions. Long-term monitoring of genetic diversity and inbreeding in reintroduced populations needs to be incorporated into field programmes, to provide the data and necessary statistical power to look for – and understand – the consequent genetic effects.

References

Alcaide, M., Negro, J. J., Serrano, D. *et al.* (2010) Captive breeding and reintroduction of the lesser kestrel *Falco naumanni*: a genetic analysis using microsatellites. *Conservation Genetics*, 11, 331–338.

Allendorf, F.W. & Luikart, G. (2007) *Conservation and the Genetics of Populations.* Blackwell, Oxford, UK.

Ardern, S.L. & Lambert, D.M. (1997) Is the black robin in genetic peril? *Molecular Ecology*, 6, 21–28.

Armbruster, P. & Reed, D. H. (2005) Inbreeding depression in benign and stressful Environments. *Heredity*, 95, 235–242.

Ball, S.J., Adams, M., Possingham, H.P. *et al.* (2000) The genetic contribution of single male immigrants to small, inbred populations: a laboratory study using *Drosophila melanogaster. Heredity*, 84, 677–684.

Balloux, F., Amos, W. & Coulson, T. (2004) Does heterozygosity estimate inbreeding in real populations? *Molecular Ecology*, 13, 3021–3031.

Beaumont, M. A. (2003) Estimation of population growth or decline in genetically monitored populations. *Genetics*, 164, 1139–1160.

Beck, B.B., Rapaport, L.G., Stanley Price, M. *et al.* (1994) Reintroduction of captive-born animals. In *Creative Conservation: interactive management of wild and captive animals*, eds P.J.S. Olney, G. Mace & A.T.C. Feistner, pp. 265–286. Chapman & Hall, London, UK.

Bekessy, A. D., Ennos, R. A., Burgman, M. A. *et al.* (2003) Neutral DNA markers fail to detect genetic divergence in an ecologically important trait. *Biological Conservation*, 110, 267–275.

Biebach, I. & Keller, L. F. (2009) A strong genetic footprint of the re-introduction history of Alpine ibex (Capra ibex ibex). Molecular Ecology, 18, 5046–5058

Biebach, I. & Keller, L. F. (2010) Inbreeding in reintroduced populations: the effects of early reintroduction history and contemporary processes. *Conservation Genetics*, 11, 527–538

Bouzat, J.L., Johnson, J.A., Toepfer, J.E. *et al.* (2010) Beyond the beneficial effects of translocations as an effective tool for the genetic restoration of isolated populations. *Conservation Genetics*, 10, 191–201.

Brekke, P., Bennett, P.M., Santure, A.W. *et al.* (2011) High genetic diversity in the remnant island population of hihi and the genetic consequences of re-introduction. *Molecular Ecology*, 20, 29–45.

Brekke, P., Bennett, P.M., Wang, J. *et al.* (2010) Sensitive males: inbreeding depression in an endangered bird. *Proceedings of the Royal Society of London B*, 277, 3677–3684.

Bristol, R.M., Tucker, R. & Groombridge, J.J. (2011) Genetic consequences of reintroduction of the Seychelles paradise flycatcher (*Terpsiphone corvina*) (in preparation).

Brouwer, L., Barr, I., van dePol, M. *et al.* (2010) MHC-dependent survival in a wild population: evidence for hidden genetic benefits gained through extra-pair fertilizations. *Molecular Ecology*, 19, 3444–3455.

Brouwer, L., Eikenaar, C., Komdeur, J. *et al.* (2007) Heterozygosity-fitness correlations in a bottlenecked island species: a case study on the Seychelles warbler. *Molecular Ecology*, 16, 3134–3144.

Castric, V. & Vekemans, X. (2004) Plant self-incompatibility in natural populations: a critical assessment of recent theoretical and empirical advances. *Molecular Ecology*, 13, 2873–2889.

Charles Darwin Foundation (2008) The reintroduction of the Floreana mockingbird to its island of origin. Charles Darwin Research Foundation and Galapagos National Park, Puerto Ayora, Galapagos.

Charles Darwin Foundation (2009) Project Floreana 2009–2013: an integrated approach. Charles Darwin Research Foundation, Puerto Ayora, Galapagos.

Cohen, S. (2002) Strong positive selection and habitat-specific amino acid substitution patterns in *Mhc* from an estuarine fish under intense pollution stress. *Molecular Biology and Evolution*, 19, 1870–1880.

Collar, N.J. & Stuart, S.N. (1985) *Threatened Birds of Africa and Related Islands. ICBP/IUCN Red Data Book, Part 1.* International Council for Bird Preservation, Cambridge, UK.

Collins, M.S., Smith, T.B., Seibels, R.E. *et al.* (1998) Approaches to the Reintroduction of the Bali Mynah. *Zoo Biology*, 17, 267–284.

DeRose M. A. & Roff, D. A. (1999) A comparison of inbreeding depression in life-history and morphological traits in animals. *Evolution*, 53, 1288–1292.

Doherty, P. & Zinkernagel, R. (1975) Enhanced immunological surveillance in mice heterozygous at the H-2 gene complex. *Nature*, 256, 50–52.

Duffy, K. (1993) Echo parakeet project – progress report August 1992–April 1993. Mauritian Wildlife Foundation.

Earnhardt, J.M. (1999) Reintroduction programmes: genetic trade-offs for populations. *Animal Conservation*, 2, 279–286.

Edwards, S.V. & Hedrick, P.W. (1998) Evolution and ecology of MHC molecules: from genomics to sexual selection. *Trends in Ecology and Evolution*, 13, 305–311.

Ellegren, H., Hartman, G., Johansson, M. *et al.* (1993) Major histocompatibility complex monomorphism and low levels of DNA fingerprinting variability in a reintroduced and rapidly expanding population of beavers. *Proceedings of the National Academy of Sciences USA*, 90, 8150–8153.

England, P.R., Osler, G.H.R., Woodworth, L.M. *et al.* (2003) Effects of intense versus diffuse population bottlenecks on microsatellite genetic diversity and evolutionary potential. *Conservation Genetics*, 4, 595–604.

Ewen, J.G., Acevedo-Whitehouse, K., Alley, M. *et al.* (2011) Empirical consideration of parasites and health in reintroduction. In *Reintroduction Biology: integrating science and management*, eds J.G. Ewen, D.P. Armstrong, K.A. Parker & P.J. Seddon, Chapter 9. Wiley-Blackwell, Oxford, UK.

Ewing, S.R., Nager, R.G., Nicoll, M.A.C. *et al.* (2008) Inbreeding and loss of genetic variation in a reintroduced population of Mauritius kestrel. *Conservation Biology*, 22, 395–404.

Fischer, J. & Lindenmayer, D.B. (2000) An assessment of the published results of animal relocations. *Biological Conservation*, 96, 1–11.

Frankham, R. (2005) Genetics and extinction. *Biological Conservation*, 12, 131–140.

Frankham, R. (2008) Genetic adaptation to captivity in species conservation programs. *Molecular Ecology*, 17, 325–333.

Frankham, R., Ballou, J.D. & Briscoe, D.A. (2002) *Introduction to Conservation Genetics*, 1st edn. Cambridge University Press, Cambridge, UK.

Frankham, R., Ballou, J.D. & Briscoe, D.A. (2010) *Introduction to Conservation Genetics*, 2nd edn. Cambridge University Press, Cambridge, UK.

Frankham, R., Lees, K., Montgomery, M.E. *et al.* (1999) Do population size bottlenecks reduce evolutionary potential? *Animal Conservation*, 2, 255–260.

Frederickson, R.J., Siminski, P., Woolf, M. *et al.* (2007) Genetic rescue and inbreeding depression in Mexican wolves. *Proceedings of the Royal Society B*, 22, 2365–2371.

Gautschi, B., Müller, J.P., Schmid, B. *et al.* (2003) Effective number of breeders and maintenance of genetic diversity in the captive bearded vulture population. *Heredity*, 91, 9–16.

Gautschi, B., Widner, A., Joshi, J. *et al.* (2002) Increased frequency of scale abnormalities and loss of genetic variation in serially bottlenecked populations of the dice snake (*Natrix tessellate*). *Conservation Genetics*, 3, 235–245.

Gaymer, R., Blackman, R.A.A., Dawson, P.G. *et al.* (1969) The endemic birds of Seychelles. *Ibis*, 111, 157–176.

Groombridge, J. (2007) Genetics and extinction of island endemics: the importance of historical perspectives. *Animal Conservation* 10: 147–148.

Groombridge, J.J., Bruford, M.W., Jones, C.G. *et al.* (2001) Estimating the severity of the population bottleneck in the Mauritius kestrel *Falco punctatus* from ringing records using MCMC estimation. *Journal of Animal Ecology*, 70, 401–409.

Groombridge, J.J., Dawson, D.A., Burke, T. *et al.* (2009) Evaluating the demographic history of the Seychelles kestrel (*Falco araea*): genetic evidence for recovery from a population bottleneck following minimal conservation management. *Biological Conservation*, 142, 2250–2257.

Groombridge, J.J., Jones, C.G., Bruford, M.W. *et al.* (2000) 'Ghost' alleles of the Mauritius kestrel. *Nature*, 403, 616.

Groombridge, J.J., Massey, J.G., Bruchm, J.C. *et al.* (2004) An attempt to recover the po'ouli by translocation and the appraisal of recovery strategy for bird species of extreme rarity. *Biological Conservation*, 118, 365–375.

Haig, S.M., Ballou, J.D. & Derrickson, S.R. (1990) Management options for preserving genetic diversity: reintroduction of Guam rails to the wild. *Conservation Biology*, 4, 290–300.

Hansson, B. & Richardson, D.S. (2005) Genetic variation in two endangered *Acrocephalus* species compared to a widespread congener: estimates based on functional and random loci. *Animal Conservation*, 8, 83–90.

Harmon, S.A. & Van DenBussche, R.A. (2000) Genetic attributes of a declining population of reintroduced eastern wild turkeys (*Meleagris gallopavo silvestris*). *Southwestern Naturalist*, 45, 258–266.

Hedrick, P.W. (2002) Pathogen resistance and genetic variation at MHC loci. *Evolution*, 56, 1902–1908.

Hedrick, P. (2005) 'Genetic restoration': a more comprehensive perspective than 'genetic rescue'. *Trends in Ecology and Evolution*, 20, 109.

Hedrick, P., Fredrickson, R. & Ellegren, H. (2001) Evaluation of d2, a microsatellite measure of inbreeding and outbreeding in wolves with a known pedigree. *Evolution*, 55, 1256–1260.

Hedrick, P.W., Lee, R.N. & Parker, K.M. (2000) Major histocompatibility complex (MHC) variation in the endangered Mexican wolf and related canids. *Heredity*, 85, 617–624.

Hedrick, P.W., Parker, K.M., Miller, E.L. *et al.* (1999) Major histocompatibility complex variation in the endangered Przewalski's horse. *Genetics*, 152, 1701–1710.

Hoeck, P.E.A., Bollmer, J.L., Parker, P.G. *et al.* (2009a) Microsatellite primers for the four Galapagos mockingbird species (*Mimus parvulus*, *M. macdonaldi*, *M. melanotis* and *M. trifasciatus*). *Molecular Ecology Resources*, 9, 1538–1541.

Hoeck, P.E.A., Bollmer, J.L., Parker, P.G. *et al.* (2009b). Differentiation with drift: a spatio-temporal genetic analysis of Galapagos mockingbird populations (*Mimus* spp.). *Philosophical Transactions of the Royal Society B*, 365, 1127–1138.

Hogg, J.T., Forbes, S.H., Steele, B.M. *et al.* (2006) Genetic rescue of an insular population of large mammals. *Proceedings of the Royal Society of London B*, 273, 1491–1499.

Hooson, S. & Jamieson, I.G. (2003) The distribution and current status of New Zealand saddlebacks *Philesturnus carunculatus carunculatus*. *Bird Conservation International*, 13, 79–95.

Hughes, A. (1991) MHC polymorphism and the design of captive breeding programs. *Conservation Biology*, 5, 249–251.

Hughes, A.L. & Yeager, M. (1998) Natural selection at major histocompatibility complex loci of vertebrates. *Annual Review of Genetics*, 32, 415–435.

Ingvarsson, P.K. (2001) Restoration of genetic variation lost – the genetic rescue hypothesis. *Trends in Ecology and Evolution*, 16, 62–63.

Jamieson, I.G. (2011). Founder effects, inbreeding and loss of genetic diversity in four avian reintroduction programs. *Conservation Biology*, 25, 115–123.

Jamieson, I.G. & Lacy, R.C. (2011) Managing genetic issues in reintroduction biology. In *Reintroduction Biology: integrating science and management*, eds J.G. Ewen, D.P. Armstrong, K.A. Parker & P.J. Seddon, Chapter 13. Wiley-Blackwell, Oxford, UK.

Jamieson, I.G., Grueber, C.E., Waters, J.M. *et al.* (2008) Managing genetic diversity in threatened populations: a New Zealand perspective. *New Zealand Journal of Ecology*, 32, 130–137.

Jamieson, I.G., Tracy, L.N. Fletcher, D. *et al.* (2007) Moderate inbreeding depression in a reintroduced population of North Island robins. *Animal Conservation*, 10, 95–102.

Jeffery, K.J.M. & Bangham, C.R.M. (2000) Do infectious diseases drive MHC diversity? *Microbes and Infection*, 2, 1335–1341.

Jenkins, M. (1979) Natural History of the Guam rail. *Condor*, 81, 404–408.

Jones, C.G. & Merton, D.V. (2011) A tale of two islands: the rescue and recovery of endemic birds in New Zealand and Mauritius. In *Reintroduction Biology:*

integrating science and management, eds J.G. Ewen, D.P. Armstrong, K.A. Parker & P.J. Seddon, Chapter 2. Wiley-Blackwell, Oxford, UK.

Jones, C.G., Heck, W., Lewis, R.E. *et al.* (1995) The restoration of the Mauritius kestrel *Falco punctatus* population. *Ibis*, 137, S173–S180.

Kalinowski, S. T. & Hedrick, P. W. (2006) Detecting inbreeding depression is difficult in captive endangered species. *Animal Conservation*, 2, 131–136

Kay, S., Millet, J., Watson, J. *et al.* (2002) Status of the Seychelles kestrel *Falco araea*: a reassessment of the populations on Mahé and Praslin 2001–2002. *Report by BirdLife Seychelles*, Victoria, Mahé, Republic of Seychelles.

Kay, S., Millet, J., Watson, J. *et al.* (2004) Status of the Seychelles kestrel *Falco araea* on Praslin: an assessment of a re-introduced population on Praslin 2002–2003. Report by BirdLife Seychelles, Victoria, Mahé, Republic of Seychelles.

Keller, K.F. & Waller, D.M. (2002) Inbreeding effects in wild populations. *Trends in Ecology and Evolution*, 17, 230–241.

Keller, L.K., Biebach, I., Ewing, S.R. *et al.* (2011) The genetics of reintroductions: inbreeding and genetic drift. In *Reintroduction Biology: integrating science and management*, eds J.G. Ewen, D.P. Armstrong, K.A. Parker & P.J. Seddon, Chapter 11. Wiley-Blackwell, Oxford, UK.

King, H.D. (1939) Life processes in gray Norway rats during fourteen years in captivity. *American Anatomical Memoires*, 17, 1–77.

Komdeur, J. (1994) Conserving the Seychelles warbler *Acrocephalus sechellensis* by translocation from Cousin Island to the islands of Aride and Cousine. *Biological Conservation*, 76, 143–152.

Komdeur, J. (1998) Long-term fitness benefits of egg sex modification by the Seychelles warbler. *Ecology Letters*, 1, 56–62.

Komdeur, J., Kappe, A.L. & van deZande, L. (1998) Influence of population isolation on genetic variation and demography in Seychelles warblers; a field experiment. *Animal Conservation*, 1, 203–212.

Lambert, D.M., King, T., Shepherd, L.D. *et al.* (2005) Serial population bottlenecks and genetic variation: translocated populations of the New Zealand saddleback (*Philesturnus carunculatus rufusater*). *Conservation Genetics*, 6, 1–14.

Latch, E.K. & Rhodes, O.E. (2005) The effects of gene flow and population isolation on the genetic structure of reintroduced wild turkey populations: are genetic signatures of source populations retained? *Conservation Genetics*, 6, 981–997.

Leberg, P.L. (1993) Strategies for population reintroduction: effects of genetic variability on population growth and size. *Conservation Biology*, 7, 194–199.

Leberg, P.L. & Firmin, B.D. (2008) Role of inbreeding depression and purging in captive-breeding and restoration programmes. *Molecular Ecology*, 17, 334–343.

Le Gouar, P., Rigal, F., Boisselier-Dubayle, M.C. *et al.* (2008) Genetic variation in a network of natural and reintroduced populations of griffon vulture (*Gyps fulvus*) in Europe. *Conservation Genetics*, 9, 349–359.

Lewis, O.T. & Thomas, C.D. (2001) Adaptation to captivity in the butterfly *Pieris brassicae* (L.) and implications for ex situ conservation. *Journal of Insect Conservation*, 5, 55–63.

Ljungqvist, M., Åkesson, M. & Hansson, B. (2010) Do microsatellites reflect genome-wide genetic diversity in natural populations? A comment on Väli *et al.* (2008). *Molecular Ecology*, 19, 851–855.

Lousteau-Lalanne, P. (1968) The Seychelles, Cousin Island nature reserve. International Council for Bird Preservation.

Lovegrove, T.G., Nieuwland, A.B. & Green, S. (1995) Interim report on the echo parakeet conservation project, February 1995. Mauritian Wildlife Foundation.

Lynch, M. (1988) Estimation of relatedness by DNA fingerprinting. *Molecular Biology and Evolution*, 5, 584–599.

Lynch, M. & Ritland, K. (1999) Estimation of pairwise relatedness with molecular markers. *Genetics*, 152, 1753–1766.

Madsen, T., Shine, R., Olsson, M. *et al.* (1999) Restoration of an inbred adder population. *Nature*, 402, 34–35.

Markert, J.A., Grant, P.R., Grant, B.R. *et al.* (2004) Neutral locus heterozygosity, inbreeding and survival in Darwin's ground finches (*Geospiza fortis* and *G. scandens*). *Heredity*, 92, 306–315.

Marshall, T.C. & Spalton, J.A. (2000) Simultaneous inbreeding and outbreeding depression in reintroduced Arabian oryx. *Animal Conservation*, 3, 241–248.

Marshall, T.C., Coltman, D.W., Pemberton, J.M. *et al.* (2002) Estimating the prevalence of inbreeding from incomplete pedigrees. *Proceedings of the Royal Society of London Series B – Biological Sciences*, 269, 1533–1539.

Miller, H.C. & Lambert, D.M. (2004) Genetic drift outweighs balancing selection in shaping post-bottleneck major histocompatibility complex variation in New Zealand robins (*Petroicidae*). *Molecular Ecology*, 13, 3709–3721.

Miller, K.A., Nelson, N.J., Smith, H.G. *et al.* (2009) How do reproductive skew and founder group size affect genetic diversity in reintroduced populations? *Molecular Ecology*, 18, 3792–3802.

Nader, W., Werner, D. & Wink, M. (1999) Genetic diversity of scarlet macaws *Ara macao* in reintroduction studies for threatened populations in Costa Rica. *Biological Conservation*, 87, 269–272.

Negro, J.J. & Torres, M.J. (1999) Genetic variability and differentiation of two bearded vulture *Gypaetus barbatus* populations and implications for reintroduction projects. *Biological Conservation*, 87, 249–254.

Nichols, R.A., Bruford, M.W. & Groombridge, J.J. (2001) Sustaining genetic variation in a small population: evidence from the Mauritius kestrel. *Molecular Ecology*, 10, 593–602.

Nicoll, M.A.C., Jones, C.G. & Norris, K. (2004) Comparison of survival rates of captive-reared and wild-bred Mauritius kestrels (*Falco punctatus*) in a re-introduced population. *Biological Conservation*, 118, 539–548.

Ouborg, N.J., Pertoldi, C., Loeschcke, V. *et al.* (2010) Conservation genetics in transition to conservation genomics. *Trends in Genetics*, 26, 177–187.

Overall, A.D.J., Byrne, K.A., Pilkington, J.G. *et al.* (2005) Heterozygosity, inbreeding and neonatal traits in Soay sheep on St Kilda. *Molecular Ecology*, 14, 3383–3393.

Paxinos, E.E., James, H.F., Olson, S.L. *et al.* (2002) Prehistoric decline of genetic diversity in the nene. *Science*, 296, 1827.

Pemberton, J. (2004) Measuring inbreeding depression in the wild: the old ways are the best. *Trends in Ecology and Evolution*, 19, 613–615.

Pemberton, J.M. (2008) Wild pedigrees: the way forward. *Proceedings of the Royal Society of London B*, 275, 613–621.

Penn, D., Damjanovich, K. & Potts, W. (2002) MHC heterozygosity confers a selective advantage against multiple-strain infections. *Proceedings of the National academy of Science. USA*, 99, 11260–11264.

Pimm, S.L., Dollar, L. & Bass Jr, O.L. (2006) The genetic rescue of the Florida panther. *Animal Conservation*, 9, 115–122.

Pritchard, J., Stephens, M. & Donnelly, P. (2000) Inference of population structure using multilocus genotype data. *Genetics*, 155, 945–959.

Queller, D.C. & Goodnight, K.F. (1989) Estimating relatedness using genetic markers. *Evolution*, 43, 258–275.

Raisin, C., Frantz, A.C., Kundu, S. *et al.* (2011) Genetic homogenisation following intensive conservation management of the endangered Mauritius parakeet. *Conservation Genetics* (submitted).

Reed, D.H. & Frankham, R. (2001) How closely correlated are molecular and quantitative measures of genetic variation? A meta-analysis. *Evolution*, 55, 1095–1103.

Richards, H. & Tollington, S. (2010) Management of the echo parakeet in the wild 2010. Unpublished report to Mauritian Wildlife Foundation, Vacoas, Mauritius.

Richardson, D.S. & Westerdahl, H. (2003) MHC diversity in two *Acrocephalus* species: the outbred great reed warbler and the inbred Seychelles warbler. *Molecular Ecology*, 12, 3523–3529.

Richardson, D.S., Bristol, R. & Shah, N.J. (2006) Translocation of Seychelles warbler *Acrocephalus sechellensis* to establish a new population on Denis Island, Seychelles. *Conservation Evidence*, 3, 54–57.

Richardson, D.S., Jury, F.L., Dawson, D.A. *et al.* (2000) Fifty Seychelles warbler (*Acrocephalus sechellensis*) microsatellite loci polymorphic in Sylviidae species and their cross-species amplification in other passerine birds. *Molecular Ecology*, 9, 2226–2231.

Richardson, D.S., Komdeur, J. & Burke, T.A. (2004) Inbreeding in the Seychelles warbler: environment-dependent maternal effects. *Evolution*, 58, 2037–2048.

Richardson, D.S., Komdeur, J., Burke, T. *et al.* (2005) MHC-based patterns of social and extra-pair mate choice in the Seychelles warbler. *Proceedings of the Royal Society of London B*, 272, 759–767.

Robert, A. (2009) Captive breeding genetics and reintroduction success. *Biological Conservation*, 142, 2915–2922.

Rocamora, G. & Richardson, D.S. (2003) Genetic and morphological differentiation between the two remnant populations of Seychelles white-eye *Zosterops modestus*. *Ibis*, 145, 34–44.

Ruiz-Lopez, M.J., Roldan, E.R.S., Espeso, G. *et al.* (2009) Pedigrees and microsatellites among endangered ungulates: what do they tell us? *Molecular Ecology*, 18, 1352–1364.

Saccheri, I., Kuussaari, M., Kankare, M. *et al.* (1998) Inbreeding and extinction in a butterfly metapopulation. *Nature*, 392, 491–494.

Seddon, P.J., Armstrong, D.P. & Maloney, R.F. (2007) Developing the science of reintroduction. *Conservation Biology*, 21, 303–312.

Seddon, P.J., Strauss, W.M. & Innes, J. (2011) Animal translocations: what are they and why do we do them? In *Reintroduction Biology: integrating science and management*, eds J.G. Ewen, D.P. Armstrong, K.A. Parker & P.J. Seddon, Chapter 1. Wiley-Blackwell, Oxford, UK.

Slate, J., David, P., Dodds, K.G. *et al.* (2004) Understanding the relationship between the inbreeding coefficient and multilocus heterozygosity: theoretical expectations and empirical data. *Heredity*, 93, 255–265.

Sommer, S. (2005) The importance of immune gene variability (MHC) in evolutionary ecology and conservation. *Frontiers in Zoology*, 2, 16.

Spurgin, L.G. & Richardson, D.S. (2010) How pathogens drive genetic diversity: MHC, mechanisms and misunderstandings. *Proceedings of the Royal Society of London B*, 277, 979–988.

Tallmon, D.A., Luikart, G. & Waples, R.S. (2004) The alluring simplicity and complex reality of genetic rescue. *Trends in Ecology and Evolution*, 19, 489–496.

Taylor, S.S. & Jamieson, I.G. (2008) No evidence for loss of genetic variation following sequential translocations in extant populations of a genetically depauperate species. *Molecular Ecology*, 17, 545–556.

Taylor, S.S., Jamieson, I.G. & Wallis, P. (2007) Historic and contemporary levels of genetic variation in two New Zealand passerines with different histories of decline. *Journal of Evolutionary Biology*, 20, 2035–2047.

Thevenon, S. & Couvet, D. (2002) The impact of inbreeding depression on population survival depending on demographic parameters. *Animal Conservation*, 5, 53–60.

Tucker, R. (2010) Conservation genetics of the Seychelles paradise flycatcher (*Terpsiphone corvina*). Unpublished MSc Thesis, Imperial College, London.

Vali, Ü., Einarsson, A., Waits, L. & Ellegren, H. (2008) To what extent do microsatellite markers reflect genome-wide genetic diversity in natural populations? *Molecular Ecology*, 17, 3808–3817.

VanderWerf, E.A., Groombridge, J.J., Fretz, J.S. *et al.* (2006) Decision-analysis to guide recovery of the po'ouli, a critically endangered Hawaiian honeycreeper. *Biological Conservation*, 129, 383–392.

van Tienderen, P.H., deHaan, A.A., van derLinden, C.G. *et al.* (2002). Biodiversity assessment using markers for ecologically important traits. *Trends in Ecology and Evolution*, 17, 577–582.

Veillet, A.C. (2007) Investigation of inbreeding and inbreeding depression in nene (*Branta sandvicensis*) using microsatellite DNA fingerprinting and pedigree information. Research Thesis, University of Hawaii, Hilo, USA.

Vila, C., Sundqvist, A.-K., Flagstad, O. *et al.* (2003) Rescue of a severely bottlenecked wolf (*Canis lupus*) population by a single immigrant. *Proceedings of the Royal Society of London B*, 270, 91–97.

Vincent, J. (1966) *Red Data Book – Aves*. International Union for Conservation of Nature and Natural Resources, Morges.

VonHoldt, B.M., Stahler, D.R., Smith, D.W. *et al.* (2008) The genealogy and genetic viability of reintroduced Yellowstone grey wolves. *Molecular Ecology*, 17, 252–274

Waller, D.M. (1993) The statics and dynamics of mating system evolution. In *The Natural History of Inbreeding and Outbreeding*, ed. N.W. Thornhill, pp. 97–117. University of Chicago Press, Chicago, USA.

Wang, J. (2007a) Triadic IBD coefficients and applications to estimating pairwise relatedness. *Genetics Research*, 89, 135–153.

Wang, J. (2007b) Parentage and sibship exclusions: higher statistical power with more family members. *Heredity*, 99, 205–217.

Watson, J. (1981) *Population ecology, food and conservation of the Seychelles kestrel Falco araea on Mahé*. PhD Thesis, Aberdeen University, Aberdeen, Scotland, UK.

Watson, J. (1989) Successful translocation of the endemic Seychelles kestrel *Falco araea* to Praslin. In *Raptors in the Modern World*, eds B.-U. Meyburg & R.D. Chancellor, pp. 363–367. World Working Group on Birds of Prey and Owls, Berlin, Germany.

Wisely S.M., McDonald, D.B. & Buskirk, S.W. (2003) Evaluation of the species survival plan and captive breeding program for the black-footed ferret. *Zoo Biology*, 22, 287–298.

Wisely, S.M., Santymire, R.M., Livieri, T.M. *et al.* (2008) Genotypic and phenotypic consequences of reintroduction history in the black- footed ferret (*Mustela nigripes*). *Conservation Genetics*, 9, 389–399.

Wolf, C.M., Griffith, B., Reed, C. *et al.* (1996) Avian and mammalian translocations: update and re-analysis of 1987 survey data. *Conservation Biology*, 10, 1142–1154.

Worley, K., Collet, J., Spurgin, L.G. *et al.* (2010) MHC heterozygosity and survival in red junglefowl. *Molecular Ecology*, 19, 3064–3075.

Zhang, B., Fang, S.-G. & Xi, Y.-M. (2006) Major histocompatibility complex variation in the endangered crested ibis *Nipponia nippon* and implications for reintroduction. *Biochemical Genetics*, 44, 113–123.

(13)

Managing Genetic Issues
in Reintroduction Biology

Ian G. Jamieson[1] and Robert C. Lacy[2]

[1]Department of Zoology, University of Otago, New Zealand
[2]Department of Conservation Science, Chicago Zoological Society,
United States of America

'Conservation geneticists need to not only talk-the-talk but also walk-the-walk with managers involved in reintroductions. Once genetic factors are placed in their proper context, then conservation genetics can play an important role in managing long-term genetic viability of reintroduced populations.'

Page 445

Encouraging managers to embrace the genetics of reintroductions

There has been a longstanding debate in the conservation literature about the relative importance of genetic issues in increasing the risk of species extinction (Caro & Laurenson, 1994; Caughley, 1994; Hedrick *et al.*, 1996). Debates of this sort are potentially useful in conservation biology because they help to clarify the issues and explore the relative importance of genetic factors in species declines (Boyce, 2002). For example, we now recognize that genetic factors do not normally work in isolation, but instead often interact with factors that cause populations to decline in the first place (e.g. introduced

Reintroduction Biology: Integrating Science and Management. First Edition.
Edited by John G. Ewen, Doug P. Armstrong, Kevin A. Parker and Philip J. Seddon.
© 2012 Blackwell Publishing Ltd. Published 2012 by Blackwell Publishing Ltd.

predators or disease) (Reed, 2010). Even in the case of island endemics, where the overwhelming evidence points to invasive pest and predators as the major cause of historical extinction (Blackburn *et al.*, 2004), genetics could still play an important role for the remaining subset of island endemics whose populations are small but which are currently protected from introduced predators (Groombridge, 2007; Jamieson, 2007a, 2007b).

Nowadays, conservation geneticists often take an active role in working alongside threatened species managers in helping to steer policy, design guidelines and implement best-practice procedures. Nevertheless, genetic issues tend to have lower priority primarily because of the long time frame over which genetic factors act, relative to other agents of decline such as habitat loss or introduced predators (Jamieson *et al.*, 2008). Therefore, genetic problems are often secondary consequences of the primary causes of demographic decline – i.e. genetic depletion can become very important to long-term viability, but it is a secondary, delayed, effect.

Take, for example, the documented case of inbreeding depression in a reintroduced population of New Zealand North Island robins (*Petroica longipes*) and its potential effect on population viability. North Island robins, like many New Zealand birds, have shown a marked decline in range and numbers in areas where they co-occur with introduced mammalian predators such as Norway rats (*Rattus norvegicus*) and stoats (*Mustela erminea*) (Higgins & Peter, 2002; Innes *et al.*, 2010). Although not as susceptible as some endemic species, robins nevertheless suffer intense nest predation and high juvenile and female mortality in areas inhabited by rats and stoats (Powlesland, 1983; Brown, 1997; Armstrong *et al.*, 2006). Because they are relatively easy to catch and translocate, and because of their tameness around humans, robins are typically one of the first species to be reintroduced in recovery programmes on forested islands or fenced sanctuaries in New Zealand (www.rsg-oceania.squarespace.com/).

In the reintroduced robin population on Tiritiri Matangi Island, Jamieson *et al.* (2007) provided evidence of moderate inbreeding depression in the form of an inbreeding load (or lethal equivalents per diploid organism) of 8.4 in juvenile survival (for definition of inbreeding load and lethal equivalent, see Keller *et al.*, this volume, Chapter 11). This meant that highly inbred juvenile robins (e.g. products of sib–sib matings, or $f = 0.25$) had a 65% lower chance of surviving their first year than did outbred robins ($f = 0$). This level of inbreeding depression is clearly problematic as the population is geographically closed, and inbreeding inevitably increases in closed populations (Frankham *et al.*, 2010; Keller *et al.*, this volume, Chapter 11). A preliminary population

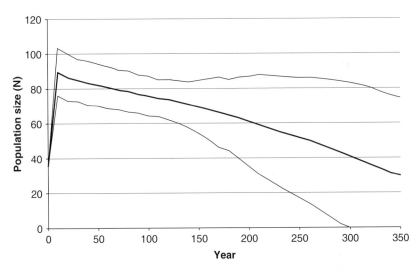

Figure 13.1 **Population projections for reintroduced North Island robins on Tiritiri Matangi Island if there was no increase in carrying capacity due to forest regeneration. The model was implemented in Vortex 9.72 (Lacy *et al.*, 2007) and included density dependence and inbreeding depression (with parameter uncertainty) based on estimates from Jamieson *et al.* (2007). Inbreeding depression was modelled with the effect of purging, with 50% of lethal equivalents due to recessive lethals. The graph shows the mean size of the adult breeding population with 95% CI, based on 500 model repetitions (I. Jamieson, D. Hegg & D. Armstrong, unpublished).**

viability model that takes into account the reduced juvenile survival rate associated with the projected rate of inbreeding, along with other demographic factors such as density dependence, indicates that the population would decline slowly if the carrying capacity of the island does not increase (Figure 13.1) (I. Jamieson, D. Hegg & D. Armstrong, unpublished data). In reality, a rapid rate of decline is unlikely to happen because the small (about 40 pairs) population is currently confined to about 20 ha of forest, and over time are likely to expand to a much larger portion of the 220 ha island, which has largely been replanted. Nevertheless, models such as these clearly indicate that inbreeding depression can lead to an increased risk of extinction.

Island managers, on the other hand, could argue that the risk to reintroduced populations due to inbreeding depression is low relative to the risk of reinvasion of rats, for example. Rates of incursion of rats to islands can be considerable, but the ability to detect and remove rats present at low densities is limited (Russell *et al.*, 2005, 2008). Therefore, given the constraints on a manager's time and funding sources, far more attention will be focused on protecting an island population against reinvasion and re-establishment of rats than towards managing for inbreeding depression. The manager's priorities are quite reasonable, but it is still important to address *all* of the threats to a population, so that the suite of planned actions are sufficient to achieve the conservation goals for the species over the long term (Armstrong & Seddon, 2008), as recommended in the IUCN Species Survival Commission guidelines for species conservation planning (IUCN/SSC, 2008), and as we illustrate later in this chapter.

Another case in point is one of the best-known and highly cited examples of 'genetic rescue'. A long-term study of an isolated population of greater prairie-chicken (*Tympanuchus pinnatus*) identified that a decline in numbers and in hatching success were associated with a reduction in genetic diversity and that both declines were dramatically reversed when birds from another genetically diverse population were introduced, resulting in the apparent rescue of the population from possible extinction (Bouzat *et al.*, 1998; Westemeier *et al.*, 1998). However, subsequent monitoring 10 years on has shown that the isolated population has once again started to decline, presumably as a result of unaddressed poor quality habitat (Bouzat *et al.*, 2009). The message from the authors of this study was that even though follow-up translocations can be effective management strategies for genetic restoration of critically endangered species, demographic recovery and long-term viability would not be realized unless there are complementary management strategies for restoring adequate habitat. Of course, in many cases a mix of both ecological and genetic factors may be responsible for the failure of reintroductions (see the Mauritius and Seychelles kestrel examples in Groombridge *et al.*, this volume, Chapter 12).

One of the first take-home messages from this chapter therefore is that the genetic consequences of reintroduction need to be placed in the perspective of more immediate concerns for managing and controlling the main agents of decline that manifested the need for a translocation in the first place. Furthermore, both managers and geneticists can sometimes overlook the fact that often the most important management action to prevent or minimize genetic problems is to maximize population growth to a high carrying capacity. Thus, the same actions that managers are taking to achieve other goals will also be the best hedge against genetic decay. Some conservation geneticists are questioning the so-called 'detrimental' paradigm of inbreeding depression, where it is assumed that inbreeding depression presupposes a direct and unambiguous relationship between genetic diversity, fitness and extinction, even though empirical evidence for such an assumption is somewhat conflicting (Bouzat, 2010). There is increasing emphasis on how genetics can be integrated into overall management strategies, rather than seeing genetic factors in isolation and as direct and immediate risks to population extinction (Jamieson *et al.*, 2008). Conservation geneticists need to not only talk-the-talk but also walk-the-walk with managers involved in reintroductions. Once genetic factors are placed in their proper context, then conservation genetics can play an important role in managing long-term genetic viability of reintroduced populations (Armstrong & Seddon, 2008).

Genetic implications for reintroductions

Assuming the original causes of local extinction have been removed or neutralized, one of the primary aims of reintroductions in a conservation context is to re-establish a population at a site or in a habitat where the species previously existed (Seddon *et al.*, this volume, Chapter 1). For pragmatic reasons, reintroductions of birds and mammals often involve the release of small numbers of individuals in habitats or sites of restricted size, such as offshore islands. Both of these factors can cause population bottlenecks and, as we have seen in previous chapters, bottlenecks can lead to the loss of genetic diversity and/or inbreeding depression (Keller *et al.*, this volume, Chapter 11). In cases where isolated, remnant populations have already lost considerable genetic diversity or experience high levels of inbreeding, introductions of new genetic variants can prevent the negative consequences of disrupted gene-flow and isolation, as in the case of genetic rescue (Tallmon *et al.*, 2004; Hedrick & Fredrickson, 2010). On the other hand, deliberate out-crossing can also lead to unintentioned genetic consequences in the form of outbreeding depression and disruption of local (adaptive) gene complexes (Edmands, 2007, but also see below).

The theory behind the processes that can lead to loss of genetic diversity, inbreeding depression, outbreeding depression, introgression, etc., is well understood and examples from the conservation genetics literature abound (see Chapters 11 and 12 in this volume), but guidelines to aid conservation managers in decision making are generally lacking (see Hedrick & Fredrickson,

2010, for guidelines on reintroductions associated with 'genetic rescue'). One reason for this might be the perception that each reintroduction is unique. While this may be the case, reintroductions should be no different from any other aspect of conservation biology, where general rules or theory can be, and are, applied (Armstrong & Seddon, 2008). For example, genetic rules for establishing breeding populations in captivity, such as maximizing founder size, equalizing founder representation and avoiding close inbreeding, are utilized regularly by zoos and captive breeding facilities around the world (Ballou & Lacy, 1995; Ballou et al., 2010a). Surely such rules could also be applied to reintroductions. However, because reintroductions are often about establishing natural, free-ranging or 'hands-off' populations, the more direct 'hands-on' management of the gene pool as practised in captive breeding facilities, such as selective breeding of certain individuals or lineages to minimize inbreeding and maximize genetic diversity, is either not desirable or impractical in a wild setting. This does not mean, however, that the desired effects cannot be accomplished with more subtle forms of management, some of which we outline in the following sections.

The same processes that affect small populations in captivity (i.e. inbreeding and genetic drift) apply also to the small populations often associated with reintroductions (Keller et al., this volume, Chapter 11). Although undoubtedly true, the nature and priority of problems that face managers of captive populations do not necessarily coincide with the nature and priority of problems of managing animals in re-established wild populations. For example, some might not perceive inbreeding as a long-term problem for their island-based reintroduction because the species in question existed on the island historically or occurred naturally on many offshore islands. Of course, inbreeding may have been occurring historically on these islands, but infrequent recolonization events from the mainland or nearby large islands may have enabled population persistence. However, the mainland may no longer act as a periodic source of dispersal or gene flow. In some regions, large populations have declined or have been extirpated on the mainland and, consequently, without any nearby source to facilitate gene flow, reintroduced island populations are now essentially geographically closed, and thus subject to ever increasing rates of inbreeding (Frankham et al., 2010).

In summary, if there are practical means of reducing the risks of inbreeding and loss of genetic diversity, then assessing these during the planning stage of any translocation involving wild populations would seem appropriate. We therefore first summarize the main guidelines that are regularly employed when managing genetic diversity in small captive populations, and then ask if and how these guidelines can be adapted to translocations and reintroductions among wild populations. We tend to focus more on managing genetic diversity than inbreeding per se because, as Keller *et al.* (this volume, Chapter 11) point out, the need to retain evolutionary potential in reintroduced populations requires larger effective populations sizes than those needed to avoid inbreeding, and therefore will often subsume problems arising from managing inbreeding and inbreeding depression. Finally, we focus on the genetic questions and issues associated with reintroductions, some of which have been outlined previously and placed into a broader reintroduction framework (Armstrong & Seddon, 2008).

Goals for management of genetic variation and inbreeding

One of the main disadvantages of reintroducing individuals from a captive rather than from a wild source population is that captive-bred founder animals may have been subject to selection in captivity. Such selection may reduce the fitness of free-ranging animals and can therefore seriously undermine the ultimate goal of restoring a species in its natural environment (Frankham, 2008). Otherwise, establishing a population with adequate genetic variation and then minimizing the loss of that variation over the medium and long terms are problems that both captive and wild reintroductions share (Earnhardt, 1999). One of the first attempts to address this problem employed aspects of population genetics theory to derive the so-called 50/500 rule, where the aim is to maintain an effective population size above 50 in the short term to avoid the immediate harmful effects of inbreeding and above 500 in the longer term to balance the loss of genetic variation through drift with the introduction of new mutations (Franklin, 1980). Because the effective population size will usually be much less than the mean census population size (Lande & Barrowclough, 1987), the minimum population size for management will generally be much larger than 50/500. A 50/500 goal may be unrealistic for managing a species at a single site, especially vertebrate species with large home ranges or breeding territories, but it can become more achievable when several sites are managed as a meta-population (Lacy, 1987; Caballero *et al.*, 2010).

However, with environments changing radically and rapidly, the concern is more with the ability of the population to adapt readily than with long-term maintenance in regenerated genetic diversity. Therefore, the main focus of management of genetic diversity for both captive breeding and reintroduction is not so much to maintain a balance between drift and mutation but to provide new populations with the best possible start by maximizing the propagule size (without significantly depleting the source population) and thus minimizing the loss of genetic diversity (Lacy, 1994). (For further details on how genetic diversity is categorized and measured in managed populations, see Box 13.1.) When designing guidelines for sustaining genetically viable populations in captivity over the immediate and long terms, Lacy (1994) approached the problem by dividing the demographic history of a captive population into three phases: founding, growth and capacity. We adopt the same approach here for reintroductions, but use the terms establishment, growth and regulation, following the rest of this volume and the recent reintroduction literature (Seddon *et al.*, 2007; Armstrong & Seddon, 2008). The reason for focusing on the different demographic phases here is because each presents a somewhat different genetic problem, and therefore may require different management strategies (Lacy, 1994). Assuming that a release site has a suitable habitat (Osborne & Seddon, this volume, Chapter 3) and there are no immediate issues with dispersal or the likely need for follow-up releases (Gouar *et al.*, this volume, Chapter 5), we address several genetic questions that are relevant to each phase of the reintroduction. Later we discuss the additional issue of whether to re-establish gene flow in isolated populations. We restrict much of our discussion to wild-to-wild reintroductions, as research on maintaining genetic diversity when reintroductions are sourced from captive populations has already been covered in detail elsewhere (e.g. Kleiman, 1989; Haig *et al.*, 1990; Lacy, 1994; Earnhardt, 1999; Fraser, 2008; Leus & Lacy, 2009; Ballou *et al.*, 2010b; Frankham *et al.*, 2010).

Box 13.1 **Measures of genetic diversity used in population management**

Geneticists have a variety of concepts and measures of 'genetic diversity', with different applicability and usefulness for different situations. Measures from the field of population genetics are mostly derived from allele frequency distributions at one or several loci (Wright, 1969;

Crow & Kimura, 1970), while those developed within quantitative genetics usually focus on partitioning the variance in observed traits into genetic and non-genetic components (Falconer & Mackay, 1996). The measures of genetic diversity most commonly used in conservation (Frankham *et al.*, 2010) come from population genetics, although our ultimate concern is with the preservation of adaptive variation in quantitative traits and the ability of populations to evolve in response to changing selective pressures (Frankel & Soulé, 1981). Fortunately, the theoretical expectation is that the most common population genetic measure of diversity (expected heterozygosity or gene diversity; see Nei, 1973) and the most common measure of quantitative genetic variation (heritability) are expected to be depleted in small populations at about the same proportional rate (Fisher, 1958). Unfortunately, other measures of genetic variation do not change linearly with gene diversity and heritability, so it is important to realize that genetic goals such as 'retaining 90% of gene diversity' would have a different meaning if applied to other measures of variation.

Within the field of population management, a handful of interrelated measures of genetic variation are commonly used (Lacy, 1995; Ballou *et al.*, 2010a; Ivy & Lacy, 2010). These terms can be confusing, so a few key terms are defined here. Note that several of the terms are applied either at the population level or at the individual level.

Observed heterozygosity (H_o, sometimes just termed 'heterozygosity'): the average proportion of loci at which an individual (or the average across individuals in a population) is heterozygous. Note that even the term 'heterozygous' depends on how finely the genetic locus is examined. An individual can be homozygous if assayed at the level of products from the locus (as by protein electrophoresis), but heterozygous when examined at the level of the DNA sequence.

Expected heterozygosity or *gene diversity* (H_e or G or GD, sometimes also termed 'heterozygosity'): the heterozygosity expected from the allele frequencies if the population were in Hardy–Weinberg equilibrium. This is often a more useful measure of population diversity than observed heterozygosity, because it better characterizes the diversity inherent in the gene pool, and, regardless of the particular distribution of alleles among current individuals, its theoretical expectation is more easily defined and it is proportional to heritability. Because it is often the loss of this gene diversity, proportional to the starting (source or founder) population that is of greatest concern, many studies report gene diversity as a per cent of a measured or estimated value, but will often neglect to identify clearly that reported values are proportional, not absolute measures of gene diversity. Indeed, in pedigree analyses and population modelling, the true gene diversity (of founders or the descendants) is often not known, but analyses based on probability theory can provide the proportional retention of G.

Founder genome equivalents (*fge*): the number of equally contributing founders with no random loss of founder alleles in descendants that would be expected to produce the same genetic diversity as in the population under study. The *fge* can be thought of as the number of unrelated animals in the source population that would contain the same gene diversity as the descendant population, and is directly related to the proportional G by the formula $1 - [1/(2\,fge)]$. Thus, $fge = 5$ implies that 90% of G has been retained.

Allele diversity (A or n_a): the average number of distinct alleles per locus in the population. Allele diversity (like heterozygosity) is dependent on the level at which the locus is examined, it varies among kinds of loci and it is hard to estimate accurately unless sampling of individuals is extensive.

Inbreeding coefficient (F, also termed the 'coefficient of consanguinity'): the probability that, for any given locus, an individual is homozygous due to identity of the two alleles by descent from a single copy in a common ancestor of both parents. Within a population, the observed heterozygosity of an individual is expected to be proportional to $1 - F$. Note that the relationship between parents and the existence or not of common ancestors, and thus the level of inbreeding, depends on how many generations back are considered in the pedigree analysis.

Mean kinship (*mk*, also termed the *average co-ancestry*): the average kinship (or co-ancestry) of an individual to all individuals in the extant population, including itself. The average *mk* across the entire population is $1 - G$, in which G is expressed proportional to the source population. Thus, in a pedigreed population, loss of proportional G can be easily estimated as the average of all pairwise kinships.

Establishment phase

Many factors can affect the success of the crucial establishment phase of a reintroduction, as discussed in the earlier chapters of this volume, before any genetic issues would come into play. Assuming that these factors are adequately addressed, one of the main genetic goals for the establishment phase is for the released individuals that survive and breed (i.e. the genetic founders) to be as representative a sample of the source population as possible. In the ideal situation, the founders would be fully representative genetically of the source population, but this is unlikely to be the case. This raises two important questions concerning the genetic composition of the founding population for managers to consider:

1. Should more than one population be used for sourcing reintroductions?
2. How many individuals should be caught and released?

The first question is often irrelevant for highly threatened species that exist only in a single, remnant wild population. If there is more than one potential source population, then it is difficult to answer this question without knowing whether there is significant genetic structure among the potential source populations. Recent studies where DNA samples from historic museum specimens have been available indicated that when genetic structuring among potential source populations is present, it may have arisen only recently and is not a reason in itself to avoid mixing source populations (Hoeck *et al.*, 2010; Reding *et al.*, 2010; Tracy & Jamieson, 2011). However, in many cases the molecular data necessary to test for genetic structure are simply unavailable. Frankham *et al.* (2011) have developed practical guidelines for managing fragmented populations based on assessing the risk of outbreeding depression (see Box 13.2). In a similar vein, we provide four general rules of thumb for deciding whether two or more source populations should be merged for releases in a single reintroduction project:

1. If the potential source populations fragmented only recently (less than ~ 150 years), then there may not be significant genetic divergence, especially in allelic diversity, and so sourcing individuals from the largest or closest source population is probably a safe strategy (see the Florida panther example below).

2. If two or more populations have been separated historically (more than ~ 1 000 years), they are likely to show significant genetic divergence and possibly locally adapted gene complexes if their local environments and habitat are significantly different. If so, then it would be best to source from the population in the habitat that is most similar to that at the reintroduction site. However, there could be a need to preserve the genetic differences in the source populations for their own right, which could be accomplished by reintroducing individuals from each population to separate sites, preferably in the same region and with a similar habitat as their respective source population. Local adaptations can originate rapidly, even in mobile species such as wolves, *Canis spp.*, and otters, *Lutra spp.*, and generate genetically differentiated ecotypes (Musiani *et al.*, 2007; Mucci *et al.*, 2010). Therefore caution is required when planning translocations among different habitat types.

3. If, on the other hand, the reintroduction is to occur to an area that was well outside the current range of the available source populations, or if the habitat is highly altered relative to the environment in which the source population had evolved, then it would often be better to release a genetically mixed population that has the best chance to adapt to that altered habitat. A good example is the restoration of peregrine falcons (*Falco peregrinus*) in eastern USA where the local subspecies had been extirpated and could not be restored in its pure form. The released birds were of mixed subspecies ancestry (mostly from captive stock), which may have aided the success of the reintroductions, given that the environment into which the population was reintroduced – primarily urban centres – was quite different from the cliff-nesting habitat that they had occupied before extirpation (Tordoff & Redig, 2001). This example shows that mixed genotypes of adaptable species may thrive when introduced into new habitats, although probably few bird species match the peregrine falcon in adaptability (Tordoff & Redig, 2001).

4. For the most part, mixing of separate subspecies in a single reintroduction should normally be avoided because it could lead to outbreeding depression (Frankham *et al.*, 2011). Unfortunately, subspecies designation is often not supported by good evidence, but instead is based on minor morphological variants that do not reflect isolated geographic populations with separate evolutionary histories. For example, Culver *et al.* (2000) found no molecular genetics support for the Florida panther (*Puma concolor coryi*) being designated a separate subspecies from the rest of the pumas in the USA, as

the isolation of the population occurred only recently due to the connecting populations having been eliminated through hunting. Moreover, the remnant Florida population was so damaged genetically that it could not have been recovered as an isolated population. Therefore eight females from the nearest population in Texas (*P. concolor stanleyana*) were introduced in 1995, resulting in a dramatic decline in observed frequency of deleterious genetic traits in animals with Texan ancestry as well as a rapid recovery of numbers (for a good review of the genetic management of the Florida panther population, see Hedrick & Fredrickson, 2010).

Box 13.2 Guidelines on avoiding outbreeding depression

Recently, Frankham *et al.* (2011) have argued that the costs and associated risks of outbreeding depression when re-establishing gene flow between isolated populations may have been overstated (e.g. Edmands, 2007), relative to the benefits of augmenting gene flow and reducing the risk of inbreeding depression and loss of genetic diversity. They review the evidence of the mechanisms and processes associated with outbreeding depression, develop a predictive theory for estimating the risk of outbreeding depression and estimate from empirical data the number of generations required for outbreeding depression to manifest itself. They conclude that, except for the mixing of different species, populations with fixed chromosomal differences or populations with a long history of adaptation to different environments, the risks of outbreeding depression are generally low.

Frankham *et al.* (2011) further develop a decision tree to guide managers in deciding whether outbreeding depression is likely, and therefore whether populations should be kept separate. The decision tree has three components. First, where the taxonomy of the species in question is resolved, crosses between two 'good' species should always be avoided. Second, if there are fixed chromosomal differences between two subspecies or populations, then there is a high risk of outbreeding depression and merging the two should again be avoided. Third, information about historical gene flow is invaluable for assessing the risk of outbreeding depression. Frankham *et al.* (2010) suggest that populations that have been isolated for 500 years or more, or if there are significant habitat/environment differences and the groups have been in their respective habitat types for more than 20 generations, then a higher risk of outbreeding depression is likely and merging should be avoided or should proceed on an experimental basis. Otherwise, the decision tree indicates that augmenting gene flow could safely proceed.

Managers need to be aware that Frankham *et al.*'s final recommendation – that augmentation should proceed even if there is no evidence of inbreeding depression or population decline – will be seen by some as controversial. They base their argument on the assumption that many small and isolated populations are likely to be extirpated while waiting for the necessary data on the effects of inbreeding to be collected. Furthermore, such data collection requires resources not generally available in the developing world. They therefore recommend a pro-active approach to the problem of genetic isolation. Many managers and scientists alike would not oppose a pro-active management strategy, especially when the risks of outbreeding depression are relatively low or non-existent. However, translocations themselves, including those designed to re-establish gene flow, require significant resources and labour (Parker *et al.*, this volume, Chapter 4), and therefore will need to be prioritized on a case-by-case basis alongside a number of competing management actions. Whether in developing or developed countries, managers are less likely to have budgets approved for translocations when populations are not in obvious decline, especially given that not all isolated populations necessarily exhibit inbreeding depression (Bouzat, 2010).

If the goal is to acquire the best possible genetic representation of the source population, then managers are likely to ask our second question: how many individuals should be released? Releasing enough individuals is often not an issue for amphibians and invertebrates as thousands of individuals can sometimes be released. For example, for a reintroduction of Puerto Rican crested toads (*Peltophryne lemur*) thousands of captive-hatched tadpoles were released at each site (Johnson, 1999). This question is more of a concern for large vertebrate species where the number of individuals translocated might normally be less than 100. In these cases, some geneticists might simply answer, 'the more the better', but that response ignores other factors that will constrain the optimal number released. From experience, managers will know they are likely to catch a certain number of individuals over a specific period of time or have the capacity to transfer and release a certain number of individuals during a translocation event. For reintroducing threatened passerine birds to offshore islands of New Zealand, for example, this number has tended to be around 30–40 individuals (Armstrong & McLean, 1995). The most fundamental constraint on releasing large numbers of individuals is impact on the source population(s). In addition, managers often have the ability to catch more individuals, but this will normally take longer and incur additional costs, and hence they are reluctant to do so unless there are sound biological reasons for catching, transferring and releasing more animals.

The genetic issue of how many individuals are needed to found a population is an area where there has been considerable research for captive populations, but one that requires much more attention when it comes to reintroductions in the wild. Retention of at least 90% of the genetic variation of the source population for 100–200 years is considered a realistic aim (Soulé et al., 1986). It is recommended that 20 genetic founders (i.e. the subset of original founding stock that contributes genes in the form of independent offspring to the new population) are adequate for establishing a captive stock, as these should contain 97.5% of the genetic variability as measured by heterozygosity or gene diversity (GD) present in a wild population (Lacy, 1989), based on the equation:

$$GD = 1 - \frac{1}{2N_f} \qquad (13.1)$$

where N_f typically refers to the number of genetic founders (i.e. the number of individuals that contribute genes (in the form of offspring) to the new population). However, Lacy (1995) has argued that N_f should refer to the number of 'founder genome equivalents', which is defined as the number of equally contributing genetic founders with no random loss of founder alleles in descendants that would be expected to produce the same genetic diversity as in the population under study. Clearly, the use of founder genome equivalents rather than genetic founders increases the required number of individuals released (Lacy, 1989, 1995). More importantly, genetic guidelines for establishing captive breeding stock are not directly transferable to reintroductions in the wild because of post-release mortality (Armstrong &

Reynolds, this volume, Chapter 6) and dispersal (Le Goaur *et al.*, this volume, Chapter 5). The actual founder group will often be much smaller than the number of individuals released and may also differ significantly in genetic composition. These limitations are somewhat offset by the potential for greater capacity for growth of the population in the wild setting.

The recommendations for captive studies tend to focus on heterozygosity as a measure of genetic diversity. Heterozygosity is important for short-term success of captive or reintroduced populations, but is an overly optimistic estimate of the effects of a bottleneck in the long term, as little heterozygosity is expected to be lost through even major bottlenecks of short duration (Nei *et al.*, 1975; Allendorf, 1986). Loss of alleles in such bottlenecks is a more appropriate measure of genetic diversity when considering long-term effects, because loss of allelic diversity will have a significant effect on future adaptability and survival of species in the wild (Allendorf & Luikart, 2007).

All of the above arguments suggest the need for a new framework for estimating the number of released individuals that translates into the number of genetic founders that is necessary to retain alleles of specified frequency through the establishment phase. Such an approach has recently been applied to guide reintroductions of the endangered New Zealand forest passerine, the yellowhead/mohua (*Mohoua ochrocephala*) (Box 13.3). In the case of the mohua, a goal of releasing ~ 100 individuals to maintain alleles with an initial frequency of 0.05 (see Tracy *et al.*, 2011), when it is known that 30 individuals is normally sufficient to establish a population in the short term (i.e. to obtain positive growth), could be considered excessive by some managers. However, the extra effort and cost is justifiable if larger release numbers increase the chances of preserving long-term viability and evolutionary potential, especially when protected but isolated sites may be the only safe refuges for a species for a long time to come. In many cases the extra cost involved might translate to an additional catching team or one or two additional days of catching effort and possibly additional transportation capacity. This could be regarded as a small price to pay when considering the preservation of genetic diversity that has been generated over thousands of generations. Additionally, post-release monitoring is strongly recommended (Nichols & Armstrong, this volume, Chapter 7), especially in the initial generations after release, so that low population growth rates can be identified and the reintroduced population supplemented with more individuals if necessary.

Box 13.3 Estimating the number of individuals to release to maintain genetic diversity in reintroductions: a case study of mohua (*Mohoua ochrocephala*)

Even when the habitat is suitable and dispersal is not a major issue, only a portion of the released individuals in reintroductions will normally survive to breed and contribute genetically to the next generation (Jamieson, 2011). Managers can compensate for the inevitable loss of individuals during the establishment phase by releasing more individuals, but how many more individuals should be released in order to minimize the loss of genetic diversity? Although 'more is usually better' from a genetics perspective, managers may prefer to translocate the minimum number necessary to maintain genetic variability because of the impact to source populations as well as financial and logistic constraints. In other words, managers require a means of calculating the number of individuals to release in reintroductions to retain alleles of specified frequency through the establishment phase, beyond which increased effort in capturing and releasing more individuals provides only marginal gains in genetic diversity (i.e. a curve of diminishing returns). Studies have modelled the demographic and genetic consequences of varying sex and age composition of release groups with respect to detrimental effects of inbreeding depression (e.g. Sarrazin & Legendre, 2000; Robert *et al.*, 2004, 2007), but there are surprisingly no general rules available for calculating optimal release numbers (as opposed to genetic founders) for preserving genetic diversity.

The problem of how many individuals to release to maintain genetic diversity was raised by managers involved in reintroductions of the mohua (*Mohoua ochrocephala*) (Figure 13.2), a threatened New Zealand forest passerine that is commonly translocated to offshore island refuges where exotic predators (rats and mustelids) are absent (Tracy, 2009; Tracy *et al.*, 2011). The solution was to model genetic drift both during and following the founding event of a typical reintroduction by simulating the probabilities of retaining genetic diversity (allele retention) for varying numbers of released individuals, taking into account the allele frequency in the source population, the initial survival rate after reintroduction, the population growth rate and the carrying capacity.

Not surprisingly, the initial modelling results indicated that the probability of allele retention was sensitive to both the number of individuals released and the growth rate of the population (Tracy *et al.*, 2011). However, managers did not have a mechanism to increase growth rates of the reintroduced populations and therefore the modelling results focused more on the effects of changing the release number.

Figure 13.2   **Mohua (photo: Ian Jamieson).**

The results indicated that increasing the number of individuals released from 30 (the typical number released) to 60 individuals would increase the probability of retaining alleles at an initial frequency of 0.05, by 18% if the growth rate is moderate ($\lambda = 1.3$) and by 31% if growth rate is low ($\lambda = 1.1$). Doubling the catching effort beyond 60 individuals leads to a disproportionately small benefit in terms of allele retention. Sourcing more than 60 individuals would, however, buffer against increased initial mortality or slower than expected population growth. For example, under a slow population growth rate ($\lambda = 1.1$) approximately 102 individuals need to be released to achieve 95% certainty that alleles at an initial frequency of 0.05 will be retained after 20 years. Subjecting simulated populations to increased annual turnover, environmental stochasticity and a lag in the initial population growth resulted in a lower probability of allele retention, although retention rates appear to be more sensitive to environmental stochasticity and lags in initial growth (Tracy *et al.*, 2011).

The above model was based on the retention of neutral alleles only and does not consider alleles under selection (see Groombridge *et al.*, this volume, Chapter 12), although selection has to be quite strong to overcome the dispersive force of drift in small populations (Allendorf & Luikart, 2007). Although all models have their limitations, the authors see this approach as an improvement over the current guidelines, which recommend that 'adequate' numbers should be released to minimize loss of diversity (e.g. Frankham *et al.*, 2010; Allendorf & Luikart, 2007).

Attempts to model release numbers for maintaining genetic diversity for reintroductions in the wild, by taking into account post-release mortality and population growth rate as well as estimates of the carrying capacity of the release site, have parallel situations in the management of captive populations. For example, the 'goals' modelling tool can be used within *PM2000* (Pollak *et al.*, 2002) software to consider the interacting effects of the number of founders, population growth rate and intensity of genetic management possible (N_e/N) to determine what carrying capacity is needed and how many more founders might be needed to reach genetic goals of maintenance of sufficient heterozygosity over time. The just-released *PMx* (Ballou *et al.*, 2010a) provides additional tools for examining genetic effects of moving animals between populations (in its 'Management Set' module) or releasing animals from one population or adding animals. However, these *PM2000* and *PMx* analyses are focused on gene diversity in intensively managed pedigrees with tracking of individuals and do not project the rates of loss of alleles nor the genetic consequences in unmanaged populations that might be highly unstable demographically. More work on integrating such analyses for captive and reintroduced populations would be valuable.

Other approaches using analysis based on microsatellite diversity have also been tried. Taylor & Jamieson (2008) calculated that 30 genetic founders (=15 breeding pairs) of the South Island saddleback *(Philesturnus c. carunculatus)* are needed to maintain alleles currently present in a number of island reintroductions based on six variable microsatellite loci. Miller *et al.* (2009) used a similar approach for investigating the effect of founder group sizes on loss of genetic diversity in reintroduced populations of a long-lived reptile, the tuatara (*Sphenodon* spp.). In reality, however, estimates of molecular diversity will not exist for most threatened species and the collection and analysis of such data are often time-consuming and expensive.

The above discussion focuses on minimizing the loss of genetic diversity and not the frequency of close inbreeding. The incidence of close inbreeding (e.g. sib–sib pairings) in reintroductions can be more frequent in the first cohort of first-time breeders, but the frequency of such events tends to decline as the population grows (Jamieson, 2011). This can be the case even if there is no evidence of active inbreeding avoidance, as in many socially monogamous birds (Jamieson *et al.*, 2009). Nevertheless, growth rate and final population size may play a greater role in determining the overall rate of inbreeding in reintroduced populations than the initial founder size (Nei *et al.*, 1975; Robert *et al.*, 2007; Biebach & Keller, 2010; Jamieson, 2011).

Finally, although a single release of a large number of individuals would reduce the risk of an initial bottleneck, releasing fewer individuals per translocation over several consecutive seasons could achieve similar results as long as the population grows as quickly as possible (see below). Overall, there are a number of interacting and competing factors to consider when it comes to determining how many individuals to release. Adoption of a quantitative genetics framework like the one outlined above could be included with a structured decision-making (SDM) approach (Gregory & Long, 2009; Nichols & Armstrong, this volume, Chapter 7) whereby utility is assigned to different potential outcomes and then the optimal management action is worked out based on these and projections from the genetic modelling. Such an approach could help support a manager's case to allow for adequate resourcing of the reintroduction including subsequent monitoring.

Growth phase

We saw in the case of the mohua (Box 13.3) that the probability of allele retention is sensitive to both the number of individuals released (and the subset of genetic founders) and the growth rate of the population, but managers tend to have more control over the former than the latter in free-ranging reintroduced populations. Nevertheless, it is worth noting that rapid population growth immediately after a translocation event (potentially managed via supplementary feeding or predator control) will minimize loss of founder alleles (Lacy, 1994). If each generation is considered as an incomplete sampling of the founders' genes, then a slow growth rate means fewer progeny are produced and the number of offspring capable of passing on each founder allele to the next generation is less. Although 12 or more progeny are required to ensure with high probability that all alleles of a founder are transmitted to at least one offspring (Thompson, 1994), more than 99% of the alleles of a founder will persist if it leaves seven offspring. Problems arise, however, if population growth derives primarily from a small subset of the founders, with much of the contribution of others being lost; hence, in the ideal situation it is important that all founders produce equal numbers of progeny.

In reality, however, the 'ideal' situation rarely occurs and it becomes important to know how much the reintroduced population has diverged from the ideal situation, especially during the initial generations. Hence the importance of not only monitoring the survival of released individuals but also monitoring their relative breeding success. For example, Ewing *et al.* (2008) documented a high level of inbreeding and loss of gene diversity over a 17-year period in a reintroduced island population of Mauritius kestrel (*Falco araea*) where 12 individuals were released but yielded only six founder genome equivalents. Jamieson (2011) identified a similar pattern in reintroduced island populations of three New Zealand passerines. With the number of released individuals ranging from 20 to 58, the number of founder genome equivalents (6–11) after only seven breeding seasons was a small fraction of the release numbers and far fewer than the 20 founder genome equivalents recommended for captive breeding (Lacy, 1995). In these three reintroduced populations, the relatively small numbers of founder genome equivalents were a result of founders' biased sex ratio and skewed breeding success as well as close inbreeding between first generation offspring (Jamieson, 2011).

What should happen when skewed mating success is a natural part of the breeding system of a threatened species? For example, as predicted from its lek mating system, the highly endangered kakapo (*Strigops habroptilus*) has high variance in male mating success (Robertson, 2006). Using microsatellite markers to determine paternity, only four adult males (of a possible 30) produced offspring over an 8-year period in a reintroduced population on Codfish Island, and one male ('Felix') fathered 7 of the 13 offspring produced (Miller *et al.*, 2003; Robertson, 2006). The impact of the skewed breeding success, plus unequal sex ratio, was that the effective population size was reduced by 79% of the census population size (Robertson, 2006). To try to increase the effective population size by minimizing skew in male breeding success and to provide better mating opportunity for another male kakapo, Richard Henry, the last surviving kakapo from a distinct lineage, Felix and two of his breeding-age sons were banished to an isolated island. More recently, artificial insemination has been successfully trialled in kakapo to manage male mating success and to increase representation from under-represented lineages (R. Moorhouse, Department of Conservation, personal communication).

Although these tactics make sense in terms of managing genetic diversity in this bottlenecked population of kakapo, mating success is naturally skewed in lek-breeding animals, raising the question of whether such management action is really necessary. Lek-breeding systems such as that of the kakapo evolved under conditions where populations were naturally large and outbred, and hence able to maintain genetic diversity; indeed, active female mate-choice in lek-breeding systems is probably taking advantage of the existence of sufficient variation in male genetic quality and/or diversity in the population (Kirkpatrick, 1987). Intervention management such as that described above may be called for when populations have gone through severe bottlenecks and are exhibiting signs of inbreeding depression that is hindering population growth and potentially losing unique genetic diversity (Jamieson *et al.*, 2006; Robertson, 2006).

Regulation phase

Although it might be more appropriate to rectify imbalances in founder representation only once a population approaches its carrying capacity (because priority should be given to maximizing growth of the population; see Lacy, 1994), this may not be practical for many reintroduced populations. This is because selective or targeted breeding – the main mechanism by which founder representation is managed in a captive population – is much less feasible in a wild population. Furthermore, once the population is at carrying capacity, it may be difficult to introduce new founders effectively into the breeding population (for an example of managing reintroduced takahe on islands at carrying capacity, see Box 13.4). At the very least, specific individuals or members of certain lineages would need to be translocated off the island (or culled) to create vacant breeding spaces before new founders could be introduced, which is logistically difficult and costly or, in the case of culling, often unacceptable. Managing founder representation also assumes that a pedigree exists for the population. If so, then there are a number of recommended strategies for identifying genetically important individuals (Haig *et al.*, 1990; Ballou & Lacy, 1995), which are regularly applied to captive populations. If there are no plans to intensively monitor the reintroduced population to maintain a pedigree, or the pedigree has too many gaps to be of much use by the time the population reaches carrying capacity, then one option for genetic management during the regulation phase is to periodically introduce new genetic stock, perhaps through reciprocal translocations with another isolated population (see below). Exactly when or at what stage managers would need to embark on such translocations to mediate potential loss of genetic diversity is an area that requires further research, although any signs of significant inbreeding depression that could affect population fitness and long-term viability would be obvious indicators to proceed with such management action.

Box 13.4 **How to facilitate gene flow when an isolated population is at carry capacity: a case study of island takahe**

The takahe (Figure 13.3) is a large, flightless rail that had been reduced to a single remnant population in Murchison Mountains of Fiordland, New Zealand. To safeguard the species from the threat of extinction due to further losses from introduced predators or catastrophic events such as a disease outbreak, 25 birds (mostly juveniles) were translocated over several years to four offshore islands, from which introduced predators had been removed (for maps and details of the translocations see Grueber & Jamieson, 2008; Lee & Jamieson, 2001). The 'island' population has since grown to 83 birds (Wickes *et al.*, 2009) and has generally been managed as a closed metapopulation, with occasional translocations among islands (Jamieson & Wilson, 2003). Following a pedigree analysis that indicated that the island population as a whole was highly inbred, had lost substantial genetic diversity and showed evidence of inbreeding depression after just three generations (Jamieson, 2003; Jamieson *et al.*, 2003; Grueber & Jamieson, 2008), it was recommended that new genetic stock be transferred from the source population (Fiordland) to the islands, to effect a recruitment rate of two breeding adults every 4–5 years (Grueber & Jamieson, 2008).

Figure 13.3   **Takahe (photo: Ian Jamieson).**

Before carrying out the translocations, it was important to first determine whether the newly introduced birds were likely to recruit into the island's breeding population. The presence of both large numbers of non-breeding adult takahe as well as multimale/female breeding groups (takahe typically breed in pairs) on each of the islands suggested that the population was close to or at carrying capacity (Wickes *et al.*, 2009). Subsequent analysis indicated that breeding capacity of three of the four offshore islands had been maximized. Moreover, implementation of the genetic restocking programme would be ineffective if new arrivals were unable to establish on breeding territories. Increasing the area of suitable breeding habitat on the islands was one possible approach for increasing carrying capacity, although the takahe's large breeding territories and home ranges (Ryan & Jamieson, 1998) suggested this strategy was likely to be limited. Instead, the Takahe Recovery Group opted to identify individuals of lineages that were over-represented in the island pedigree (i.e. target individuals with high mean kinship; see Ballou & Lacy, 1995) and translocate these birds back to the original source population in Fiordland, after successfully completing a thorough disease screening programme. The immediate aims of these translocations are to eliminate close inbreeding ($f \geq 0.125$) and equalize founder contribution among breeding pairs. Once these objectives are achieved, low-level translocations off the islands will continue in order to limit future inbreeding and facilitate the introduction of new genetic stock (i.e. increase gene flow) into the island breeding population, while at the same time augmenting the growth rate of the expanding wild population in Fiordland. Low-level gene flow might additionally limit or slow adaptation of takahe to the islands' temperate climate and pasture grassland habitat (Jamieson & Ryan, 2000, 2001; Jamieson, 2003), but their small effective population sizes indicate that the islands will be constantly subject to strong genetic drift and inbreeding.

Re-establishing gene flow

In addition to trying to restore a population to its former range or to specific habitat sites, managers are often faced with the question of whether to re-establish gene flow between isolated and fragmented populations. Assisted gene flow is a means of introducing new genetic variation to an isolated population and hence reducing the rate of inbreeding and slowing the loss of genetic diversity through drift (Keller *et al.*, this volume, Chapter 11). Deliberate mixing of populations consisting of different species or subspecies is normally avoided because of the dilution of gene pools and the dangers of outbreeding depression (see Box 13.2). Natural versus human-induced hybridization and introgression is a controversial topic in conservation and approaches to dealing with these processes are covered in detail by others (e.g. Allendorf *et al.*, 2001). Here, we limit ourselves with intraspecific crosses where two populations share alleles and introgression is less likely to cause outbreeding depression.

If two populations have become isolated only recently (<150 years) due to anthropogenic factors, but the populations are large (>500 individuals), then there may be no immediate need to re-establish gene flow, especially if the two populations appear to be stable and healthy. If one or both populations are small (<500 individuals) but stable, then periodic assisted gene flow may be an effective means of slowing the inevitable increase in rate of inbreeding and decline in heterozygosity (see above). Most theoretical analyses suggest that the 'one migrant per generation' rule would be sufficient to offset genetic deterioration within subpopulations, but account should be taken of the fact that the rule applies only to 'ideal' populations with random mating and is based on other unrealistic assumptions: 1–10 migrants per generation or more is preferable (Mills & Allendorf, 1996; Vucetich & Waite, 2000). Generation times are likely to range from 1–4 years for small birds and mammals to 10–20 years for long-lived species. Note that the number of migrants refers to genetic founders – individuals that recruit into the breeding population – so, again, managers would have to determine how many individuals would need to be released to account for post-release mortality.

Although genetic deterioration is seen as an important component of extinction risk for small, isolated island populations (Frankham, 2005), the number of known cases of island populations threatened by genetic malfunction is few (Jamieson, 2007a, 2007b). Furthermore, we are not aware of any species reintroductions currently under an ongoing management regime of introducing one or more migrants per generation. The need for maintaining gene flow will vary from species to species, particularly in relation to the current size of the population, and also needs to be put into context with other conservation priorities. Furthermore, it is expected that a group of reintroduced island populations are likely to show genetic differentiation and divergence over time, especially in allele frequency distributions. This in itself does not constitute a reason to transfer individuals between populations to try to maintain similar allele frequencies, especially when weighed up against the expense and increased risk of translocating undesirable pathogens (Taylor & Jamieson, 2008; Hedrick & Fredrickson, 2010). Given enough time, and if the populations are large enough for natural selection to outweigh random drift, divergences can involve local adaptation through gene complexes, which could be lost when isolated populations are mixed (Allendorf et al., 2001). The risk of losing important local adaptations needs to be weighed against the immediate risk that inbreeding depression poses or the reduced ability of genetically depauperate populations to adapt to future changes in the environment. We therefore recommend that the requirements for re-establishing gene flow be examined on a case-by-case basis and that priority be given to reintroduced populations that are limited in number and size but that play an important role in a species recovery programme. This is yet another area in the emerging field of reintroduction biology where further research and data are required to test existing theory and to determine how robust that theory is to the variation among cases and species (Seddon et al., 2007; Armstrong & Seddon, 2008).

Concluding comments

Even in cases where translocation leads to a large population, considerable genetic diversity can still be lost during the establishment phase of a reintroduction and beyond, depending on the founder size and population growth rate. Perhaps more worryingly, though, is the increasing number of reintroductions occurring on much smaller scales than those recommended here. For example, many community sponsored reintroduction programmes throughout the mainland of New Zealand involve predator-controlled or predator-resistant fenced sanctuaries of relatively small scale (<100 ha) (www.sanctuariesnz.org; see Burns et al., 2012). These community-led projects raise funds for the initial translocations to seed the sanctuary, but are unlikely to have the resources for the ongoing transfers that are likely to be required to minimize the loss of genetic variation over the long term. Although these initiatives are a great sign of the community embracing population restoration, they may require a commitment of funds and resources for genetic restocking from source populations and/or frequent transfers between sanctuaries (in much the same way that captive breeding facilities manage threatened species) that has possibly neither been appreciated nor budgeted for. More generally, the many potential genetic issues that we have addressed here and in the previous chapters suggest a need for better integration of reintroductions into long-term management plans. Overall, the development of long-term strategies and guidelines would be consistent with a goal of minimizing inbreeding and maintaining genetic diversity in reintroduced populations.

Acknowledgements

The authors thank Richard Frankham for allowing them to read unpublished manuscripts. Catherine Grueber, along with the book's editors, read drafts of the manuscript and provided many helpful comments. Ian Jamieson's research in the area of conservation genetics of reintroductions is funded by the Department of Conservation (Contract 3576), Landcare Research (Contract C09X0503) and University of Otago. Robert Lacy's research in conservation genetics is funded by the Chicago Zoological Society and grants from the Institute of Museum and Library Services (USA) and the Association of Zoos and Aquariums.

References

Allendorf, F.W. (1986) Genetic drift and the loss of alleles versus heterozygosity. *Zoo Biology*, 5, 181–190.
Allendorf, F.W. & Luikart, G. (2007) *Conservation and the Genetics of Populations.* Blackwell, Oxford, UK.
Allendorf, F.W., Leary, R.F., Spruell, P. *et al.* (2001) The problem with hybrids: setting conservation guidelines. *Trends in Ecology and Evolution*, 16, 613–622.
Armstrong, D.P. & McLean, I.G. (1995) New Zealand translocations: theory and practice. *Pacific Conservation Biology*, 2, 39–54.
Armstrong, D.P. & Reynolds, M.H. (2011) Modelling reintroduced populations: the state of the art and future directions. In *Reintroduction Biology: integrating science and management*, eds J.G. Ewen, D.P. Armstrong, K.A. Parker & P.J. Seddon, Chapter 6. Wiley-Blackwell, Oxford, UK.
Armstrong, D.P. & Seddon, P.J. (2008) Directions in reintroduction biology. *Trends in Ecology and Evolution*, 23, 20–25.
Armstrong, D.P., Raeburn, E.H., Lewis, R.M. *et al.* (2006) Estimating the viability of a reintroduced New Zealand robin population as a function of predator control. *Journal of Wildlife Management*, 70, 1028–1036.

Ballou, J.D. & Lacy, R.C. (1995) Identifying genetically important individuals for management of genetic variation in pedigreed populations. In *Population Management for Survival and Recovery*, eds J.D. Ballou, M. Gilpin & T.J. Foose, pp. 76–111. Columbia University Press, New York, USA.

Ballou, J.D., Lees, C., Faust, L.J. *et al.* (2010a) Demographic and genetic management of captive populations. In *Wild Mammals in Captivity: principles and techniques for zoo management*, 2nd edn, eds D.G. Kleiman, K.V. Thompson & C.K. Baer, pp. 219–252. University of Chicago Press, Chicago, Illinois, USA.

Ballou, J.D., Pollak, J.P. & Lacy, R.C. (2010b) *PMx: software for demographic and genetic analysis and management of pedigreed populations*. Chicago Zoological Society, Brookfield, Illinois, USA.

Biebach, I. & Keller, L.F. (2010) Inbreeding in reintroduced populations: the effects of early reintroduction history and contemporary processes. *Conservation Genetics*, 11, 527–538.

Blackburn, T.M., Cassey, P., Duncan, R.P. *et al.* (2004) Avian extinctions and mammalian introductions on oceanic islands. *Science*, 305, 1955–1958.

Bouzat, J.L. (2010) Conservation genetics of population bottlenecks: the role of chance, selection, and history. *Conservation Genetics*, 11, 463–478.

Bouzat, J.L., Johnson, J.A., Toepfer, J.E. *et al.* (2009) Beyond the beneficial effects of trans-locations as an effective tool for the genetic restoration of isolated populations. *Conservation Genetics*, 10, 191–201.

Bouzat, J.L., Lewin, H.A. & Paige, K.N. (1998) The ghost of genetic diversity past: historical DNA analysis of the greater prairie chicken. *American Naturalist*, 152, 1–6.

Boyce, M.S. (2002) Reconciling the small-population and the declining-population paradigms. In *Population Viability Analysis*, eds S.R. Bessinger & D.R. McCullough, pp. 41–49. University of Chicago Press, Chicago, Illinois, USA.

Brown, K.P. (1997) Predation at nests of two New Zealand endemic passerines: implications for bird community restoration. *Pacific Conservation Biology*, 3, 91–98.

Burns, B., Innes, J. & Day, T. (2012) The use and potential of pest-proof fencing for ecosystem restoration and fauna conservation in New Zealand. In *Fencing for Conservation*, eds M.J. Somers & M.W. Hayward. Springer-US, New York, USA.

Caballero, A., Rodríguez-Ramilo, S.T., Ávila, V. *et al.* (2010) Management of genetic diversity of subdivided populations in conservation programmes. *Conservation Genetics*, 11, 409–419.

Caro, T.M. & Laurenson, M.K. (1994) Ecological and genetic factors in conservation: a cautionary tale. *Science*, 263, 485–486.

Caughley, G. (1994) Directions in conservation biology. *Journal of Animal Ecology*, 63, 215–244.

Culver, M., Johnson, W.E., Pecon-Slattery, J. *et al.* (2000) Genomic ancestry of the American puma (*Puma concolor*). *Journal of Heredity*, 91, 186–197.

Crow, J.F. & Kimura, M. (1970) *An Introduction to Population Genetics Theory*. Harper & Row, New York, USA.

Earnhardt, J.M. (1999) Reintroduction programmes: genetic trade-offs for populations. *Animal Conservation*, 2, 279–286.

Edmands, S. (2007) Between a rock and a hard place: evaluating the relative risks of inbreeding and outbreeding for conservation and management. *Molecular Ecology*, 16, 463–475.

Ewing, S.R., Nager, R.G., Nicoll, M.A. *et al.* (2008) Inbreeding and loss of genetic variation in a reintroduced population of Mauritius kestrel. *Conservation Biology*, 22, 395–404.

Falconer, D.S. & Mackay, T.F.C. (1996) *Introduction to Quantitative Genetics*, 4th edn. Longman, Harlow, UK.

Fisher, R.A. (1958) *The Genetical Theory of Natural Selection*, 2nd edn. Dover, New York, USA.

Frankel, O.H. & Soulé, M.E. (1981) *Conservation and Evolution*. Cambridge University Press, Cambridge, UK.

Frankham R. (2005) Genetics and extinction. *Biological Conservation*, 12, 131–140.

Frankham R. (2008) Genetic adaptation to captivity in species conservation programs. *Molecular Ecology*, 12, 131–140.

Frankham, R., Ballou, J.D. & Briscoe, D.A. (2010) *Introduction to Conservation Genetics*, 2nd edn. Cambridge University Press, Cambridge, UK.

Frankham, R., Ballou, J.D., Eldridge, M.D.B. *et al.* (2011) Predicting probability of outbreeding depression. *Conservation Biology*, doi: 10.1111/j.1523-1739.2011. 01662.x.

Franklin, I.R. (1980) Evolutionary change in small populations. In *Conservation Biology: an evolutionary-ecological perspective*, eds M.E. Soulé & B.A. Wilcox, pp. 135–149. Sinauer Associates, Sunderland, Massachusetts, USA.

Fraser, D.J. (2008) How well can captive breeding programs conserve biodiversity? A review of salmonids. *Evolutionary Applications*, 1, 535–586.

Gregory, R. & Long, G. (2009) Using structured decision making to help implement a precautionary approach to endangered species management. *Risk Analysis*, 29, 518–532.

Groombridge, J. (2007) Genetics and extinction of island endemics: the importance of historical perspectives. *Animal Conservation*, 10, 147–148.

Groombridge, J.J., Raisin, C., Bristol, R. *et al.* (2011) Genetic consequences of reintroductions and insights from population history. In *Reintroduction Biology: integrating science and management*, eds J.G. Ewen, D.P. Armstrong, K.A. Parker & P.J. Seddon, Chapter 12. Wiley-Blackwell, Oxford, UK.

Grueber, C.E. & Jamieson, I.G. (2008) Quantifying and managing the loss of genetic variation through the use of pedigrees in a non-captive endangered species. *Conservation Genetics*, 9, 645–651.

Haig, S.M., Ballou J.D. & Derrickson, S.R. (1990) Management options for preserving genetic diversity: reintroduction of Guam rails to the wild. *Conservation Biology*, 4, 290–300.

Hedrick, P.W. & Fredrickson, R. (2010) Genetic rescue guidelines with examples from Mexican wolves and Florida panthers. *Conservation Genetics*, 11, 615–626.

Hedrick, P.W., Lacy, R.C., Allendorf, F.W. *et al.* (1996) Directions in conservation biology: comments on Caughley. *Conservation Biology*, 10, 1312–1320.

Higgins, P.J. & Peter, J.M. (2002). *Handbook of Australian, New Zealand and Antarctic Birds*, vol. 6, *Pardalotes to Shrike-thrushes*. Oxford University Press, Melbourne, Australia.

Hoeck, P.E.A., Beaumont, M.A., James, K.E. *et al.* (2010) Saving Darwin's muse: evolutionary genetics for the recovery of the Floreana mockingbird. *Biology Letters*, 6, 212–215.

Innes, J., Kelly, D., Overton, J.M. *et al.* (2010) Predation and other factors currently limiting New Zealand forest birds. *New Zealand Journal of Ecology*, 34, 86–114.

IUCN/SSC (2008) *Strategic Planning for Species Conservation: a handbook*. IUCN Species Survival Commission, Gland, Switzerland.

Ivy, J.A. & Lacy, R.C. (2010) Using molecular methods to improve the genetic management of captive breeding programs for threatened species. In *Molecular Approaches in Natural Resource Conservation and Management*, eds J.A. DeWoody, J.W. Bickham, C.H. Michler, K.N. Nicols, O.E. Rhodes & K.E. Woeste, pp. 267–295. Cambridge University Press, Cambridge, UK.

Jamieson, I.G. (2003) No evidence that dietary nutrient deficiency is related to poor reproductive success of translocated takahe. *Biological Conservation*, 115, 165–170.

Jamieson, I.G. (2007a) Has the debate over genetics and extinction of island endemics truly been resolved? *Animal Conservation*, 10, 139–144.

Jamieson, I.G. (2007b) Role of genetic factors in extinction of island endemics: complementary or competing explanations? *Animal Conservation*, 10, 151–153.

Jamieson, I.G. (2011) Founder effects, inbreeding and loss of genetic diversity in four avian reintroduction programs. *Conservation Biology*, 25, 115–123.

Jamieson, I.G. & Ryan, C.J. (2000) Increased egg infertility associated with translocating inbred Takahe (*Porphyrio hochstetteri*) to island refuges in New Zealand. *Biological Conservation*, 94, 107–114.

Jamieson, I.G. & Ryan, C.J. (2001) Closure of the debate over the merits of translocating takahe to predator-free islands. In *The Takahe: fifty years of conservation management and research*, eds W.G. Lee & I.G. Jamieson, pp. 96–113. University of Otago Press, Dunedin, New Zealand.

Jamieson, I.G. & Wilson, G.C. (2003) Immediate and long-term effects of translocations on breeding success in the takahe *Porphyrio hochstetteri*. *Bird Conservation International*, 13, 299–306.

Jamieson, I.G., Grueber, C.E., Waters, J.M. *et al.* (2008) Managing genetic diversity in threatened populations: a New Zealand perspective. *New Zealand Journal of Ecology*, 32, 130–137.

Jamieson, I.G., Roy, M.S. & Lettink, M. (2003) Sex-specific consequences of recent inbreeding in an ancestrally inbred population of New Zealand takahe. *Conservation Biology*, 17, 708–716.

Jamieson, I.G., Taylor, S.S., Hegg, L. *et al.* (2009) Why some species of birds do not avoid inbreeding: insights from New Zealand robins and saddlebacks. *Behavioral Ecology*, 20, 575–584.

Jamieson, I.G., Tracy, L.N., Fletcher, D. *et al.* (2007) Moderate inbreeding depression in a reintroduced population of North Island robins. *Animal Conservation*, 10, 95–102.

Jamieson, I.G., Wallis, G.P. & Briskie, J.V. (2006) Inbreeding and endangered species management: is New Zealand out-of-step with the rest of the world? *Conservation Biology*, 20, 38–47.

Johnson, B. (1999). Recovery of the Puerto Rican crested toad. *Endangered Species Bulletin*, 24, 8–9.

Keller, L.K., Biebach, I., Ewing, S.R. *et al.* (2011) The genetics of reintroductions: inbreeding and genetic drift. In *Reintroduction Biology: integrating science and management*, eds J.G. Ewen, D.P. Armstrong, K.A. Parker & P.J. Seddon, Chapter 11. Wiley-Blackwell, Oxford, UK.

Kirkpatrick, M. (1987) Sexual selection by female choice in polygynous animals. *Annual Review Ecology and Systematics*, 18, 43–70.

Kleiman, D.G. (1989) Reintroduction of captive mammals for conservation. *Bioscience*, 39, 152–161.

Lacy, R.C. (1987) Loss of genetic diversity from managed populations: interacting effects of drift, mutation, immigration, selection, and population subdivision. *Conservation Biology*, 1, 143–158.

Lacy, R.C. (1989) Analysis of founder representation in pedigrees: founder equivalents and founder genome equivalents. *Zoo Biology*, 8, 111–113.

Lacy, R.C. (1994) Managing genetic diversity in captive populations of animals. In *Restoration of Endangered Species*, eds M.L. Bowles & C.J. Whelan, pp. 63–89. Cambridge University Press, Cambridge, UK.

Lacy, R.C. (1995) Clarification of genetic terms and their use in the management of captive populations. *Zoo Biology*, 14, 565–578.

Lacy, R.C., Borbat, M. & Pollak, J.P. (2007) *VORTEX: a stochastic simulation of the extinction process*. Version 9.72. Chicago Zoological Society, Brookfield, Illinois, USA.

Lande, R. & Barrowclough, G.F. (1987) Effective population size, genetic variation, and their use in population management. In *Viable Populations for Conservation*, ed. M.E. Soulé, pp. 87–123. Cambridge University Press, Cambridge, UK.

Lee, W.G. & Jamieson, I.G. (2001) *The Takahe: fifty years of conservation management and research*. University of Otago Press, Dunedin, New Zealand.

Le Gouar, P., Mihoub, J.-B. & Sarrazin, F. (2011) Dispersal and habitat selection: behavioural and spatial constraints for animal translocations. In *Reintroduction biology: integrating science and management*, eds J.G. Ewen, D.P. Armstrong, K.A. Parker & P.J. Seddon, Chapter 5. Wiley-Blackwell, Oxford, UK.

Leus, K. & Lacy, R.C. (2009) Genetic and demographic management of conservation breeding programs oriented towards reintroduction. In *Iberian Lynx ex situ Conservation: an interdisciplinary approach*, eds A. Vargas, C. Breitenmoser & U. Breitenmoser, pp. 74–84. Fundación Biodiversidad/IUCN Cat Specialist Group, Madrid, Spain.

Miller, H.C., Lambert, D.M., Millar, C.D. *et al.* (2003) Minisatellite DNA profiling detects lineages and parentage in the endangered kakapo (*Strigops habroptilus*) despite low microsatellite DNA variation. *Conservation Genetics*, 4, 265–274.

Miller, K.A., Nelson, N.J., Smith, H.G. *et al.* (2009) How do reproductive skew and founder group size affect genetic diversity in reintroduced populations? *Molecular Ecology*, 18, 3792–3802.

Mills, L.S. & Allendorf, F.W. (1996). The one-migrant-per-generation rule in conservation and management. *Conservation Biology*, 10, 1509–1518.

Mucci, N., Arrendal, J., Ansorge, H. *et al.* (2010) Genetic diversity and landscape genetic structure of otter (*Lutra lutra*) populations in Europe. *Conservation Genetics*, 11, 583–599.

Musiani, M., Leonard, J.A., Cluff, H.D. *et al.* (2007) Differentiation of tundra/taiga and boreal coniferous forest wolves: genetics, coat colour and association with migratory caribou. *Molecular Ecology*, 16, 4149–4170.

Nei, M. (1973) Analysis of gene diversity in subdivided populations. *Proceedings of the National Academy of Science USA*, 70, 3321–3332.

Nei, M., Maruyama, T. & Chakraborty, R. (1975) The bottleneck effect and genetic variability in populations. *Evolution*, 29, 1–10.

Nichols, J.D. & Armstrong, D.P. (2011) Monitoring for reintroductions. In *Reintroduction Biology: integrating science and management*, eds J.G. Ewen, D.P. Armstrong, K.A. Parker & P.J. Seddon, Chapter 7. Wiley-Blackwell, Oxford, UK.

Osborne, P.E. & Seddon, P.J. (2011) Selecting suitable habitats for reintroductions: variation, change and the role of species distribution modelling. In *Reintroduction*

Biology: integrating science and management, eds J.G. Ewen, D.P. Armstrong, K.A. Parker & P.J. Seddon, Chapter 3. Wiley-Blackwell, Oxford, UK.

Parker, K.A., Dickens, M.J., Clarke, R.H. *et al.* (2011). The theory and practice of catching, holding, moving and releasing animals. In *Reintroduction Biology: integrating science and management*, eds J.G. Ewen, D.P. Armstrong, K.A. Parker & P.J. Seddon, Chapter 4. Wiley-Blackwell, Oxford, UK.

Pollak, J.P., Lacy, R.C. & Ballou, J.D. (2002) *Population Management 2000*. Chicago Zoological Society, Brookfield, Illinois, USA.

Powlesland, R.G. (1983) Breeding and mortality of the South Island robin in Kowhai Bush, Kaikoura. *Notornis*, 30, 265–282.

Reding, D.M, Freed, L.A., Cann, R.L. *et al.* (2010) Spatial and temporal patterns of genetic diversity in an endangered Hawaiian honeycreeper, the Hawaii Akepa (*Loxops coccineus coccineus*). *Conservation Genetics*, 11, 225–240.

Reed, D.H. (2010) Albatross, eagles and newts, Oh my!: expectations to the prevailing paradigm concerning genetic diversity and population viability? *Animal Conservation*, 13, 448–457.

Robert, A., Couvet, D. & Sarrazin, F. (2007). Integration of demography and genetics in population restorations. *Ecoscience*, 14, 463–471.

Robert, A., Sarrazin, F., Couvet, D. *et al.* (2004) Releasing adults versus young in reintroductions: interactions between demography and genetics. *Conservation Biology*, 18, 1078–1087.

Robertson, B.C. (2006) The role of genetics in kakapo recovery. *Notornis*, 53, 173–183.

Russell, J.C., Beaven, B.M., MacKay, J.W.B. *et al.* (2008) Testing island biosecurity systems for invasive rats. *Wildlife Research*, 35, 215–221.

Russell, J.C, Towns, D.R., Anderson, S.H. *et al.* (2005) Intercepting the first rat ashore. *Nature*, 437, 1107.

Ryan, C.J. & Jamieson, I.G. (1998) Estimating the home range and carrying capacity for takahe (*Porphyrio mantelli*) on predator-free offshore islands: implications for future management. *New Zealand Journal of Ecology*, 22, 17–24.

Sarrazin, F. & Legendre, S. (2000). Demographic approach to releasing adults versus young in reintroductions. *Conservation Biology*, 14, 1–14.

Seddon, P.J, Armstrong, D.P. & Maloney, R.R. (2007) Developing the science of reintroduction biology. *Conservation Biology*, 21, 303–312.

Seddon, P.J., Strauss, W.M. & Innes, J. (2011) Animal translocations: what are they and why do we do them? In *Reintroduction Biology: integrating science and management*, eds J.G. Ewen, D.P. Armstrong, K.A. Parker & P.J. Seddon, Chapter 1. Wiley-Blackwell, Oxford, UK.

Soulé, M., Giplin, M., Conway, W. *et al.* (1986) The millennium ark: how long a voyage, how many staterooms, how many passengers? *Zoo Biology*, 5, 111–114.

Tallmon, D.A., Luikart, G. & Waples, R.S. (2004) The alluring simplicity and complex reality of genetic rescue. *Trends Ecology and Evolution*, 19, 489–496.

Taylor, S.S. & Jamieson, I.G. (2008) No evidence for loss of genetic variation following sequential translocations in extant populations of a genetically depauperate species. *Molecular Ecology*, 17, 545–556.

Thompson, E.A. (1994) Genetic importance and genomic descent. In *Population Management for Survival and Recovery: Proceedings of the 1989 Front Royal Workshop on analytical methods for population viability analyses*, eds J.D. Ballou, M. Gilpin & T.J. Foose, pp. 112–123. Columbia University Press, New York, USA.

Tordoff, H.B. & Redig, P.T. (2001) Role of genetic background in the success of reintroduced peregrine falcons. *Conservation Biology*, 15, 528–532.

Tracy, L.N. (2009) *Historic and contemporary population genetics and their management implications for an endangered New Zealand passerine, the mohua (Mohoua ochrocephala)*. MSc thesis, University of Otago, Dunedin, New Zealand.

Tracy, L.N. & Jamieson, I.G. (2011) Historic DNA reveals contemporary population structure results from anthropogenic effects, not pre-fragmentation patterns. *Conservation Genetics*, 12, 517–526.

Tracy, L.N., Wallis, G., Efford, M. *et al.* (2011) Preserving genetic diversity in threatened species reintroductions: how many individuals should be released? *Animal Conservation*, 14, 439–446.

Vucetich, J.A. & Waite, T.A. (2000) Is one migrant per generation sufficient for the genetic management of fluctuating populations? *Animal Conservation*, 3, 261–266.

Westemeier, R.L., Brawn, J.D., Simpson, S.A. *et al.* (1998) Tracking the long-term decline and recovery of an isolated population. *Science*, 282, 1695–1698.

Wickes, C., Crouchley, D. & Maxwell, J. (2009) *Takahe Recovery Plan: 2007–2012*. Department of Conservation, Wellington, New Zealand.

Wright, S. (1969) *Evolution and the Genetics of Populations*, vol. II, *The theory of gene frequencies*. University of Chicago Press, Chicago, Illinois, USA.

Summary

Philip J. Seddon[1], Doug P. Armstrong[2], Kevin A. Parker[3] and John G. Ewen[4]

[1]Department of Zoology, University of Otago, New Zealand
[2]Massey University, Palmerston North, New Zealand
[3]Massey University, Auckland, New Zealand
[4]Institute of Zoology, Zoological Society of London, United Kingdom

On the face of it this reintroduction business seems very straightforward doesn't it? Take some animals or plants from one place and put them in another area where the species used to be and start up a new population. How hard can that be, especially when you think about all those introduced species that have become invasive pests? It was perhaps this thinking in the past that resulted in a proliferation of ill-conceived reintroduction attempts (Lyles & May, 1987; Stanley Price & Soorae, 2003), doomed to fail because of lack of planning and poor implementation. To a large extent the IUCN Reintroduction Specialist Group (RSG) was created to provide a more rigorous basis for reintroduction planning and guidance for reintroduction implementation, and certainly the most influential output of the RSG has been the 1998 Reintroduction Guidelines (IUCN, 1998).

Much has changed in the two-plus decades since the guidelines were first disseminated. There has been a huge increase in the number of reintroduction projects worldwide and in the taxonomic breadth of reintroduction programmes; there has been a concurrent increase in the number of peer-reviewed outputs resulting from post-release monitoring, and this expansion

Reintroduction Biology: Integrating Science and Management. First Edition.
Edited by John G. Ewen, Doug P. Armstrong, Kevin A. Parker and Philip J. Seddon.
© 2012 Blackwell Publishing Ltd. Published 2012 by Blackwell Publishing Ltd.

of knowledge has burgeoned into the fledgling discipline of reintroduction biology. There is now a good understanding that reintroduction project success depends on a multitude of factors, not least of which are public and legislative support, adequate funding, rigorous selection of founders, appropriate species-specific methods for capture, transport, handling and release, sensible post-release monitoring and planning for interventions, and most importantly that the reintroduction site has a suitable habitat. A successful reintroduction will not be an academic exercise, and not purely a management action, nor solely a community led initiative – researchers, managers, NGOs and community groups cannot work in isolation with great hopes of success; those reintroduction projects held up as examples to be emulated are ones in which public support has combined with the expertise of wildlife managers and the analytical and theoretical knowledge of scientific researchers. Far from being few and infrequent, examples of high-quality reintroductions now abound; it is an exciting period for both the discipline of reintroduction biology and the use of reintroduction biology to restore populations of endangered species. We are grateful to be part of this period of growth and development, and privileged to be able to make a contribution with this edited volume.

What we have attempted to do here is to compile a state-of-the-art for reintroduction practice, with chapters written by those working at the 'coal face' of conservation translocations, often with one shoed foot in the academic arena and one booted foot firmly in the field. Our intended readership is that hybrid being, the 'reintroduction practitioner', whether they find their home in an NGO, a community conservation group, a government conservation agency or a research institution. We have also made an effort to step away from a collection of case studies. While these are informative, their messages are often case specific, with clear links to general approaches and an overarching theory difficult to see. An exception is Jones & Merton (Chapter 2), where a collection of case studies is presented by two leading reintroduction practitioners, but a key feature of these cases is the careful selection of management options based on careful a priori thinking. This text then covers the theory, justification and general application of issues clearly relevant to successful reintroduction. It includes methodology for the selection of appropriate release sites (Osborne & Seddon, Chapter 3) and the management of individuals (often linked to stress) during the translocation process (Parker et al., Chapter 4). Post-release dispersal and when, why and how it may occur, and how it could be managed, are reviewed in Le Gouar et al. (Chapter 5). Armstrong & Reynolds (Chapter 6), Nichols & Armstrong (Chapter 7) and

McCarthy *et al.* (Chapter 8) present an important framework within which to cast any reintroduction programme and for dealing with any perceived management problems or options. Pathogen and health awareness is reviewed in Ewen *et al.* (Chapter 9) and available disease risk assessment procedures are presented in Sainsbury *et al.* (Chapter 10). Finally, Keller *et al.* (Chapter 11), Groombridge *et al.* (Chapter 12) and Jamieson & Lacy (Chapter 13) discuss important genetic considerations and how best to manage them.

Books can have a limited useful shelf life, and to some extent we have built-in planned obsolescence for this volume by asking our contributors to not only review the background to the different issues but to attempt to chart a way forward to guide future directions and work. One measure of success at achieving our aim could be that this book becomes redundant in the near future as knowledge gaps are filled, key questions are addressed, a carefully considered and refined process with clear justification and transparency becomes the norm, and increased reintroduction success the result.

What's in a name?

Collectively we've invested some time and energy into this emerging discipline of reintroduction biology – the science and practice of restoring populations of animals and plants through translocations. However, strictly speaking, reintroduction, the re-establishment of a population within vacant parts of the species historical range, is only one tool in a conservation translocation spectrum that encompasses re-enforcement of extant populations at one end, through to the creation of novel ecosystems with new species assemblages at the other. Should the discipline, the associated IUCN specialist group, and indeed this book, be called Conservation Translocation? We don't think so. Reintroduction is a term that has been well branded, is well understood globally and describes an action that will remain the 'pointy end' of the conservation translocation stick. Having said that, however, we do now need to consider future developments at the, arguably, more risky and controversial end of the spectrum – what will be the role of conservation introductions?

The future of conservation translocations

As we adjust to living in the Anthropocene and acknowledge that humans have and continue to modify the global environment, the notion of novel ecosystems may become more acceptable. We will then have to start to make some tough choices about which species we wish to see in what places. Increasingly besieged protected areas will be islands of semi-natural conditions in seas of highly modified human-influenced landscape – effectively small fragments in which ongoing intervention will be necessary to sustain the wildlife populations within. Such conservation management interventions will include the addition of new individuals to increase genetic diversity or abundance of extant species or the addition of individuals of species new to the area to fulfil specific missing ecological functions that will enhance ecosystem resilience. We can also appreciate a strong case being made for the more risky conservation introductions, including reactive or even proactive translocation of populations to areas outside the species' historical range in order to protect them from human-mediated habitat change – so-called Assisted Colonizations. This is not as new as some might imagine; for example in New Zealand conservation management, species have sometimes been marooned on offshore islands outside their probable historic ranges to protect them from almost certain extinction in mainland areas overrun with introduced mammalian predators.

We can also envisage an even more extreme conservation introduction, whereby populations are established in areas in which the species has never been, in species assemblages that have no historical analogues. This is a de facto consequence of anthropogenic habitat change anyway, and the battle to somehow preserve an arbitrary target system replicating a specific past ecological condition cannot be won everywhere. There are serious debates currently raging over the desirability of actively creating and engineering novel ecosystems for the benefit of both nature and humans. It is against this background that the RSG has undertaken to review and revise the Reintroduction Guidelines, to ensure that assisted colonization and other conservation introductions are not used lightly or even dishonestly to usher in a new era of ill-conceived translocations, dumping unwanted or excess animals under the guise of poorly defined conservation objectives.

Regardless of the degree to which historical range is used as a basis for release site selection, we believe that the principles advocated in this book will always apply, and it is our intent and our hope that the information compiled here will help you contribute to project success, whatever part you have to play in it.

References

Armstrong, D.P. & Reynolds, M.H. (2011) Modelling reintroduced populations: the state of the art and future directions. In *Reintroduction Biology: integrating science and management*, eds J.G. Ewen, D.P. Armstrong, K.A. Parker & P.J. Seddon, Chapter 6. Wiley-Blackwell, Oxford, UK.

Ewen, J.G., Acevedo-Whitehouse, K., Alley, M. *et al.* (2011) Empirical consideration of parasites and health in reintroduction. In *Reintroduction Biology: integrating science and management*, eds J.G. Ewen, D.P. Armstrong, K.A. Parker & P.J. Seddon, Chapter 9. Wiley-Blackwell, Oxford, UK.

Groombridge, J.J., Raisin, C., Bristol, R. *et al.* (2011) Genetic consequences of reintroductions and insights from population history. In *Reintroduction Biology: integrating science and management*, eds J.G. Ewen, D.P. Armstrong, K.A. Parker & P.J. Seddon, Chapter 12. Wiley-Blackwell, Oxford, UK.

IUCN (World Conservation Union) (1998) *Guidelines for Re-introductions.* IUCN/SSC Re-introduction Specialist Group, IUCN, Gland, Switzerland, and Cambridge, United Kingdom.

Jamieson, I.G. & Lacy, R.C. (2011) Managing genetic issues in reintroduction biology. In *Reintroduction Biology: integrating science and management*, eds J.G. Ewen, D.P. Armstrong, K.A. Parker & P.J. Seddon, Chapter 13. Wiley-Blackwell, Oxford, UK.

Jones, C.G. & Merton, D.V. (2011) A tale of two islands: the rescue and recovery of endemic birds in New Zealand and Mauritius. In *Reintroduction Biology: integrating science and management*, eds J.G. Ewen, D.P. Armstrong, K.A. Parker & P.J. Seddon, Chapter 2. Wiley-Blackwell, Oxford, UK.

Keller, L.K., Biebach, I., Ewing, S.R. *et al.* (2011) The genetics of reintroductions: inbreeding and genetic drift. In *Reintroduction Biology: integrating science and management*, eds J.G. Ewen, D.P. Armstrong, K.A. Parker & P.J. Seddon, Chapter 11. Wiley-Blackwell, Oxford, UK.

Le Gouar, P., Mihoub, J.-B. & Sarrazin, F. (2011) Dispersal and habitat selection: behavioural and spatial constraints for animal translocations. In *Reintroduction Biology: integrating science and management*, eds J.G. Ewen, D.P. Armstrong, K.A. Parker & P.J. Seddon, Chapter 5. Wiley-Blackwell, Oxford, UK.

Lyles, A.M. & May, R.M. (1987) Problems in leaving the ark. *Nature*, 326, 245–246.

McCarthy, M.A., Armstrong, D.P. & Runge, M.C. (2011) Adaptive management of reintroduction. In *Reintroduction Biology: integrating science and management*, eds J.G. Ewen, D.P. Armstrong, K.A. Parker & P.J. Seddon, Chapter 8. Wiley-Blackwell, Oxford, UK.

Nichols, J.D. & Armstrong, D.P. (2011) Monitoring for reintroductions. In *Reintroduction Biology: integrating science and management*, eds J.G. Ewen, D.P. Armstrong, K.A. Parker & P.J. Seddon, Chapter 7. Wiley-Blackwell, Oxford, UK.

Osborne, P.E. & Seddon, P.J. (2011) Selecting suitable habitats for reintroductions: variation, change and the role of species distribution modelling. In *Reintroduction Biology: integrating science and management*, eds J.G. Ewen, D.P. Armstrong, K.A. Parker & P.J. Seddon, Chapter 3. Wiley-Blackwell, Oxford, UK.

Parker, K.A., Dickens, M.J., Clarke, R.H. *et al.* (2011). The theory and practice of catching, holding, moving and releasing animals. In *Reintroduction Biology: integrating science and management*, eds J.G. Ewen, D.P. Armstrong, K.A. Parker & P.J. Seddon, Chapter 4. Wiley-Blackwell, Oxford, UK.

Sainsbury, A.W., Armstrong, D.P. & Ewen, J.G. (2011) Methods of disease risk analysis for reintroduction programmes. In *Reintroduction Biology: integrating science and management*, eds J.G. Ewen, D.P. Armstrong, K.A. Parker & P.J. Seddon, Chapter 10. Wiley-Blackwell, Oxford, UK.

Stanley Price, M.R. & Soorae, P.S. (2003) Reintroductions: whence and whither? *International Zoo Yearbook*, 38, 61–75.

Index